NATO ASI Series

Advanced Science Institutes Series

A series presenting the results of activities sponsored by the NATO Science Committee, which aims at the dissemination of advanced scientific and technological knowledge, with a view to strengthening links between scientific communities.

The Series is published by an international board of publishers in conjunction with the NATO Scientific Affairs Division

A	Life Sciences	Plenum Publishing Corporation
B	Physics	London and New York
C	Mathematical and Physical Sciences	Kluwer Academic Publishers
D	Behavioural and Social Sciences	Dordrecht, Boston and London
E	Applied Sciences	
F	Computer and Systems Sciences	Springer-Verlag
G	Ecological Sciences	Berlin Heidelberg New York
H	Cell Biology	London Paris Tokyo Hong Kong
I	Global Environmental Change	Barcelona Budapest

PARTNERSHIP SUB-SERIES

1. Disarmament Technologies	Kluwer Academic Publishers
2. Environment	Springer-Verlag/Kluwer Acad. Publishers
3. High Technology	Kluwer Academic Publishers
4. Science and Technology Policy	Kluwer Academic Publishers
5. Computer Networking	Kluwer Academic Publishers

The Partnership Sub-Series incorporates activities undertaken in collaboration with NATO's Cooperation Partners, the countries of the CIS and Central and Eastern Europe, in Priority Areas of concern to those countries.

NATO-PCO DATABASE

The electronic index to the NATO ASI Series provides full bibliographical references (with keywords and/or abstracts) to about 50 000 contributions from international scientists published in all sections of the NATO ASI Series. Access to the NATO-PCO DATABASE compiled by the NATO Publication Coordination Office is possible in two ways:

- via online FILE 128 (NATO-PCO DATABASE) hosted by ESRIN, Via Galileo Galilei, I-00044 Frascati, Italy.

- via CD-ROM "NATO Science & Technology Disk" with user-friendly retrieval software in English, French and German (© WTV GmbH and DATAWARE Technologies Inc. 1992).

The CD-ROM can be ordered through any member of the Board of Publishers or through NATO-PCO, Overijse, Belgium.

Series G: Ecological Sciences, Vol. 39

Springer
Berlin
Heidelberg
New York
Barcelona
Budapest
Hong Kong
London
Milan
Paris
Santa Clara
Singapore
Tokyo

Biological Fixation of Nitrogen for Ecology and Sustainable Agriculture

Edited by

Andrzej Legocki

Polish Academy of Sciences
Institute of Bioorganic Chemistry
Noskowskiego 12/14
PL-61-704 Poznan, Poland

Hermann Bothe

Universität zu Köln, Botanisches Institut
Gyrhofstraße 15
D-50923 Köln, Germany

Alfred Pühler

Universität Bielefeld
Lehrstuhl für Genetik
Postfach 8640
D-35501 Bielefeld, Germany

With 53 Figures (1 colour plate) and 28 Tables

 Springer

Proceedings of the NATO Advanced Research Workshop "Biological
Fixation of Nitrogen for Ecology and Sustainable Agriculture", held in
Poznan, Poland, September 10–14, 1996

Cataloging-in-Publication Data applied for

Die Deutsche Bibliothek - CIP-Einheitsaufnahme

**Biological fixation of nitrogen for ecology and sustainable
agriculture** : with 28 tables ; [proceedings of the NATO
Advanced Research Workshop "Biological Fixation of Nitrogen
for Ecology and Sustainable Agriculture", held in Poznań,
Poland, September 10 - 14, 1996] / ed. by Andrezej Legocki ... -
Berlin ; Heidelberg ; New York ; Barcelona ; Budapest ; Hong
Kong ; London ; Milan ; Paris ; Santa Clara ; Singapore ;
Tokyo : Springer, 1997
  (NATO ASI series : Ser. G, Ecological sciences ; Vol. 39)
  ISBN 3-540-62056-7
NE: Legocki, Andrezej [Hrsg.]; Advanced Research Workshop Biological
  Fixation of Nitrogen for Ecology and Sustainable Agriculture <1996,
  Poznań>; NATO: NATO ASI series / G

ISBN 3-540-62056-7 Springer-Verlag Berlin Heidelberg New York

Typesetting: Camera ready by authors/editors
Printed on acid-free paper
SPIN 10532423   31/3137 - 5 4 3 2 1 0

PREFACE

For nearly thirty years there have been optimistic reports on the value of biological nitrogen fixation (BNF) in agriculture and how new research was going to revolutionise the use of BNF. However, we have not yet seen the conversion of imaginative new BNF technologies into significant application in agricultural practice. Neither have the BNF scientific community been developing workable ways in which existing agricultural practice might be more sustainable and less ecologically damaging. This is extremely frustrating because the potential exists to satisfy the needs of nearly all crop plants for fixed nitrogen through the use of nitrogen-fixing bacteria, or their genes. Indeed, it is not unreasonable to think that BNF genes could be exploited also in ruminant animals to reduce greatly their need for protein.

BNF is of great value to farmers, particularly where nodulating legumes are grown. We have extensive documentation of the amounts of nitrogen that can be fixed under different conditions and have a much better appreciation for the limitations of BNF in practice than was evident ten or more years ago. Research has led to the production of strains of *Rhizobium* with enhanced nitrogen fixing activity and these have recently become available commercially.

The NATO Advanced Research Workshop connected with the 2nd European Conference on Nitrogen Fixation brought together a large number of leading scientists as well as many young and enthusiastic researchers. This created a unique and inspiring atmosphere to discuss most interesting issues of BNF in the context of sustainability and conversion into the agricultural practice.

New horizons of nitrogen fixation research include: plant-microbe signalling, microbial ecology, carbon-nitrogen metabolism, oxygen and hormonal regulation, legume genetics, introduction of genetically modified microorganisms into the environment as well as co-evolution of symbiotic systems. The understanding of these fundamental issues should greatly contribute to further scientific, agricultural and ecological development.

The Workshop was organised by the Institute of Bioorganic Chemistry of the Polish Academy of Sciences. We wish to express our deep gratitude to the members of the local and international committees and to the administrative staff of the Science Centre of the Polish Academy of Sciences in Poznań for their enthusiasm and dedicated work.

Poznań, September 1996

Andrzej B. Legocki
Hermann Bothe
Alfred Pühler

CONTENTS

1. Structural Chemistry and Biochemistry of Nitrogenase

2. Signal Perception, Transduction and Cell Cycle Genes in Nodulation

3. Plant Genes Involved in Nodulation

4. Bacterium - Plant Surface Interaction

5. Molecular Microbial Ecology

6. Nitrogen Fixing Systems

7. Nitrogen Fixation in Sustainable Agriculture

8. Carbon-Nitrogen Metabolism in Symbiotic Systems

9. Oxygen Regulation in Nitrogen Fixation

10. Model Plants for Nitrogen Fixation and Legume Genetics

11. Coevolution of Symbiotic Systems

Part 1

Structural Chemistry and
Biochemistry of Nitrogenase

Nitrogenase: Two Decades of Biochemical Genetics

Jeverson Frazzon[1], Johan Spee[2], Jason Christiansen[3] and Dennis R. Dean[3]

[1]Dept. of Biotechnology, Universidade Federal do Rio Grande do Sul, Porto Alegre, Brazil
[2]Department of Biochemistry, Agricultural University, Wageningen, The Netherlands
[3]Department of Biochemistry, Virginia Tech, Blacksburg, Virginia, 24061, USA

Introduction. Biological nitrogen fixation is catalyzed by nitrogenase (1). This enzyme is composed of two component proteins usually designated the Fe protein and the MoFe protein. During catalysis the component proteins associate and dissociate in a MgATP-dependent manner, where each association-dissociation step involves the delivery of one electron from the Fe protein to the MoFe protein. Each intercomponent electron transfer cycle is coupled to the hydrolysis of two MgATP. Because turnover requires 6 electrons for N_2 reduction, and 2 electrons for the obligate evolution of H_2 that accompanies N_2 reduction, the minimal stoichiometry of the reaction is usually indicated as:

$$N_2 + 8e^- + 8H^+ + 16MgATP \rightarrow 2NH_3 + H_2 + 16MgADP + 16Pi$$

The Fe protein is a homodimer that contains the MgATP binding site(s) and serves as the obligate electron donor to the MoFe protein. The electron transfer unit within the Fe protein is a Fe_4S_4 cubane that is symmetrically situated between the identical subunits. The MoFe protein is an $\alpha_2\beta_2$ heterotetramer that contains two pairs of metalloclusters. One of these clusters is a linked Fe_8S_7 double-cubane, called the P cluster. Each P cluster symmetrically bridges an $\alpha\beta$ unit. The other complex metallocluster is called FeMo-cofactor and it contains an Fe_7S_9Mo core and homocitrate, as an organic constituent. FeMo-cofactor is contained entirely within the MoFe protein α-subunit. The current view is that each $\alpha\beta$ unit of the MoFe protein comprises an independent catalytic unit and that the P cluster initially accepts the electron(s) delivered by intercomponent electron transfer and then brokers their intramolecular delivery to FeMo-cofactor where substrate binding and reduction occurs (2).

The biochemical-genetic approach to biological research involves the characterization of mutant strains that are altered in a specific biochemical feature as a means to decipher genetic organization and biochemical function. In the case of nitrogenase, the goals of such a biochemical-genetic strategy can be placed in the context of the following questions: How many genes are required for the expression and assembly of nitrogenase? How are the genes associated with nitrogenase organized? How is nitrogen fixation (*nif*) gene regulation balanced and coordinated? What are the biochemical events involved in nitrogen fixation? How does nitrogenase work? In the mid 1970's serious biochemical-genetic strategies aimed at elucidating the answers to these questions were initiated both at the University of Wisconsin and at the Nitrogen Fixation Unit in the UK. Since then, many investigators scattered throughout the World have contributed to our knowledge of nitrogenase assembly, catalysis and regulation. In this brief review we first discuss our current knowledge of the biochemical events that lead to formation of an active nitrogenase. In the second section we describe how site-directed mutagenesis and gene replacement techniques are being used to probe the mechanism for nitrogenase catalysis.

Genetic organization and biochemical function of *nif* genes. In the early days, the biochemical-genetic strategy for the analysis of *nif* genes and their products involved mixing extracts from two different mutants, both of which were individually impaired in nitrogenase activity, and then asking whether or not nitrogenase activity could be reconstituted. Because of its ease of genetic manipulation, *Klebsiella pneumoniae* was the initial organism of choice for

NATO ASI Series, Vol. G 39
Biological Fixation of Nitrogen
for Ecology and Sustainable Agriculture
Edited by A. Legocki, H. Bothe, A. Pühler
© Springer-Verlag Berlin Heidelberg 1997

these studies. Thus, biochemical complementation experiments could be readily augmented by genetic complementation studies, ultimately leading to the construction of a physical, functional, and genetic map of the *K. pneumoniae nif*-specific genes and their products. Many groups contributed to this effort and there were some differences along the way. However, 18 of the 20 *nif* genes and their products, as well as the complex organization of their corresponding transcriptional units, were correctly identified in this way. The fact that these accomplishments were achieved in the pre-recombinant DNA era (for example, see ref 3) is a testament to the abilities of the investigators involved as well as to the power of traditional genetic and biochemical techniques. We now know that the all of the *K. pneumoniae nif*-genes are tightly clustered on the genome and that they are organized into eight transcriptional units. The functions of the products of these genes can be grouped in the following ways (4): **nitrogenase structural genes** (*nifH* [Fe protein], *nifD* [MoFe protein α-subunit], *nifK* [MoFe protein β-subunit]); **electron transport** (*nifJ* [pyruvate-flavodoxin-oxidoreductase], *nifF* [flavodoxin]); **positive regulatory protein** (*nifA*); **negative regulatory protein** (*nifL*); **Fe protein maturation** (*nifM* [putative cis-trans prolyl-isomerase], **FeMo-cofactor biosynthesis and insertion** (*nifHBENQVY*); **mobilization of Fe and S** (*nifUS*), **function not known** (*nifWZXT*). It is noted that products of some of the *nif*-specific genes appear to have more than one function. For example, the Fe protein, which is the product of *nifH*, appears to function in nitrogenase catalysis, FeMo-cofactor formation, and insertion of FeMo-cofactor into an apo-form of the MoFe protein.

The classical biochemical-genetic approach permitted Shah and co-workers to develop a biochemical assay for the biosynthesis of FeMo-cofactor (5). The biosynthesis of FeMo-cofactor is of particular interest because it provides the site for substrate binding and reduction. Our current understanding of FeMo-cofactor biosynthesis is as follows (for a recent review, see ref. 6): (i) FeMo-cofactor is separately synthesized and then inserted into an apo-form of the immature MoFe protein; (ii) FeMo-cofactor is preassembled on a molecular scaffold provided by an $\alpha_2\beta_2$ complex of the *nifEN* gene products; (iii) The organic constituent of FeMo-cofactor is formed by the activity of a homocitrate synthase encoded by *nifV*; (iv) At least a portion of FeMo-cofactor's Fe-S core is assembled through the activity of the *nifB* gene product prior to its donation to the *nifEN* gene products scaffold; (v) The *nifQ* gene product is involved in activation or mobilization of the Mo that is required for FeMo-cofactor assembly; (vi) The *nifY* gene product encodes a molecular prop that enforces a structure of the apo-form of the MoFe protein that is able to accept the separately synthesized FeMo-cofactor (in *A. vinelandii* an analogous protein called "gamma" appears to have this function); (vii) Fe protein is required for both the formation and insertion of FeMo-cofactor; (viii) MgATP is required for FeMo-cofactor formation.

Although the approach mentioned above led to the identification and, in some cases, the function of the overall players involved in the formation, maturation, and regulation of nitrogenase, there are also natural limitations to the approach. One specific problem is that, other than the nitrogenase structural and electron transport components, most *nif* gene products are accumulated intracellularly only in very small quantities. Thus, the sophisticated biochemical and biophysical techniques that have been applied towards the study of the nitrogenase component proteins have not been amenable for the study of the cluster assembly proteins. It is at this stage that various recombinant DNA techniques have now begun to prove valuable. The study of the *nifEN* gene products that compose the molecular scaffold upon which FeMo-cofactor is assembled provides one good example. The detailed biochemical analysis of the NifEN scaffold presents two problems. First, the scaffold is produced only in small quantities, and second, the biochemical assay for NifEN activity is somewhat laborious. These problems have been recently circumvented by replacing the weak *nifEN* gene expression elements with the much stronger nitrogenase structural gene regulatory elements and by "tagging" the N-terminal sequence of NifE with six histidines. In this way it has been possible to produce and then rapidly purify large amounts of the NifEN complex by using immobilized metal affinity chromatographic techniques. It should be pointed out that the NifEN complex produced and prepared in this way has yet to undergo rigorous biochemical scrutiny, nevertheless, recent work from our laboratory clearly demonstrate that the purified NifEN is active in *in vitro* FeMo-cofactor biosynthesis.

Another approach has involved using "reverse" biochemistry. In this strategy a particular *nif*-specific gene is hyperexpressed in *Escherichia coli*, the product purified and then examined for properties that could give clues to its biochemical function or mechanism. This approach has proven enormously successful. For example, polymerase-chain-amplification techniques were used to make gene cartridges for each of the individual *Azotobacter vinelandii* *nif*-genes so that they could be fused to the transcriptional and translational control elements present in the T7 expression system described by Tabor. In this way it was possible to produce sufficient amounts of soluble forms of both the negative (*nifL* gene product) and positive (*nifA* gene product) regulatory elements that heretofore could not be isolated in their native form. Another example is provided by the *nifS* gene that was hyperexpressed in *E. coli* and whose product was found to contain pyridoxal phosphate. This simple biochemical observation quickly led to the remarkable discovery that *nifS* encodes a cysteine desulfurase that uses cysteine as a substrate in the activation of sulfur for Fe-S cluster formation (7). Considering that the *nifS* gene product is accumulated in nitrogen fixing cells at a level of only about 0.07% of the total cellular proteinit is difficult to imagine how its function could have been determined without the intervention of recombinant DNA methods. An interesting feature of *nifS*, as well as of *nifU* (and a putative gene indicated as orf6 in *A. vinelandii*), is that they are members of a gene family that appear to be highly conserved throughout nature. We suggest that these three genes encode proteins that are players in a universal mechanism for the mobilization of Fe and S for Fe-S cluster assembly.

Nitrogenase Catalysis The primary biochemical-genetic strategy for the analysis of nitrogenase catalysis has involved targeting specific residues that might participate in a specific process, for example MgATP binding, and then substituting these residues by using site-directed mutagenesis and gene replacement techniques. Altered nitrogenase component proteins produced in this way are subsequently purified and their catalytic and biophysical properties characterized. Prior to the availability of a structural model for the component proteins this strategy depended on indirect methods that were based, primarily, on various comparisons of primary sequences (reviewed in reference 2). On this basis, some of the residues involved in nucleotide binding and hydrolysis, residues located within both the Fe protein and MoFe protein that are involved in component interaction, residues involved in ligating the Fe protein' Fe_4S_4 cluster, residues located within the P cluster binding region, and certain residues located within the FeMo-cofactor environment were correctly identified. Nevertheless, indirect methods fell far short of predicting all of the structural and functional features of the nitrogenase component proteins. The important contribution of the initial biochemical-genetic studies on nitrogenase catalysis is that reliable methods for producing and characterizing altered nitrogenase were developed prior to, rather than after, the emergence of X-ray crystallographic models.

Some of the basic questions that are associated with nitrogenase catalysis that are now being successfully addressed include the following: How do the Fe protein and MoFe protein interact? What is the role of MgATP binding and hydrolysis during nitrogenase catalysis? What are the inter- and intra-molecular electron paths? How are multiple electrons accumulated within the MoFe protein? How do the metallocluster polypeptide environments contribute to catalysis? Two of these issues, the nature of component protein interaction and role of MgATP, are briefly discussed below as examples of the biochemical-genetic analysis of nitrogenase catalysis.

Component Protein Interaction A component docking model that pairs the two-fold symmetric surface of the Fe protein homodimer with the exposed surface of a MoFe protein pseudosymmetric αβ-unit interface has been proposed (8). An important feature of this model is that it positions the Fe protein's Fe_4S_4 cluster as close as possible to the MoFe protein's P cluster. This arrangement accommodates the view that the primary electron transfer event (namely, the transfer of an electron from the Fe protein to the MoFe protein) involves delivery of an electron from the Fe_4S_4 cluster of the Fe protein to a MoFe protein P cluster. Inspection of the Fe protein structural model shows that there is a crown of positively charged residues located within the proposed docking interface that surround the Fe protein's Fe_4S_4 cluster. Similarly, the proposed MoFe protein docking surface has two patches of negatively charged aspartate residues in the complementary pseudosymmetric surface that is proposed to interact with the Fe protein. This arrangement permits one to consider that reciprocal ionic interactions between the component proteins might be critical to the initial docking event. Indeed, this situation was anticipated prior to the availability of structural models because it was known that

ADP-ribosylation of arginine-100, located on the proposed docking surface and close to the Fe protein's Fe_4S_4 cluster, prevents productive intercomponent electron transfer (9). Furthermore, it is known that nitrogenase catalysis is sensitive to high salt conditions, indicating that ionic interactions are important for productive component protein interaction (10).

The role of all of these residues has now been probed by substituting them with a variety of other residues followed by the biochemical characterization of the altered proteins. In the case of the Fe protein, residues arginine-100, arginine-140 and lysine-143 have been substituted with other amino acids. All of these substitutions result in altered Fe proteins whose interactions with the normal MoFe protein are hypersensitive to salt and also cause the uncoupling of MgATP hydrolysis from productive electron transfer (10, 11). In the case of the MoFe protein, the negatively charged aspartate residues that are located within regions of primary sequence and structural symmetry between the α- and β-subunits have been substituted by asparagines. The results of the biochemical characterization of these altered MoFe proteins was unexpected. The substitution of α-subunit residue aspartate-162, β-subunit aspartate-160, or β-subunit aspartate-161 by asparagine had very little effect on catalytic activity. In contrast, substitution of the α-subunit aspartate-160 residue by asparagine resulted in a biochemical phenotype nearly the same as the altered histidine-100 substituted Fe protein. This result shows that the conservation of primary sequence and structural symmetry for the MoFe protein subunits is not functionally conserved. This conclusion also became apparent in our analyses of MoFe proteins that have alterations in the P cluster environment. For example, to date, no substitutions placed at the α-subunit cysteine-95 and cysteine-154 P cluster ligand positions have resulted in altered MoFe proteins that are catalytically active. In contrast, the analogous β-subunit cysteine-88 and cysteine-153 P cluster ligand positions are able to tolerate some substitutions (12). These observations support the general view that the Mo Fe protein α-subunit has the dominant role in intermolecular electron transfer and catalysis. This view is consistent with the fact that FeMo-cofactor is contained entirely within the MoFe protein α-subunit.

Role of MgATP in Nitrogenase Catalysis On the basis of primary amino acid sequence comparisons Robson recognized that the Fe protein is a member of the molecular switch family of proteins (14). This suggestion was later confirmed by the structural analysis of the Fe protein. The molecular switch family includes those proteins that use nucleotide binding and hydrolysis events to drive conformational changes that are linked to signal transduction mechanisms. Upon MgATP binding the Fe protein undergoes a conformational change that results in movement of its Fe_4S_4 cluster to the polypeptide surface together with a change in the redox potential of the Fe_4S_4 cluster. This conformational change readies the Fe protein for intercomponent interaction. MgATP hydrolysis only occurs upon the docking event which ultimately leads to component dissociation. Thorneley and Lowe have suggested that component protein dissociation is the rate limiting step in nitrogenase catalysis (13). It appears that MgATP binding and hydrolysis are necessary for reciprocal communication between the Fe protein and the MoFe protein, but why is MgATP necessary for nitrogenase activity? There are a number of viable hypotheses that are being tested, but the one we find most attractive is that MgATP binding and hydrolysis are needed to ensure that electron transfer is directed only towards the MoFe protein. This point is an important one because multiple electrons must be accumulated within the MoFe protein before substrate reduction occurs. For this reason we envision that the role of MgATP binding and hydrolysis is to control an electron gate. For example, the electron gate might be opened upon MgATP binding to permit intercomponent electron transfer and then closed upon nucleotide hydrolysis to ensure the unidirectionality of electron flow.

The biochemical-genetic analysis of the role of MgATP in nitrogenase catalysis has represented some of the most exciting developments in recent nitrogenase research. Although the overall strategies involved in the analysis of Fe protein have relied primarily on similar well studied signal transduction systems, the particular advantage of nitrogenase research is that it has been possible to dissect specific events in the process because of the many biochemical and biophysical tools that are available. For example, conformational changes that are elicited within the Fe protein upon MgATP binding can be quantitatively monitored by EPR, MCD, and NMR spectroscopies. Also, the binding of MgATP and its hydrolysis are easily followed and the transfer of an electron from the Fe protein to the MoFe protein can be monitored by stopped-flow spectrometry.

Because the Fe protein's MgATP binding site and its Fe_4S_4 cluster are remote from each other, there must be a pathway by which they can communicate with each other (15). The structural model reveals that the MgATP binding site and the Fe_4S_4 cluster are connected via a short helix which includes aspartate-125 and the cluster coordinating cysteine-132 residue. Aspartate-125 is a key residue in the Walker B motif and it appears to be bridged to the Walker A motif lysine-15 residue. This arrangement suggests that the salt bridge is broken upon MgATP binding resulting in a conformational shift that is propagated to cysteine-132 leading to the observed changes in the biophysical properties of the Fe_4S_4 cluster. Such a conformational change could then result in the positioning of the catalytic base, possibly aspartate-129, needed for water assisted MgATP hydrolysis. This model has found support in a variety of amino acid substitution studies (16-20). For example, substitution of aspartate-126 by glutamate-125 results in an Fe protein that undergoes the conformational shift in the presence of either MgADP or MgATP. Substitution of the lysine-15 residue by glutamine-15 results in an altered Fe protein that cannot undergo the nucleotide induced conformational change. Substitution of glutamate for the aspartate-129 residue, which is located between lysine-15 and aspartate-125 results in an altered Fe protein that could undergo the MgATP induced conformation change and could still interact with the MoFe protein, but cannot effect primary electron transfer nor support MoFe protein-stimulated MgATP hydrolysis. These features support the proposed role for aspartate-129 as the primary catalytic base that relays the signal from the MoFe protein (unknown at this time) to effect MgATP hydrolysis. A final compelling piece of evidence that the MgATP site and the Fe protein's Fe_4S_4 cluster communicate through the proposed loop was provided by an elegant experiment recently reported by the Seefeldt group. They constructed a mutant strain where leucine-127 was deleted. This altered Fe protein exhibited properties strikingly similar to the MgATP bound form of the normal Fe protein, even in the absence of nucleotide (Biochemistry, in Press). Of particular interest is that this altered Fe protein forms a stable complex when mixed with the MoFe protein and remains able to transfer an electron to the MoFe protein even in the absence of nucleotides.

Summary Comment Over the past twenty years the biochemical-genetic approach has provided a great deal of insight concerning the assembly of nitrogenase and the mechanism for nitrogenase catalysis. However, fundamental questions remain. For example, we know very little about the details of where and how the substrate binds to the active site. We know very little about the dynamics of nitrogenase catalysis nor do we know very much about the chemical details of metallocluster assembly. In all of these cases we know quite a bit about who the players are and how they fit into their respective schemes but we don't know the details. The fact that altered forms of the Fe protein, as well as the use of nucleotide analogs (21, 22), have recently been used to successfully trap "transition state" intermediates in nitrogenase catalysis leads us to believe that further application of the biochemical-genetic strategy will continue to yield new insights.

Acknowledgement This work was supported by a grant from the National Science Foundation. Jeverson Frazzon was supported by the Brasilian National Research Council.

1.   Burris, R. H. (1991) J. Biol. Chem.. 266: 9339-9342.
2.   Peters, J. W., Fisher, K. & Dean, D. R. (1995) Ann. Rev. Micro. 49: 335-366.
3.   Roberts, G.P. & Brill, W. J. (1980) J. Bacteriol. 144: 210-216.
4.   Arnold, W., Rump, A., Klipp, W, Priefer, U. B. & Puhler, A. (1988) J. Mol. Bio. 203: 715-738.
5.   Shah, V. K., Imperial, R. A., Ugalde, R., Ludden, P. W. & Brill, W. J. (1986) Proc. Natl. Acad. Sci. USA 83: 1636-1640.
6.   Muchmore, S. W. Jack, R. F. & Dean, D. R. (1995) in Mechanism of Metallocluster Assembly pp. 111-133.
7.   Zheng, L., White, R. H., Cash, V. L., Jack, R. F. & Dean, D. R. (1993) Proc. Natl. Acad. Sci. USA 90: 2754-2758.
8.   Howard, J. B. & Rees, D. C. (1994) Ann. Rev. Biochem. 63: 235-264.
9.   Pope, M. R., Murrell, S. A. & Ludden, P. W. (1985) Proc. Natl. Acad. Sci. USA 82: 3173-3177.

10. Wolle, D., Kim, C.-H., Dean, D. R. & Howard, J. B. (1992) J. Biol. Chem. 267: 3667-3673.
11. Seefeldt, L. C. (1994) Protein Science 3: 2073-2081.
12. Dean, D. R., Setterquist, R. A., Brigle, K. E., Scott, D. J., Laird, N. F. & Newton, W. E. (1990) Molec. Microbiol. 4:1505-1512.
13. Thorneley, R. N. F. & Lowe, D. J. (1983) Biochem. J. 215: 393-403.
14. Robson, R. L. (1984) FEBS Lett. 173: 394-398.
15. Geogiadis, M. M., Komiya, Chakrabarti, P., Woo, D., Kornuc, J. J. & Rees, D. C. (1992) Science 257: 1653-1659.
16. Wolle, D., Dean, D. R., & Howard, J. B. (1992) Science 258: 992-995.
17. Seefeldt, L. C., Morgan, T. V., Dean, D. R. & Mortenson, L. E. (1992) J. Biol. Chem. 267: 6680-6688.
18. Lanzilotta, W. N. Ryle, M. J. & Seefeldt, L. C. (1995) Biochemistry 34: 10713-10723.
19. Lanzilotta, W. N., Fisher, K. & Seefeldt, L. C. (1995) Biochemistry 35: 7188-7196.
20. Ryle, M. J., Lanzilotta, W. N., Mortenson, L. E., Watt, G. D. & Seefeldt, L. C. (1995) J. Biol. chem. 270: 13112-13117.
21. Duyvis, M. G., Wassink, H. & Haaker, H. (1996) FEBS. Lett. 380: 233-236.
22. Renner, K. A. & Howard, J. B. (1996) Biochemistry 35: 5353-5358.

Molybdenum-Nitrogenase: Structure and Function

William E. Newton

Department of Biochemistry, Virginia Polytechnic Institute and State University, Blacksburg, VA 24061-0308, USA

Molybdenum (Mo) was believed, as early as 1930, to be absolutely essential for nitrogen fixation, even though vanadium (V) was found to be almost as stimulatory to bacterial growth on N_2 (1). Recently, however, it has been shown that biological nitrogen fixation can occur without Mo. Nitrogenase is now known to exist in three genetically distinct forms; the conventional Mo-based system (Mo-nitrogenase) and the two alternative systems, V-nitrogenase and nitrogenase-3 (2-4). In general, the enzymes from all N_2-fixing bacteria are very similar, comprising two component proteins. For Mo-nitrogenase, these are the MoFe protein and the Fe protein (5), neither of which has activity alone. The deduced amino-acid sequences of the subunits of the MoFe, VFe and FeFe proteins show significant identity with one another and the conservation of both the domain structures around the eight strictly conserved Cys residues and the spacing between them indicates that all three protein types will have the same general structural features. All N_2-fixing species examined so far have Mo-nitrogenase. In some species, it occurs alone, while in others it is found in all permutations with the V-based and third systems. It is not at all clear why each species has its own combination of these nitrogenases (6). This paper reviews only studies of the Mo-based nitrogenase system.

1. Requirements for Catalysis and Substrate Reduction.

All nitrogenases require MgATP, a low-potential reductant and an anaerobic environment for catalytic activity. Under these conditions, a variety of substrates (acetylene, nitrous oxide, azide and HCN), in addition to the physiological substrate, N_2, can be reduced. During catalysis, N_2 is reduced to NH_3 plus MgADP (adenosine diphosphate) and phosphate. Because MgADP is an inhibitor of nitrogenase catalysis by competing for the MgATP-binding sites on the Fe protein, *in vitro* assays recycle MgADP to MgATP. Reductants include ferredoxins and flavodoxins *in vivo* and sodium dithionite $(Na_2S_2O_4)$ *in vitro*. The optimal stoichiometry for nitrogenase function is four moles MgATP hydrolyzed for each pair of electrons transferred to substrate, this ratio is independent of the substrate reduced.

$$N_2 + 16\ MgATP^{2-} + 4\ S_2O_4^{2-} + 24\ H_2O \xrightarrow{\text{nitrogenase}}$$

$$2\ NH_3(aq) + H_2 + 16\ MgADP^- + 16\ HPO_4^{2-} + 8\ HSO_3^- + 16\ H^+$$

With the notable exception of *C. pasteurianum*, the component proteins from all Mo-based nitrogenases interact as heterologous crosses to form catalytically active

NATO ASI Series, Vol. G 39
Biological Fixation of Nitrogen
for Ecology and Sustainable Agriculture
Edited by A. Legocki, H. Bothe, A. Pühler
© Springer-Verlag Berlin Heidelberg 1997

enzymes (7). Carbon monoxide (CO) is a potent inhibitor of all nitrogenase-catalyzed substrate reductions with the exception of H^+ reduction (8) and H_2 is a specific competitive inhibitor of N_2 reduction, affecting neither reduction of any other substrate nor its own evolution (9).

Substrate reduction is accomplished by sequential association and dissociation of the two proteins and, during each cycle, two molecules of MgATP are hydrolyzed and a single electron is transferred from the Fe protein to the MoFe protein (10,11) with the dissociation step being rate limiting *in* vitro at about 6 sec^{-1} (11). Although the kinetics of all the partial reactions have been measured precisely, only recently is information being accumulated about the physical details of the mechanism, including how the two proteins interact, the pathway of electron flow within the MoFe protein, the nature of possible reaction intermediates, etc. The recently reported structures of the *Azotobacter* Fe protein (12) and MoFe protein (13) have spawned computer-generated docking models for the possible protein-protein contacts. Furthermore, mutagenesis and cross-linking studies have implicated Arg-101 (14) and Glu-113 (15) of the Fe protein and Lys-399 of the beta-subunit (15) and Asp-161 of the alpha-subunit (16) of the MoFe protein in Fe protein-MoFe protein docking.

MgATP binding causes an, as yet, uncharacterized conformational change in Fe protein that affects its redox potential, the shape of its electron paramagnetic resonance (EPR) signal (17) and the accessibility of the iron (18) in the [4Fe-4S] cluster. Further, MgATP hydrolysis occurs only when the Fe protein and MoFe protein are complexed. These observations, together with details from the crystal structure, which shows well-separated binding sites for the [4Fe-4S] cluster and the nucleotide, are the basis of an hypothesis by which MgATP binding and hydrolysis coordinates and regulates the uni-directional flow of electrons to the MoFe protein (19).

2. Structure of the MoFe Protein.

The recently available three-dimensional structure of the MoFe protein shows a central core consisting of both beta-subunits, outside of which lie the two alpha-subunits, which do not contact each other. The structure confirms the presence of two types of prosthetic group, the P-cluster and the FeMo-cofactor, each of which contains about 50% of the Fe and S^{2-} content. These clusters are distributed in pairs, which are separated by about 70A, *i.e.*, they are on opposing sides of the protein. Each P-cluster symmetrically bridges an alpha/beta-subunit interface, whereas one FeMo-cofactor is ensconsed within each alpha-subunit.

The P-cluster has been modeled in several ways - most often as two [4Fe-4S] cubes bound to each other through two cysteinyl bridges (from Cys-88 of the alpha-subunit and Cys-95 of the beta-subunit) and a disulfide bond with only two terminal Cys-Fe interactions each (13). This structure confirms the predictions made from primary sequence conservation and extrusion characteristics (20). Each FeMo-cofactor, which consists of one Mo atom, seven Fe atoms, nine sulfides and a homocitrate molecule, is the origin of the biologically unique, characteristic S=3/2 EPR spectrum of the MoFe protein. The crystal structure shows that the FeMo-cofactor is bound to the alpha-subunit polypeptide by only two direct bonds, through alpha-Cys-275 and alpha-His-442. In addition, putative hydrogen-bonding interactions occur through alpha-subunit residues, Gln-191 and Gln-440 (both to the homocitrate), Arg-359, Arg-96 and His-195. Many of these interactions were predicted through our mutagenesis studies (see below), which predated the structure determination.

3. Mutagenesis Studies.

Although the FeMo-cofactor has been implicated in substrate reduction, efforts to reduce substrates using the FeMo-cofactor extracted from the MoFe protein have been mostly equivocal. Thus, the FeMo-cofactor's polypeptide environment must play a vital role(s). Potential roles include provision of proton- or electron-transfer pathways, contributions to the electronic features of the substrate-reduction site, and proper orientation of FeMo-cofactor. These roles can be probed through site-directed amino-acid substitutions within the FeMo-cofactor's polypeptide environment and monitoring their disruption of catalytic and spectroscopic properties.

3.1. Substitutions in the FeMo-cofactor Environment. The original site-directed mutagenesis of the *A. vinelandii nifD* gene produced strong confirmatory evidence that the FeMo-cofactor is the site of substrate reduction because a *single* amino-acid substitution at either alpha-Gln-191 (by Lys or Glu) or alpha-His-195 (by Asn) resulted in simultaneous changes in *both* substrate specificity (*e.g.*, no N_2 fixation, low H_2 evolution, and both C_2H_4 and C_2H_6 produced from C_2H_2) and EPR spectrum (21). Moreover, substitution at alpha-His-195 by Gln (to give H195Q MoFe protein), while also eliminating N_2 fixation, produces a quite dissimilar phenotype (22). The H195Q-MoFe protein has almost wild-type levels of H_2 evolution and, like wild type, reduces C_2H_2 to C_2H_4 only. Both reactions are potently inhibited by N_2, even though it is not a substrate, resulting in the uncoupling of MgATP hydrolysis from substrate reduction without altering the rate of MgATP hydrolysis. This inhibition by N_2 of both H_2 evolution and C_2H_2 reduction catalyzed by the H195Q-MoFe protein suggests occupancy of a common active site.

The H195Q-MoFe protein has a "wild-type" K_m for C_2H_2 reduction, but this reduction is hypersensitive to inhibition by CO (22). CO has been reported to relieve the 50% inhibition by CN^- of total electron flux suffered by wild type when reducing HCN at pH 7.4 completely, thereby implying a common binding site (Li et al., 1982). Quite surprisingly, the H195Q-MoFe protein showed no inhibition of electron flux with 5 mM NaCN, which also had no effect on the $ATP/2e^-$ ratio.

Thus, these data suggest that both N_2 and CN^- have binding sites on the central face of FeMo-cofactor, which is involved with His-195, and that CN^- and CO do not share exactly the same binding site. Because N_2 binding inhibits C_2H_2 and H^+ reduction and because the affinity for CO is increased in the H195Q-MoFe protein, it is likely that these molecules also have close-by binding sites that may overlap. Moreover, for each of these three substrates, there may well be more than one binding site. In fact, the altered MoFe protein with alpha-Arg-277 substituted by His shows inhibitor (CO)-induced cooperativity ($n = 2$) during C_2H_2 reduction, indicating at least three simultaneously-occupiable binding sites on the MoFe protein. Furthermore, altered MoFe proteins, which result from substitution by Lys at Arg-359 or Arg-96, which hydrogen-bond with the Mo-Fe-S subcluster, suffer CO-sensitive H_2 evolution, but only by ca. 50%. However, these substitutions have no effect on the K_m's for N_2 and C_2H_2. Thus, two H_2-evolution sites are indicated.

3.2. Substitutions in the P-Cluster Environment. The obvious structural symmetry surrounding each P-cluster pair is not reflected by functional symmetry. Of the four terminally ligating Cys residues, substitution at only beta-Cys-153 is tolerated (23), even though alpha-Cys-154 occupies an almost equivalent position on the other $[Fe_4S_4]$ sub-cluster. Possibly, the location of beta-Cys-153 furthest away from FeMo-cofactor indicates a less structurally and functionally demanding role in catalysis. Also, ligation of the same Fe atom by beta-Ser-188 might be important. In the three P-cluster structural models (proposed from x-ray data), the structure around the alpha subunit-ligated [4Fe-4S] sub-cluster remains invariant; all structural variations occur around the beta-subunit subcluster, the one that also tolerates amino-acid substitutions.

Both of the bridging ligands, alpha-Cys-88 and beta-Cys-95, are part of prosthetic group-spanning helices and, therefore, of potential electron-transfer pathways. However, only substitutions at Cys-88 retain significant activities (20). Again, there is no obvious structural difference among them except that Cys-88 is further from FeMo-cofactor and, therefore, may play a less demanding catalytic role. Spectroscopic and catalytic probes indicate that the FeMo-cofactors of the Cys-88-substituted MoFe proteins are uncompromised, but the properties of their P-clusters have changed. In the dithionite-reduced state, the C88G- and C88T-MoFe proteins have P-clusters in their one-electron oxidized redox state. This change in redox properties is consistent with the changed electron-transfer and protein-protein interaction parameters of the altered MoFe proteins that contain them, supporting the concept that the P-cluster is the likely recipient of electrons from the Fe protein.

4. Acknowlegdements.

Thanks to J. Cantwell, D. Dean, K. Fisher, C.-H. Kim, J. Shen and K. Thrasher for studies supported by NIH (DK-37255) and to J. Bolin, M.K. Johnson and D.C. Rees.

5. References.

1. H. Bortels, *Arch. Mikrobiol.* **1**, 333 (1930); *Zentr. Bakt. Parasitenk. Abt. II* **87**, 476 (1933).
2. P.E. Bishop *et al.*, *PNAS, USA* **77**, 7342 (1980); *Science* **232**, 92 (1986).
3. R.L. Robson *et al.*, *Nature (London)* **322**, 388 (1986).
4. R.N. Pau, *TIBS* **14**, 183 (1989).
5. W.A. Bulen, J.R. LeComte, *Proc. Natl. Acad. Sci. USA* **56**, 979 (1966).
6. W.E. Newton in *New Horizons in Nitrogen Fixation* (R. Palacios, J. Mora, W.E. Newton, eds.) Kluwer, Dordrecht, 1993, p. 5.
7. D.W. Emerich, R.H. Burris, *J. Bacteriol.* **134**, 936 (1978).
8. R.W.F. Hardy, *et al.*, *Biochem. Biophys. Res. Commun.* **20**, 539 (1965).
9. J.L. Hwang, C.H. Chen, R.H. Burris, *Biochim. Biophys. Acta* **292**, 256 (1973).
10. R.N.F. Thorneley, D.J. Lowe, *Biochem. J.* **224**, 887 (1984).
11. R.V. Hageman, R.H. Burris, *Biochemistry* **17**, 4117 (1978).
12. M.M. Georgiadis *et al.*, *Science* **257**, 1653 (1992).
13. C. Kim, D.C. Rees, *Science* **257**, 1677 (1992).
14. R.G. Lowery *et al.*, *Biochemistry* **28**, 1206 (1989).
15. A. Willing, J.B. Howard, *J. Biol. Chem.* **265**, 6596 (1990).
16. C.-H. Kim, L. Zheng, W.E. Newton, D.R. Dean in ref. **6**, p. 105.
17. W. Zumft, G. Palmer, L. Mortenson, *Biochim. Biophys. Acta* **292**, 413 (1973).
18. G.A. Walker, L.E. Mortenson, *Biochemistry* **13**, 2382 (1974).
19. D. Wolle, D.R. Dean, J.B. Howard, *Science* **258**, 992 (1992).
20. D.R. Dean *et al.*, *Molec. Microbiol.* **4**, 1505 (1990).
21. D.J. Scott *et al.*, *Nature (London)* **343**, 188 (1990).
22. C.-H. Kim, W.E. Newton, D.R. Dean, *Biochemistry* **34**, 2798 (1995).
23. H.D. May, D.R. Dean and W.E. Newton, *Biochem. J.* **277**, 457 (1991).

Part 2

Signal Perception, Transduction and Cell Cycle Genes in Nodulation

Rhizobium Nodulation Factors in Perspective

F. Debellé[1], G. P. Yang[1], M. Ferro[2], G. Truchet[1], J. C. Promé[2] and J. Dénarié[1]

[1]Laboratoire de Biologie Moléculaire des Relations Plantes-Microorganismes, CNRS-INRA, B.P.27, 31326 Castanet-Tolosan Cedex, France

[2]Institut de Pharmacologie et de Biologie Structurale, CNRS, 205 route de Narbonne, 31077 Toulouse Cedex, France

Since the discovery of Nod factors (NFs) six years ago (Lerouge et al 1990), many studies have concentrated on the role of the various rhizobial *nod* genes in the control of NF synthesis, the diversity of NF structures, and the various types of responses that NFs elicit in plants. As recent reviews have been published in this field (Schultze et al., 1994; Spaink, 1995; Fellay et al., 1995; van Rhijn and Vanderleyden, 1995 ; Dénarié et al., 1996), we will simply discuss in this brief chapter a few topics which are of particular interest to us either because they have led to a better understanding of the molecular basis of host specificity or because they open important perspectives. Due to tight space limitation, we have referred preferentially to reviews, rather than to the original papers.

1 Beyond the apparent diversity of rhizobium-legume interactions, the pivotal role of the rhizobium *nodABC* genes

The family *Leguminosae* is one of the largest families of flowering plants, including more than 18,000 species. This family is very diverse, ranging from forest giants to tiny ephemerals, and extends to all terrestrial habitats (Young and Johnston, 1989). The symbiotic bacterial partners are also very diverse and do not form a discrete clade. Although they are collectively referred to as rhizobia, they are now classified in four genera *Rhizobium*, *Bradyrhizobium*, *Azorhizobium* and *Sinorhizobium*, that are quite distinct (Martinez-Romero and Caballero-Mellado, 1996). Some rhizobia are indeed more closely related to nonsymbiotic bacteria than they are to other rhizobia. Diversity exists in the mechanism of infection and in the anatomy of nodules, and these traits are controlled by the plant, suggesting that the role of bacteria is to activate the plant genetic programme by way of signals. Rhizobium-plant associations are specific, as a given bacterium can only nodulate a defined number of plants.

It is the genetic analysis of the rhizobial ability to infect and nodulate legumes, in a specific manner, which has led to the discovery of Nod factors (NFs). Interestingly, this analysis has revealed that only a limited number of nodulation genes are common to all rhizobial species, the regulatory *nodD* and the structural *nodABC* genes (Schultze et al., 1994). The *nodABC* genes are pivotal, since their inactivation results in the complete loss of the ability to elicit infection and nodulation, in all the systems studied.

NATO ASI Series, Vol. G 39
Biological Fixation of Nitrogen
for Ecology and Sustainable Agriculture
Edited by A. Legocki, H. Bothe, A. Pühler
© Springer-Verlag Berlin Heidelberg 1997

2 *R. meliloti nod* genes and Nod factors, and the molecular basis of specific nodulation

In *Rhizobium meliloti*, we have shown that *nodABC* genes are required for the synthesis of diffusible signal molecules, the Nod factors (Nod because their synthesis is under the control of *nodABC* genes). *R. meliloti* NFs were found to be tetrasaccharides or pentasaccharides of N-acetyl glucosamine linked in ß1-4 (= chitin oligomers), with three substitutions: the reducing sugar bears a sulphate group and the terminal non-reducing sugar is O-acetylated on the carbon 6 and N-acylated by a C16 fatty acyl chain with two double bonds in positions 2 and 9. Each substitution is under the control of specific *nod* genes and is important for infection and nodulation of alfalfa (Schultze et al., 1994; Dénarié et al., 1996). The host range gene *nodH* codes for a sulfotransferase which allows NF sulfation. Mutants in *nodH* produce non-sulfated NFs, are unable to infect and nodulate alfalfa, but have gained the ability to infect and nodulate vetch, a non-host legume. *nodFE* mutants produce NFs that are acylated by a trivial cellular fatty acid, vaccenic acid, instead of the specific poly-unsaturated C16 ones. A change in the structure of the N-acyl chain results in a strongly reduced ability to penetrate into the plant, and to elicit the formation of infection threads. *nodL* mutants, which produce NFs that are not O-acetylated, have a phenotype similar to that of *nodFE* mutants. Double *nodF/nodL* mutants are completely unable to infect and nodulate alfalfa. Thus, the three substitutions of NFs are essential for efficient infection and nodulation of alfalfa.

3 Chemical diversity of Nod factors from various rhizobial species: Variations on a theme

NFs have been purified from various rhizobial species (Dénarié et al., 1996). They all belong to the same chemical family: they are mono-N-acylated chitin oligomers, mostly pentamers and tetramers. The chitin oligomer backbone, however, is diversely substituted in the various species, on the two terminal glucosamine residues.

At the non-reducing end, two rhizobial strategies for N-substitutions have been identified. In the more common strategy, the N atom is acylated by vaccenic acid, and is also frequently methylated (by the NodS protein). Other substitutions at the non-reducing end involve carbamoylation, frequently at position 6 (by NodU) (Carlson et al., 1994; Fellay et al., 1995). This strategy, that we propose to call the *nodSU* strategy, does not leave much room for structural diversification (and thus for specific recognition) at the non-reducing end.

Another strategy, involving the host range *nodFE* genes, is used by some fast-growing rhizobia (Spaink, 1995). NodFE proteins specify the synthesis of particular polyunsaturated fatty acids. This strategy has made possible a diversification of the N-acyl moiety. For example the proteins NodFE specify the synthesis of C16:2 and C16:3 fatty acid chains in *R. meliloti*, C18:4 in *R. leguminosarum* bv. *viciae* and C20:3 and C20:4 in *R. leguminosarum* bv. *trifolii*. Poly-unsaturated fatty acids have also been found in NFs from *R. huakuii* and *R. galegae* (G.P. Yang and M. Ferro, personal communication). *nodE* genes are important host range genes indicating that subtle variations in the structure of the acyl chain are recognized by plant hosts. It has recently been found that NodA are specific acyl-transferases. For example, in contrast to *R. meliloti* NodA, the *R. tropici* NodA is unable to transfer the C16 unsaturated

Structure of Nod factors from various rhizobial species

Host plant[a]	Rhizobial species	R1	R2[d]	R3	R4	R5	n[c]
Medicago	R. meliloti	H	C16:2, C16:3 C18-C26(ω-1)OH	Ac(O-6),H	S	H	1,2,3
Astragalus	R. huakuii	H	C18:4	H	S	H	**3**
Vicia	R. l. bv viciae	H	C18:1,C18:4	Ac(O-6),H	H	H	2,3
Pisum cv. Afghanistan	R. l. bv viciae TOM	H	C18:1,C18:4	Ac(O-6)	Ac	H	2,3
Trifolium	R. l. bv trifolii	H	C18:1, C18:3 C20:3, C20:4	Ac(O-6)	H	H	1,2,3
Lotus	R. loti	Me	C18:1	Cb(O-4)	AcFuc	H	3
Phaseolus	R. etli	Me	C18:1	Cb(O-4),H	AcFuc	H	3
	R. tropici	Me	C18:1	H	S,H	H	3
Acacia	R. sp GRH2	Me,H	C18:1	H	S,H	H	2,3,4
	S. teranga	Me	C18:1	Cb	S,H	H	3
Lablab	R. sp NGR234	Me	C18:1	Cb(O-6 and O-3 or O-4), H	MeFuc AcMeFuc MeSFuc	H	3
Glycine	R. fredii	H	C18:1	H	MeFuc,Fuc	H	**1,2,3**
	B. japonicum	H	C18:1	H	MeFuc	H	3
	B. elkanii	H,Me	C18:1	Ac(O-6),H,Cb	MeFuc,Fuc	H	2,3
Sesbania	A. caulinodans	Me	C18:1	Cb(O-6)	Fuc, H	Ara	2,3
	S. saheli	Me	C18:1	Cb	Fuc, H	Ara	2,3

[a] Plant genera from which rhizobial strains were isolated
[b] R6=H in all strains except B. elkanii where R5=H, glycerol
[c] The bold numbers indicate the number of glucosamine residues of the most abundant Nod factors
[d] Selected fatty acyl substituents
Me=methyl; Ac=acetate; S=sulfate; Cb=carbamate; Fuc=fucose; Ara=arabinose

and the (ω-1) hydroxylated fatty acids onto the chitooligosaccharide backbone (Debellé et al., 1996). It thus appears that in each "*nodFE*" rhizobium species *nodA* has coevolved with the corresponding *nodFE* genes to ensure the transfer of the specific fatty acids. In fact, the "common" *nodA* genes appear to be host-range determinants in these bacteria.

Another "common" gene, *nodC*, controls the length of the chito-oligosaccharide backbone (chitin pentamers or tetramers) and is also a host-range determinant, suggesting that this structural feature is also important for recognition (Roche et al., 1996).

Substitutions at the reducing end are more or less varied. The "*nodFE*" rhizobia have only sulfate or acetate groups. In contrast, the "*nodSU*" rhizobia can exhibit sugar substitutions such as fucosyl or/and arabinosyl, and these sugars can in turn be substituted by small groups such as sulfate or acetate, enabling a large number of variations (Carlson et al., 1994; Lorquin et al., 1996; Mergaert et al., 1996). It thus appears that the "*nodSU*" bacteria mostly exhibit variations at the reducing end by sugar O-substitutions, whereas "*nodFE*" bacteria exhibit more diversification at the non-reducing end, by N-acyl substitutions.

4 NF structure is related to the taxonomy of the plant host, not to that of the bacterium

The first indication that NF structure is more related to the plant, the signal receiver, than to the rhizobia, the signal emitter, came from the study of soybean-nodulating bacteria, *B. japonicum* and *S. fredii*. Although these rhizobia are quite distant phylogenetically, they both produce similar major NFs that are fucosylated and not sulfated (Carlson et al., 1994).

More evidence has recently been obtained from the study of NFs from *Azorhizobium caulinodans* and *Sinorhizobium saheli*, rhizobia which nodulate *Sesbania rostrata*, an aquatic legume with nodules formed on both roots and stems. These rhizobia are also quite distant phylogenetically but both produce similar (and peculiar) major NFs that are doubly substituted at the reducing end, by fucose and arabinose (Lorquin et al., 1996; Mergaert et al., 1996).

Another example concerns substitutions at the other end of the molecule, the non-reducing end. As stated above, a number of rhizobial species (the "*nodFE* "rhizobia) produce NFs that are N-acylated with poly-unsaturated fatty acids which vary with the rhizobial species or biovar, that is with the host range. These species are distributed in all groups of fast-growing rhizobia, *Rhizobium* stricto sensu, *Sinorhizobium*, the *R. loti* cluster, and *R. galegae* which is closely related to *Agrobacterium*. No correlation was found between the production of polyunsaturated fatty acids and the bacterial taxonomical position. In contrast, examination of a phylogenetic tree of *Leguminosae* (Young and Johnston, 1989) reveals that all these rhizobia nodulate plants which are members of related tribes. *R. huakuii* and *R. galegae* nodulate *Astragalus* and *Galega* respectively, and these two plant genera are members of the *Galegeae* tribe. *R. meliloti* and *R. l.* bv. *trifolii* nodulate *Medicago* and *Trifolium* respectively, members of the *Trifolieae* tribe, and *R. l.* bv. *viciae* nodulate *Pisum* and *Vicia* of the *Vicieae* tribes. *Trifolieae*, *Vicieae* and *Galegeae* tribes are supposed to derive from a common galegoid ancestor. No polyunsaturated fatty acids have been found in NFs from rhizobia that nodulate plants belonging to all the other legume tribes not deriving from the galegoid complex.

Interestingly, *R. loti*, which nodulates plants of the *Loteae* tribe, is a "*nodSU*" rhizobium with vaccenic acid N-acylation. *Loteae* are believed to derive from a galegoid ancestor but recent phylogenetic studies suggest that they may have diverged early from the phylum which led to other tribes such as *Trifolieae* and *Vicieae*. The study of NFs from rhizobia nodulating plants belonging to the tribes mentionned above and to the related *Coronillieae*, *Hedysareae* and *Cicereae* tribes could clarify the phylogenetic relationships of these legumes.

Interpretting these results from an evolutionary point of view, we can hypothesize that most extant legumes, as archaic legumes, possess NF recognition mechanisms (receptors?) which do not involve a specific recognition of the N-acyl moiety. Once in legume evolution, presumably in the *Galegeae* ancestors, appeared a new mechanism (a new receptor type?) allowing the recognition of specific features of the lipid moiety such as the presence of conjugated double bonds. This "mutation" allowed a diversification in the recognition of fatty acids enabling variations in the number of double bonds and in the carbon chain lengths. Co-evolution between symbiotic partners was made possible by the allelic variation of rhizobial *nodFE* genes and allelic variation of the corresponding plant NF receptors. In the future, we expect that the study of *nodFE* and *nodA* genes and of the NFs they specify on the one hand, and the cloning and characterisation of the NF receptors of the homologous plants on the other hand, will provide a strong basis for the analysis of the molecular basis of co-evolution between the two partners. It is very likely that strong selection pressure is exerted at the level of the interaction between ligands (NFs) and receptors to allow bacteria to penetrate into a specific host.

Whereas the presence of this new type of NF fatty acyl recognition mechanism seems monophyletic in plants, the corresponding *nodFE* strategy is widespread in fast growing rhizobia suggesting that horizontal transfer has occurred among rhizobial soil populations. Obviously the two partners do not evolve at the same rate, and when one plant genotype with a given type of NF receptor(s) is introduced in a soil, it is likely that if the adequate rhizobia, producing the appropriate NFs (keys) are not present, horizontal genetic transfer occurring among soil rhizobia will generate new combinations of *nod* genes and thus the production of various NFs. Bacteria possessing the appropriate key will enter the legume and eventually multiply (Relic et al., 1994).

5 Nod factors as tools to characterise rhizobial populations

To address questions concerning for example rhizobium population dynamics and genetics, and gene transfer in the soil, ecologists have to characterize large numbers of field isolates. Modern identification and classification of strains use techniques of nucleic acid characterisation, including DNA hybridization and 16S sequencing, allowing strains to be assigned a precise taxonomical position. These approaches are presently revealing the great degree of phylogenetic diversity among nodule-forming soil bacteria (Martinez-Romero and Caballero-Mellado, 1996).

However, for plant-associated microorganisms, such as rhizobia, it is also essential to characterise, at least to some degree, their symbiotic properties. This requires testing numerous strains on various plant hosts, which is space consuming (in greenhouses) and very laborious. In addition, the choice of putative hosts is always limited, often arbitrary, and results are difficult to reproduce because the symbiotic behaviour frequently depends on growth media and culture conditions. Recent results suggest that the use of simplified methods for studying NFs could help in characterising plant-associated traits of rhizobia.

In Senegal, *Sinorhizobium teranga* isolates have two mutually exclusive host-ranges. Some isolates nodulate *Acacia* and some isolates nodulate *Sesbania*, which is also nodulated by *Sinorhizobium saheli*. It was found that the introduction of the *nodD1* gene of *R.* sp. NGR234 into most rhizobial strains resulted in an increased production of NFs that made possible their detection by thin layer chromatography (TLC) after growing rhizobia in the presence of radioactive acetate or sulfate (Lortet et al., 1996). This revealed that a number of *S. saheli* isolates that nodulate *Sesbania* have a particular NF profile, similar to the *A. caulinodans* NF profile, with no detectable sulfated *nod*-dependent compounds. *S. teranga* isolates nodulating *Sesbania* have the same TLC profile. In contrast, *S. teranga* strains nodulating *Acacia* had a different TLC profile showing the production of sulfated compounds. Further chemical studies showed that the NFs of *Acacia*-nodulating strains had structures similar to those of NFs produced by *R. tropici* strains that have a broad host range and nodulate beans and *Leucaena*. In contrast, *Sesbania*-nodulating strains had NFs similar to those of *A. caulinodans* and *S. saheli* (Lorquin et al., 1996). This is further evidence that NF structures correlate with host-range and not with the taxonomical position of the bacterium. Future research should aim at improving procedures of NF TLC identification and characterisation, by devising methods with a better resolving power which are applicable to large numbers of isolates. Cooperation between ecologists and specialists of NFs should be fruitful, for both a better symbiotic characterization of field isolates, and a better assessment of the biodiversity in Nod signalling.

6 Complexity of Nod factor perception and transduction mechanisms: The need for a genetic approach

Purified NFs elicit responses in legume hosts similar to those induced by bacteria themselves in the course of infection and nodulation. Responses are observed in several root tissues, including the epidermis and the cortex, and exhibit different structural requirements (Mylona et al., 1995; Dénarié et al., 1996). These responses will be described in a number of chapters of this book, and here we will just point out observations which suggest a great complexity in the NF perception and transduction mechanisms. A striking feature is the very low concentrations (in the picomolar range) at which NFs trigger responses in epidermal and root hair cells. Responses include a rapid membrane depolarisation and changes in calcium fluxes (Ehrhardt et al., 1996), and the induction of the expression of early nodulin genes (Mylona et al., 1995). The specificity of NFs and the low concentrations at which they are active suggest the existence of high affinity NF receptors.

The study of the symbiotic behaviour of *R. meliloti* mutants that produce modified NF signals suggests the existence of different mechanisms of NF recognition at the root hair level (Dénarié et al., 1996). Double *nodF/nodL* mutants produce NFs that are N-acylated by vaccenic acid instead of the polyunsaturated ones and are not O-acetylated. Such mutants are neither able to initiate the formation of infection threads nor to enter into the plant, indicating the presence of "entry" receptors with stringent structural requirements for NFs. However, these mutants do elicit very striking root hair and epidermal cell wall deformations, and the activation, at a distance, of cortical cells which results in the accumulation of starch granules in plastids. This suggests the presence on epidermal and root hair cells of a second type of NF receptor that has less stringent structural requirements. This model involving more than one type of NF receptor is supported by further genetic evidence. Pea mutants carrying the *sym2* gene cannot be infected and nodulated by *R.l.* bv. *viciae* European strains; however, these strains do elicit root hair deformations on these mutant plants (Kozik et al., 1995; Bisseling et al., in this book).

The double *nodF/nodL* mutant of *R. meliloti* also uncouples plant responses in deeper tissues of the roots. This mutant does not trigger the formation of nodule meristems, but does strongly induce transcription of the *enod40* gene in the pericycle (W.C. Yang, F. de Billy and T. Bisseling, personal communication). This suggests that the induction of these two types of responses has different NF structural requirements, an indication that there is more than one mechanism of NF perception or transduction.

This complexity means that in addition to the present approaches essentially based on cell biology, biochemistry and molecular biology, a genetic approach should be developed to dissect these pathways. Pea nodulation mutants are being used for this purpose (see Bisseling et al., in this book). In addition, an international network is being set up to coordinate and share efforts to develop the small seeded autogamous *Medicago truncatula* as an appropriate system for the genetic dissection of symbiosis, on the plant side. There is no doubt that plant genetics will help decipher NF perception and transduction in the near future.

7 Evolutionary origin of Nod factors: Are rhizobia mimicking fungi or plants?

The discovery of the chemical nature of NFs was a great surprise. The *nod* genes were previously believed to control cell division and the induction of nodule organogenesis by specifying phytohormone synthesis. Why do rhizobia use lipo-chitooligosaccharides to trigger these various plant responses? Are these signalling molecules unique to rhizobia or do rhizobia mimic other biological signals?

In contrast to fungi, bacteria have not been so far reported to synthesize chitin polymers. Would rhizobium NFs be signalling compounds reminiscent of ancient plant-fungal associations? Vesicular arbuscular mycorrhizae (VAM) are the most widespread and probably the most ancient plant-fungus symbiotic associations. They are found in more than 80% of extant plant taxa in pteridophytes, bryophytes, gymnosperms and mono- and dicotyledon angiosperms. Molecular clocks and fossil records suggest that these associations were established when plants first colonized terrestrial ecosystems. Very interestingly a significant proportion of the pea mutants which have been found to be defective for nodulation, are also altered for mycorrhization. Three independent plant genes have been identified that are required for allowing infection by both rhizobia and VA mycorrhizae (Duc et al., 1989). These mutations affect very early stages of both types of symbiotic infections. In contrast, these Nod$^-$/Myc$^-$ mutants are not affected in their interactions with a number of fungal pathogens. Recently, two *M. truncatula* Nod$^-$ mutants have been isolated and they are both also Myc$^-$ (Sagan et al., 1995). From this, it is reasonable to hypothesize that these two types of symbionts share some common early signalling steps. Future studies will tell whether VA mycorrhizae are producing symbiotic signals similar to the rhizobium nodulation signals.

Another hypothesis is that rhizobia trigger plant developmental responses by using signal molecules which mimic plant endogenous growth regulators. Several lines of evidence indicate that lipo-chitooligosaccharides (LCOs) are also able to elicit developmental responses in non-legume plants (Dénarié et al., 1996). The expression of *nodAB* genes in transgenic tobacco plants changes leaf morphology as well as whole plant development, suggesting that substrate(s) for NodB and NodA enzymes, probably chitin oligomers, exist in plants and that their modification influences morphogenesis (Röhrig et al, 1995). Moreover, synthetic LCOs alleviate the requirement for auxin and cytokinin to sustain growth of cultured tobacco protoplasts, and the structure of the fatty acyl moiety was found to be important for biological activity (Röhrig et al., 1995). These data indicate that LCOs are able to act as plant growth regulators in non-legumes. Future studies will tell whether NFs are part of a yet unidentified family of endogenous plant signals.

Acknowledgements

We thank Clare Gough for her careful review of this manuscript. Part of this work was supported by grants from the Human Frontier Science Programme (RG-378-92) and from the European Communities BIOTECH Programme (PTP CT93-0400).

References

Carlson, R.W., Price, N.P.J., and Stacey, G. (1994). The biosynthesis of rhizobial lipo-oligosaccharide nodulation signal molecules. Mol. Plant-Microbe Interact. *7*: 684-695.

Debellé, F., Plazanet, C., Roche, P., Pujol, C., Savagnac, A., Rosenberg, C., Promé, J.C., and Dénarié, J. (1996). The NodA proteins of *Rhizobium meliloti* and *R. tropici* specify the N-acylation of Nod factors by different fatty acids. Mol. Microbiol., in press.

Dénarié, J., Debellé, F., and Promé, J.C. (1996). *Rhizobium* lipo-chitooligosaccharide nodulation factors: Signaling molecules mediating recognition and morphogenesis. Annu. Rev. Biochem. *65:*503-535.

Duc, G., Trouvelot, A., Gianinazzi-Pearson, V., and Gianinazzi, S. (1989). First report of non-mycorrhizal plant mutants (Myc-) obtained in pea (*Pisum sativum* L.) and fababean (*Vicia faba* L.). Plant Sci. *60*:215-222.

Ehrhardt, D.W., Wais, R., and Long, S.R. (1996). Calcium spiking in plant root hairs responding to *Rhizobium* nodulation signals. Cell *85*: 673-681.

Fellay, R., Rochepeau, P., Relic, B., and Broughton, W.J. (1995). Signals to and emanating from *Rhizobium* largely control symbiotic specificity. In Pathogenesis and host specificity in plant diseases. Histopathological, biochemical, genetic and molecular bases. U.S. Singh, R.P. Singh, and K. Kohmoto, eds. (Oxford: Pergamon/Elsevier Science Ltd), pp. 199-220.

Kozik, A., Heidstra, R., Horvath, B., Kulikova, O., Tikhonovich, I., Ellis, T.H.N., Van Kammen, A., Lie, T.A., and Bisseling, T. (1995). Pea lines carrying *sym1* or *sym2* can be nodulated by *Rhizobium* strains containing *nodX*; *sym1* and *sym2* are allelic. Plant Sci. *108*: 41-49.

Lerouge, P., Roche, P., Faucher, C., Maillet, F., Truchet, G., Promé, J.C., and Dénarié, J. (1990). Symbiotic host-specificity of *Rhizobium meliloti* is determined by a sulphated and acylated glucosamine oligosaccharide signal. Nature *344*: 781-784.

Lorquin, J., Lortet, G., Ferro, M., Méar, N., Dreyfus, B., Promé, J.C., and Boivin, C. (1996). Nod factors from *Sinorhizobium saheli* and *S. teranga Sesbania*-nodulating strains are both arabinosylated and fucosylated, submitted.

Lortet, G., Méar, N., Lorquin, J., Dreyfus, B., de Lajudie, P., Rosenberg, C., and Boivin, C. (1996). Nod factor TLC profiling as a tool to characterize symbiotic specificity of rhizobial strains: Application to *Sinorhizobium saheli, S. teranga* and *Rhizobium* sp. strains isolated from *Acacia* and *Sesbania*. Mol. Plant-Microbe Interact., in press

Martinez-Romero, E. and Caballero-Mellado, J.(1996). *Rhizobium* phylogenies and bacterial genetic diversity. Critical Reviews in plant Sciences *15*: 113-140.

Mergaert, P., D'Haeze, W., Fernandez-Lopez, M., Geelen, D., Goethals, K., Promé, J.C., Van Montagu, M., and Holsters, M. (1996). Fucosylation and arabinosylation of Nod factors in *Azorhizobium caulinodans*: involvement of *nolK, nodZ* as well as *noeC* and/or downstream genes. Mol. Microbiol. *21*: 409-419.

Mylona, P., Pawlowski, K., and Bisseling, T. (1995). Symbiotic nitrogen fixation. Plant Cell *7*, 869-885.

Relic, B., Perret, X., Estrada-Garcia, M.T., Kopcinska, J., Golinowski, W., Krishnan, H.B., Pueppke, S.G., and Broughton, W.J. (1994). Nod factors of *Rhizobium* are a key to the legume door. Mol. Microbiol. *13*, 171-178.

Roche, P., Maillet, F., Plazanet, C., Debellé, F., Ferro, M., Truchet, G., Promé, J.C., and Dénarié, J. (1996). The common *nodABC* genes of *R. meliloti* are host range determinants. in press

Röhrig, H., Schmidt, J., Walden, R., Czaja, I., Miklasevics, E., Wieneke, U., Schell, J., and John, M. (1995). Growth of tobacco protoplasts stimulated by synthetic lipo- chitooligosaccharides. Science *269*, 841-843.

Sagan, M., Morandi, D., Tarenghi, E., and Duc, G. (1995). Selection of nodulation and mycorrhizal mutants in the model plant *Medicago truncatula* (Gaertn) after gamma-ray mutagenesis. Plant Sci. *111*, 63-71.

Schultze, M., Kondorosi, E., Ratet, P., Buire, M., and Kondorosi, A. (1994). Cell and molecular biology of *Rhizobium*-Plant interactions. International Review of Cytology *156*, 1-75.

Spaink, H.P. (1995). The molecular basis of infection and nodulation by rhizobia: The ins and outs of sympathogenesis. Annu. Rev. Phytopathol. *33*, 345-368.

Van Rhijn, P., and Vanderleyden, J. (1995). The *Rhizobium*-plant symbiosis. Microbiol. Rev. *59*: 124-142.

Young, J.P.W., and Johnston, A.W.B. (1989). The evolution of specificity in the legume-*Rhizobium* symbiosis. Trends Ecol. Evol. *4*, 341-349.

THE HOST-SPECIFIC ROLE OF CHEMICAL MODIFICATIONS AT THE REDUCING TERMINUS OF LIPO-CHITIN OLIGOSACCHARIDES

A.O. Ovtsyna[1,2], A. Veldhuis[1], I.M. Lopez-Lara[1], A.H.M. Wijfjes[1], C. Quinto[3], R. Geurts[4], .T.Bisseling[4] , D. B. Scott[5], I.A. Tikhonovich[2], B.J.J. Lugtenberg[1], and H.P. Spaink[1*]

[1] Leiden University, Institute of Molecular Plant Sciences, Wassenaarseweg 64, 2333 AL Leiden, The Netherlands
[2] All-Russia Research Institute for Agricultural Microbiology, St. Petersburg, 189620, Russia
[3] Instituto de Biotecnologia, U.N.A.M., Apartado postal 510-3, Cuernavaca Morelos 62271, Mexico.
[4] Wageningen Agricultural University, Department of Molecular Biology, 6703 HA Wageningen, the Netherlands
[5] Molecular Genetics Unit, Massey University, Palmerston North, New Zealand
[*]Author for correspondence. FAX: 3171-5275088
Email: Spaink@Rulsfb.Leidenuniv.Nl

INTRODUCTION

Signal molecules which determine the host specificity of rhizobia for particular leguminous plants include flavonoids secreted by the host plant roots and the bacterial lipo-chitin oligosaccharides (LCOs). This presentation will mainly focuss on the role of the LCOs. For instance, a highly unsaturated fatty moiety of LCOs plays a major role in the specificity of rhizobia which associate with indeterminate-nodulating plants. The genes *nodF* (acyl carrier protein), *nodE* (β-ketoacylsynthase) and *nodA* (acyl transferase) are involved in the production of LCOs containing such a special fatty acids. For determinate nodulating plants a fucosyl residue (located at the reducing *N*-acetylglucosamine residue of LCOs) plays a major role in determining specificity. The *nodZ* gene which is present in various rhizobia such as *Bradyrhizobium japonicum*, *Rhizobium loti* and *R.etli*, is a fucosyltransferase involved in the addition of this fucose (Lopez-Lara *et al*, 1996). The role of other determinants of host specificity, such as *nodX* and *nolL* will be discussed.

THE *NODX* GENE

Firmin *et al* (1993) have shown that the *NodX* gene of the *R.leguminosarum* biovar viciae strain TOM is involved in the acetylation of LCOs. The position of the acetyl group of the LCOs of this strain was determined to be at the C6 position of the reducing *N*-acetylglucosamine unit. The presence of this additional acetyl substituent

NATO ASI Series, Vol. G 39
Biological Fixation of Nitrogen
for Ecology and Sustainable Agriculture
Edited by A. Legocki, H. Bothe, A. Pühler
© Springer-Verlag Berlin Heidelberg 1997

appears to be very important for rhizobia to be able to efficiently nodulate various primitive pea cultivars such as cultivar Afganistan. Two allelic loci which are called *Sym1* and *sym2* have been studied in detail by the group of Bisseling (e.g. see Kozik, 1996). The presence of these loci in pea introgression lines is completely correlated with the restricted host range for strains which contain the *nodX* gene. A gene-for-gene correlation has therefore been proposed. In this paper we present experiments which were aimed at testing the specificity of the additional acetyl group for nodulation of the pea line Afghanistan. For this purpose an isogenic set of derivatives of *R.leguminosarum* biovar viciae strain 248 was produced which differs only in the gene present on a broad host range plasmid cloning vector. In the case that the *nodX* gene from the strain TOM was introduced the LCOs from the resulting strain were shown to be different by thin layer chromatography analysis. LCOs produced by the *nodX*-containing 248 derivative shows the presence of additional LCO species which probably contain an additional acetyl group. The *nodX*-containing strain was able to nodulate efficiently on two different Afghanistan pea cultivars. Since in the absence of the *nodX* gene no nodulation was observed this result confirms the important role of the additional acetyl moiety. We have also tried the effect of the presence of genes which are involved in other modifications of the LCO backbone. Surprizingly, the presence of the *nodZ* gene in strain 248 also led to efficient nodulation of the tested Afghan pea cultivars. This leads to the novel conclusion that recognition Afghan pea cultivars of modifications at the non-reducing terminal saccharide moiety is apparently not very specific. This result could be indicative for a role of the additonal acetyl group in protection against breakdown or scavenging by gene-products encoded by the *sym1/2* loci.

THE *NolL* GENE

The nucleotide sequence of the *R.loti NolL* gene was recently reported by Scott *et al.* (1996) and was shown to be homologous to the *nodX* gene of *R.leguminosarum*. Our recent results indicate that NolL plays a role in the acetylation of the NodZ-determined fucosyl residue of LCOs. This was indicated by the analysis of radiolabeled LCOs of a *nolL* mutant derived of *R.loti* strain NZP2037 (Scott *et al.*, 1996). We are curently testing the presumed transacetylase activity of NolL and NodX proteins by *in vitro* enzymatic analyses. Until know experiments have been very difficult due to the high toxicity of overproduction of the *nodX* or *nolL* proteins. Using a gain of function approach, the *nolL* gene was shown to play an important role in determination of specificity of nodulation of *Lotus* plants (Table 2). We are presently analyzing in the host-specific characteristics of the *nolL* gene in more detail by testing other plant species.

Table 1. Effect of the presence of *nodZ* and *nolL* on nodulation of *R.leguminosarum* biovar viciae on various heterologous hosts. Plasmids containing the indicated genes were introduced into strain RBL5560. Plant species tested were *Vicia sativa*, *Macroptilium atropurpureum*, *Vigna unguiculata* and *Lotus preslii*. For details on the *nodDFITA* (flavonoid independent transcription activation) gene we refer to Spaink *et al* (1989). +, 70-100% nodulation; +/- 20-40% nodulation; -, 0% nodulation. N.t.: not tested.

| Introduced gene | nodulation on plant species | | | |
	Vicia	*Macroptilium*	*Vigna*	*Lotus*
control vector	+	-	-	-
nodDFITA	+	-	-	-
nodDFITA plus *nodZ*	+	+	+	-
nodDFITA plus *nolL*	+	-	*n.t.*	-
nodZ plus *nolL*	+	*n.t.*	*n.t.*	+/-

ACKNOWLEDGEMENTS

This work was supported by the Netherlands Organisation for Scientific Research (NWO), by a Marie Curie fellowship awarded to C.Q., by the PTP project of the European Union (BIO2-CT93-0400) and an INTAS support for the work of A.O.O.

REFERENCES

Firmin, J.L., Wilson, K.E., Carlson, R.W., Davies, A.E., and Downie, J.A. (1993) Resistance to nodulation of cv Afghanistan peas is overcome by *nodX* which mediates an *O*-acetylation of of the *Rhizobium leguminosarum* lipo-oligosaccharide nodulation factor. *Mol. Microbiol.* **10**: 351-360.

Kozik, A. (1996) Fine mapping of the *sym2* locus of pea linkage group I. Thesis Wageningen Agricultural University, ISBN 90-5485-543-6

López-Lara, I.M., Blok-Tip, L., Quinto, C., Garcia, M.L., Stacey, G., Bloemberg, G.V., Lamers, G.E.M., Lugtenberg, B.J.J., Thomas-Oates, J.E., and Spaink, H.P. (1996) NodZ of *Bradyrhizobium* extends the nodulation host range of *Rhizobium* by adding a fucosyl residue to nodulation signals. *Mol. Microbiol.* 21:397-408.

Scott, D.B., Young, C.A., Collins-Emerson, J.M., Terzaghi, E.A., Rockman, E.S., Lewis, P.E., and Pankhurst, C.E. (1996) Novel and complex chromosomal arrangement of *Rhizobium loti* nodulation genes. *Mol. Plant-Microbe Int.* 9: 187-197.

Spaink, H.P. (1996) Regulation of plant morphogenesis by lipo-chitin oligosaccharides. *Crit. Reviews Plant Sciences* in press

Spaink, H.P., Okker, R.J.H., Wijffelman, C.A., Tak, T., Goosen-deRoo, L., Pees, E., van Brussel, A.A.N., and Lugtenberg, B.J.J. (1989) Symbiotic properties of rhizobia containing a flavonoid-independent hybrid *nodD* product. *J. Bacteriol.* 171: 4045-4053.

The Complex Interaction between *Bradyrhizobium japonicum* and its Symbiotic Host Plants.

Gary Stacey[1,2], Jonathan Cohn[1], Brad Day[1], John Loh[1], Miniviluz Garcia[1], Joonhyun Paik[1], and Joyce Yuen[1]. [1]Center for Legume Research, [1]Dept. of Microbiology, [2]Dept. of Ecology and Evolutionary Biology, University of Tennessee, Knoxville, TN, USA, 37996-0845

1 Introduction

In any research area, initial research findings usually generate excitement and the rapid subsequent development of simple models to explain the results. Inevitably these initial models require modification and it is the unique case where the beginning model survives future research results. Such model building is extremely helpful in explaining results to the scientific community, focusing thinking about a problem, and also helping to design future experiments. However, there is also the danger that such models may lead to a dogmatic view of the way a biological system may work. Such dogma is the antithesis of creative thinking and can also discourage researchers from pursuing a research area. In other words, theresearcher may feel that the model answers the question so why pursue it any further. The study of symbiotic nitrogen fixation has not been immune to such model building and its associated good and bad points.

2 Regulation of *nod* gene expression in *Bradyrhizobium japonicum*

Early models to explain *nod* gene regulation in *Rhizobium* species stated that the NodD protein bound to the *nod* gene promoter, interacted with host-produced flavonoids, and activated transcription. We now know that the situation is much more complex than these models describe and involves more regulatory proteins than just NodD. However, in our own investigations, the dogma generated by these earlier models gave us some problems. It was not uncommon for scientific colleagues to discourage us from pursuing studies of *nod* gene regulation in *B. japonicum*. We were told that *B. japonicum nod* gene regulation would be similar to other rhizobia and so why spend the time working on this system.

2.1 Nod V and NodW

Figure 1 describes our current understanding of the regulatory pathways controlling *nod* gene expression in *B. japonicum* strain USDA110. Our earlier work in this area has been reviewed (Stacey, 1995) so we will describe here only our recent findings. An unique feature of *nod* gene regulation in *B. japonicum* is the involvement of the two-component regulatory system, NodV and NodW. Göttfert et al. (1990) were the first to

NATO ASI Series, Vol. G 39
Biological Fixation of Nitrogen
for Ecology and Sustainable Agriculture
Edited by A. Legocki, H. Bothe, A. Pühler
© Springer-Verlag Berlin Heidelberg 1997

report these genes and to show their importance to the ability of *B. japonicum* to nodulate certain host plants (e.g., cowpea), but not soybean. We subsequently showed that NodVW are required for the induction of *nod* gene expression by isoflavones

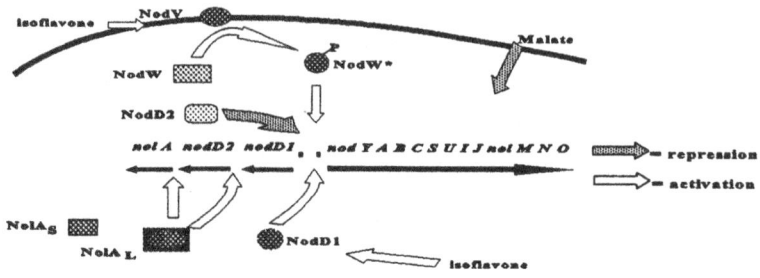

Figure 1. Model of *nod* gene regulation in *B. japonicum.*

(Sanjuan et al., 1994). Indeed, in strains deleted for *nodD1*, *nodD2*, and *nolA*, the NodVW regulatory system was found to be sufficient and essential for wild-type levels of *nod* gene expression.

Based on studies of other two-component regulatory proteins (Charles et al., 1992), one would predict that NodV is a sensor kinase that activates NodW by phosphorylation. Indeed, we found that NodV can autophosphorylate in vitro in the presence of Υ^{32}P-ATP. Subsequent incubation of labeled NodV with purified NodW results in transfer of the ^{32}P to NodW. A NodW mutant was constructed in which the aspartate subject to phosphorylation was replaced by asparagine. This mutant was incapable of being phosphorylated by NodV in vitro and, moreover, was unable to activate the expression of a *nodY-lacZ* fusion in vivo. The mutant NodW was also unable to complement the nodulation defect of a *nodW* mutant. Therefore, phosphorylation of NodW is required for biological activity. Immunoprecipitation experiments showed that NodW is phosphorylated in vivo only in the presence of isoflavones known to induce *nod* gene expression. These data clearly show that NodV and NodW comprise a second isoflavone recognition system (the first being NodD) in *B. japonicum*. Since NodD alone suffices in other rhizobia, one wonders why *B. japonicum* requires NodD and NodVW.

2.2 NolA

The role of NodV and NodW as positive activators of *nod* gene expression was discovered in a strain deleted of *nodD1*, *nodD2*, and *nolA* (Sanjuan et al., 1994). In this strain, the NodVW system allows for full *nod* gene expression. However, strains lacking only NodD1 are unable to induce *nod* gene transcription. These results led us to postulate that NolA was repressing gene activation by NodVW in the *nodD1* mutant strain (Dockendorff et al., 1994). Indeed, when NolA was expressed from a multicopy plasmid, *nod* gene expression was significantly reduced.

These earlier studies were limited due to the lack of *nolA* mutant strains. More recently, we constructed these mutants, but found no affect on *nodYABC* expression (Garcia et al., 1996). Thus, our original hypothesis that NolA directly represses *nod* gene expression appears to be incorrect. (Another good model destroyed.) However, further studies of the *nolA* mutant strains suggest that NolA is a positive activator of transcription, autoregulating its own expression and activating *nodD2* expression. Expressing NodD2 from a multicopy plasmid significantly reduces *nodYABC* expression suggesting that NolA may mediate its repressive effects via NodD2. However, previous experience cautions us from making this conclusion too strongly.

NolA and TipA are members of the MerR-family of transcriptional regulators. The *tipA* gene of *Streptomyces lividans* has been shown to encode two proteins, $TipA_L$ and $TipA_S$ (Holmes et al. 1993). We now have evidence that *nolA* also encodes two proteins, $NolA_L$ and $NolA_S$. $NolA_L$ has an N-terminal helix-turn-helix motif and likely acts as a transcriptional regulator. Data suggest that $NolA_L$ and $NolA_S$ expression is controlled transcriptionally by the use of two transcriptional start sites. The most 3' site (P2) is located only 7 bp away from ATG1 needed for $NolA_L$ translation. We postulate that translation of a mRNA beginning at P2 is likely to be inefficient and, therefore, ATG2 is used, leading to the synthesis of $NolA_S$. C-terminal *lacZ* fusions to *nolA* require NolA (presumably $NolA_L$) for expression. However, N-terminal fusions, prior to ATG2, do not require NolA for expression and, indeed, are inducible by soybean seed extract, but not by isoflavones that induce *nodYABC* expression. These data suggest that an unidentified component in soybean seed extract is inducing *nolA* transcription.

Figure 1 and the data discussed above should convince any reader that the regulation of *nod* gene expression in *B. japonicum* has many unique features that distinguish it from other rhizobia. We feel fortunate that we did not take previous models as the last word in *nod* gene regulation and pursued our work in this system.

3 Nod signal recognition

It is now known that most of the *nod* gene products are enzymes that synthesize lipo-chitin molecules that act as a signal to the host (reviewed in Dénarié et al., 1996). Addition of these molecules at low levels ($\leq$ nM) to host roots can result in the formation of a structure nearly identical to the nodule formed by bacterial inoculation. Early experiments led to the development of a model where modification of the basic tetrameric or pentameric chitin backbone structure by substituents, such as sulfate or 2-O-methylfucose, led to a molecule that was specifically recognized by the suitable legume host. Thus, host specificity was easily explained by nod signal-receptor specificity. This model is still widely quoted but does not match the complexity suggested by current information. For example, sulfation of the *R. meliloti* nod signal is required for activity on alfalfa roots (Dénarié et al., 1996). However, *R. meliloti nodH* mutants that cannot sulfate the nod signal can still nodulate as much as 40% of the alfalfa plants inoculated (Ogawa et al., 1991). Likewise, we reported that 2-O-methylfucosylation of the *B. japonicum* nod signal was required for biological activity on soybean roots (Stacey et al., 1994). However, *B. japonicum nodZ* mutants can still nodulate soybean. We now have an explanation for these results in that, although a non-fucosylated pentameric nod signal is inactive, a non-fucosylated tetramer has activity on soybean (Stokkermans et al., 1995).

In two recent papers (Minami et al., 1996a,b), we reported work suggesting that two recognition events are likely involved in nod signal action on soybean roots. The first recognition event shows limited chemical specificity. For example, even a chitin pentamer can induce the rapid, transient expression of the early nodulin, ENOD40, in soybean roots. However, a specific lipo-chitin nod signal is required to induce prolonged expression of ENOD40. Single nod signals added to soybean roots were unable to induce expression of the early nodulin ENOD2. In this case, it was found that a mixture of at least two nod signals was required to induce ENOD2 expression. We postulate that these two nod signals are recognized by at least two, distinct receptors, only one of which requires a structurally specific nod signal. However, as in other models, this idea probably understates the true biological complexity.

4 Conclusions

There is an inherent tendency among researchers to explain their research results in the form of simple models that can be easily explained and convey the basic ideas of their discoveries. However, all such models are simply hypotheses and, by the scientific method, should be vigorously tested by experimentation. Dogma arising from a too literal interpretation of models can be damaging to the scientific endeavor. Such thinking may lead to the idea that the question is answered discouraging researchers from vigorously pursuing a research area. Such dogmatic thinking may also lead researchers to ignore or gloss over research findings that do not fit the model. It is our feeling that some of this has happened in the study of rhizobia-legume interactions. Critical thinking should be applied to all models, including those that we have described above.

References

Charles, T.C., et al. 1992. Ann. Rev. Phytopath. 30: 463-484.
Dénarié, J., et al. 1996. Annu. Rev. Biochem. 65: 503-535.
Dockendorff, T., et al. 1994. Mol. Plant-Microbe Int. 7:596-602.
Garcia, M.L., et al. (1996) Mol. Plant-Microbe Int. (In press).
Göttfert et al., 1990. Proc. Natl. Acad. Sci. (USA) 87: 2680-2684.
Holmes, D., et al. 1993. EMBO J. 13: 138-146.
Minami, E., et al. 1996a. Plant J. 10: 23-32.
Minami, E., et al. 1996b. Mol. Plant-Microbe Int. 9: (In press).
Sanjuan, J., et al. 1992.. Proc. Natl. Acad. Sci. USA 89: 8789-8793.
Sanjuan, J., et al. 1994. Mol. Plant-Microbe Int. 7: 364-369.
Stacey, G., et al. 1994. J. Bacteriol. 176: 620-633.
Stacey, G. 1995. Microbiol. Lett. 127: 1-9.
Stokkermans, T.J.W., et al. 1995.. Plant Physiol. 108: 1587-1595.

Studies on *MsENOD40* gene expression in alfalfa (*Medicago sativa* L.) and white sweetclover (*Melilotus alba* Desr.)

Hirsch, A.M.[1,2], Y. Fang[1], P. van Rhijn[1], S. Galili[3], O. Shaul[3], N. Atzmon[3], S. Wininger[3], Y. Eshead[3], Y. Li[1], V. To[1], and Y. Kapulnik[3]

[1] Department of Molecular, Cell and Developmental Biology and [2] Molecular Biology Institute, University of California, Los Angeles, CA USA 90095-1606

[3] Institute of Field and Garden Crops ARO, The Volcani Center, Bet Dagan 50-250, Israel

Keywords. Early nodulin gene, *MsENOD40*, cytokinin, nodule development, mycorrhizae

1. Introduction

Plant hormones have been postulated to be involved with nodule development for at least 60 years (see Hirsch and Fang, 1994 for review). One of the most enigmatic of the plant hormones is cytokinin. Little is known about its biosynthesis in the plant or its mechanism of action. Numerous investigations have shown that there is an interaction between cytokinin and auxin, but exactly how these two phytohormones mediate development, especially in the root, is unclear.

A chronological list of some of the literature indicating the involvement of cytokinins in nodulation and nodulation-related events in legumes and non-legumes as well as reports measuring increased levels of cytokinin in nodules is presented in Table 1.

Table 1. Cytokinins and Nodulation

Activity	Reference
Cytokinins induce pseudonodules on tobacco roots	Arora et al., 1959
Kinetin triggers cell divisions in mature cortical cells	Torrey, 1961
Cytokinins induce pseudonodules on *Alnus glutinosa* roots	Rodriguez-Barrueco and Bermudez de Castro, 1973
Cortical cell proliferation occurs in devascularized pea root explants	Libbenga et al., 1973
Cytokinins detected in nodules of *Phaseolus vulgaris*	Puppo et al., 1974
Cytokinins detected in nodules of *Vicia faba*	Henson and Wheeler, 1976
Identification of cytokinins in pea nodules	Syono and Torrey, 1976
Cytokinins detected in nodules of *Phaseolus mungo*	Jaiswal et al., 1981
High levels of cytokinins measured in pea nodules	Badenoch-Jones et al., 1987
Cytokinins induce *ENOD2* expression in *Sesbania rostrata*	Dehio and deBruijn, 1992
Nod⁻ *R. meliloti* carrying *Agrobacterium trans*-zeatin synthase gene induce nodules and *ENOD2* gene expression in alfalfa	Cooper and Long, 1994
Cytokinins induce pseudonodules on siratro roots	Relic et al., 1994
Cytokinins induce *ENOD12* expression in alfalfa	Bauer et al., 1996

2. Results and Discussion

2.1. Cytokinins elicit *MsENOD40* gene expression in uninoculated roots.

We have previously determined that the *MsENOD40* gene is expressed in a diversity of cell types in alfalfa in addition to the tissues of the nodule: emerging lateral root primordia, leaf marginal meristems, stem procambium, and floral tissue (Asad et al., 1994). Using northern analysis, we have found that the *MsENOD40* gene is induced not only by *Rhizobium meliloti* inoculation, but also in uninoculated roots by the addition of purified Nod factor or the cytokinin 6-benzylaminopurine (BAP) (P. van Rhijn, Y. Fang, S. Galili, O. Shaul, N. Atzmon, S. Wininger, Y. Eshead, Y. Kapulnik, Y. Li, V. To, and A.M. Hirsch, manuscript in preparation).

To analyze this response further, we screened an alfalfa genomic library with the full-length cDNA clone to obtain phage DNAs that contained putative promoter elements. Our previous Southern analysis indicated that there were one or two *MsENOD40* genes in alfalfa (Asad et al., 1994). Subsequent analysis of the four different genomic clones that were isolated from the library showed that there were clones containing two distinct promoters, differing from one another in their restriction patterns (data not shown). The promoter regions were characterized further and then individually cloned into pBluescript. Sequence comparisons showed that the 5'-proximal sequences of the two promoters were identical. However, the 5'-distal ends of the two promoters varied significantly; they were only 40% similar to each other. The two different *MsENOD40* promoters were transcriptionally fused to the reporter gene *uidA* (*GUS*) and introduced into *Agrobacterium tumefaciens* by electroporation. These constructs were then used to transform alfalfa (*Medicago sativa* cv. Regen) as previously described (Hirsch et al., 1995).

We found that the expression patterns of the two different promoters varied under non-symbiotic conditions (Y. Fang and A.M. Hirsch, manuscript in preparation). However, under symbiotic conditions, the expression patterns correlated with those previously found for alfalfa by using *in situ* hybridization methods (Asad et al., 1994). A mature nodule showing GUS localization in the nodule meristem is illustrated in Fig. 1A.

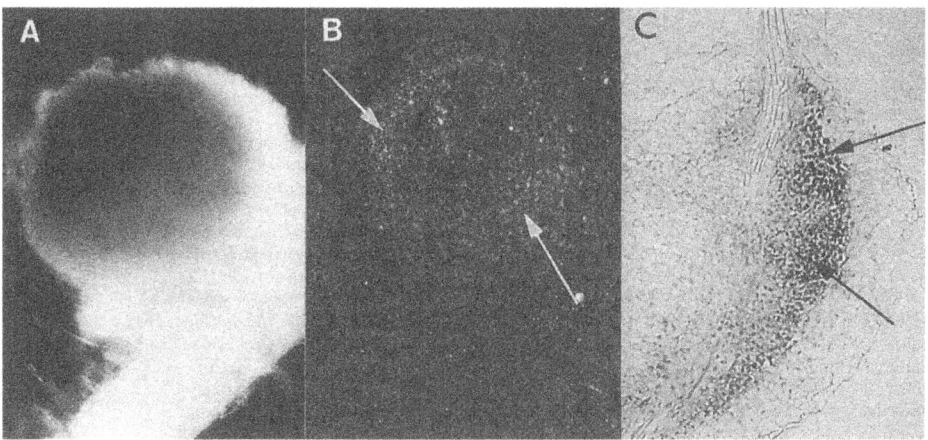

Fig. 1. Sites of expression of *MsENOD40*. (A) Mature alfalfa nodule from a transgenic plant containing the full-length *MsENOD40* promoter linked to *GUS*. ((B) *In-situ* hybridization localization of *MsENOD40* transcripts (shown here in dark-field as white dots) in pericyclic derivatives of an NPA-induced alfalfa nodule-like structure; oblique section. (C) Longitudinal section of a sweetclover pseudonodule induced by NPA. The digoxigenin-labeled probe hybridizes to *ENOD40* transcripts in derivatives of the pericycle.

We then investigated GUS activity levels of the different promoter-*GUS* constructs in response to BAP. We examined ten different plant lines for each of the two full-length promoters. Although the individual lines varied, the five shown here exhibited increased GUS activity in the BAP-treated roots compared to the untreated controls (Fig. 2).

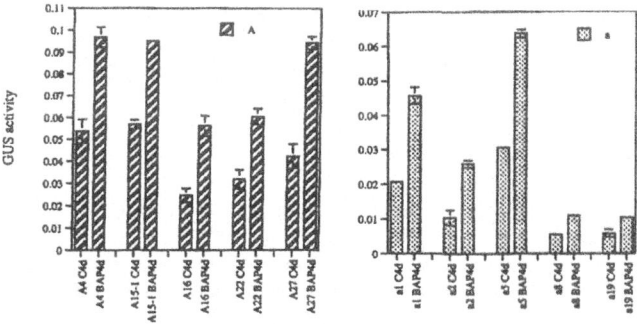

Fig. 2. Measurement of GUS activity in five different transgenic alfalfa lines contaning the full-length *MsENOD40-1* promoter and five other plants with the full-length *MsENOD40-2* promoter linked to *uidA*. A colorimetric assay was used after 4 days of treatment with 10^{-6} M benzylaminopurine (BAP).

2.2. *MsENOD40* is expressed in alfalfa and sweetclover roots treated with NPA.

Another way to perturb hormone balance is to treat uninoculated roots with the auxin transport inhibitor *N*-(1-naphthyl)phthalamic acid (NPA). We had previously shown that such treatment resulted in the formation of nodule-like structures that contain transcripts for the early nodulin genes *ENOD2* and *ENOD12* (Hirsch et al., 1989; Scheres et al., 1992). Similarly, alfalfa roots treated with 10^{-5} M NPA form pseudonodules that express *MsENOD40* transcripts (Fig. 1B).

Recently, we have initiated a research program on white sweetclover (*Melilotus alba* Desr.), a small-seeded, dwarfed annual with a life cycle of approximately 8 weeks. Like alfalfa, sweetclover is nodulated by *R. meliloti*. In addition, sweetclover is readily mutable; a number of single-gene mutants have already been produced and investigated (Kneen and LaRue, 1988). For example, Wu et al. (1996) have shown that five different nodulation defective mutants of white sweetclover as well as their wild-type parent form nodule-like structures in response to NPA and that these pseudonodules contain *ENOD2* transcripts. We treated an anthocyanin mutant (BT65), which was determined to be deficient in luteolin (T.A. LaRue, unpublished observations) with NPA. *In situ* hybridization analysis of *MsENOD40* gene expression shows that this gene is expressed in the pericycle of the NPA-induced pseudonodule (Fig. 1C).

2.3. *MsENOD40* is expressed in mycorrhizal roots.

LaRue and Weeden (1994) proposed that nodulation evolved from an association between roots and arbuscular mycorrhizal fungi. To test that hypothesis, we have probed total RNA isolated from mycorrhizal roots of alfalfa with a probe derived from *MsENOD40-2* (Asad et al., 1994). *MsENOD40* is expressed in these roots as early as 10 days after inoculation with the fungus. In addition, mycorrhizal roots accumulate significantly more *trans*-zeatin riboside than control roots (P. van Rhijn, Y. Fang, S. Galili, O. Shaul, N. Atzmon, S. Wininger, Y. Eshead, Y. Kapulnik, Y. Li, V. To, and A.M. Hirsch, manuscript in preparation).

3. Conclusion

We have shown that the early nodulin gene *ENOD40* is expressed in alfalfa roots under various conditions which lead to a hormonal perturbation in the root. This finding raises several questions as to how the expression of this gene is regulated downstream of Nod factor perception. For further evaluation of the responses of the *MsENOD40* promoter to cytokinin, we are examining transgenic alfalfa carrying constructs of nested deletions linked to *GUS*.

4. References

Arora, N., F. Skoog, and O.N. Allen. 1959. Kinetin-induced pseudonodules on tobacco roots. Amer. J. Bot. 46:610-613.

Asad, S., Y. Fang, K. Wycoff, and A.M. Hirsch. 1994. Isolation and characterization of cDNA and genomic clones of *MsENOD40*; transcripts are detected in meristematic cells. Protoplasma. 183:10-23.

Badenoch-Jones, J., C.W. Parker, and D.S. Letham. 1987. Phytohormones, *Rhizobium*-mutants, and nodulation in legumes VII. Identification and quantification of cytokinins in effective and ineffective pea root nodules using radioimmunoassay. J. Plant Growth Regul. 6:97-111.

Bauer, P., P. Ratet, M.D. Crespi, M. Schultze, and A. Kondorosi. 1996. Nod factors and cytokinins induce similar cortical cell division, amyloplast deposition and *MsENOD12A* expression pattterns in alfalfa roots. Plant J. 10:91-105.

Cooper, J.B. and S.R. Long. 1994. Morphogenetic rescue of *Rhizobium meliloti* nodulation mutants by *trans*-zeatin secretion. Plant Cell. 6:215-225.

Dehio, C. and F.J. deBruijn. 1992. The early nodulin gene *SrENOD2* from *Sesbania rostrata* is inducible by cytokinin. Plant J. 2:117-128.

Henson, I.E. and C.T. Wheeler. 1976. Hormones in plants bearing nitrogen-fixing root nodules: The distribution of cytokinins in *Vicia faba* L. New Phytol. 76:433-439.

Hirsch, A.M., Bhuvaneswari, T.V., Torrey, J.G., and Bisseling, T. 1989. Early nodulin genes are induced in alfalfa root outgrowths elicited by auxin transport inhibitors. Proc. Natl. Acad. Sci. USA 86:1244-1249.

Hirsch, A.M. and Y. Fang. 1994. Plant hormones and nodulation: What's the connection? Plant Mol. Biol. 26:5-9.

Hirsch, A.M., L.M. Brill, P.O. Lim, J. Scambray, and P. van Rhijn. 1995. Steps toward defining the role of lectins in nodule development in legumes. Symbiosis. 19:155-173.

Jaiswal, V., S.J.H. Rizvi, D. Mukerji, and S.N. Mature. Cytokinins in root nodules of *Phaseolus mungo*. Ann. Bot. 48:301-305.

Kneen, B.E. and T.A. LaRue. 1988. Induced symbiosis mutants of pea (*Pisum sativum*) and sweetclover (*Melilotus alba annua*). Plant Sci. 58:177-182.

LaRue, T.A and N.F. Weeden. 1994. The symbiosis genes of the host. In: Kiss, G.B. and Endre, G. (eds.) Proc. 1st European Nitrogen Fixation Conference. Officina Press, Szeged, Hungary. pp. 147-151.

Libbenga, K.R., F. van Iren, R.J. Bogers, and M.F. Schraag-Lamers. 1973. The role of hormones and gradients in the initiation of cortex proliferation and nodule formation in *Pisum sativum* L. Planta (Berl.). 114:29-30.

Puppo, A., J. Rigaud, and P. Barthe. 1974. Sur la presence de cytokinines dans les nodules de *Phaseolus vulgaris* L. C. R. Acad. Sci. Paris, Ser. D. 279:2029-2032.

Relic, B., F. Talmont, J. Kopcinska, W. Golinowski, J.-C. Promé, and W.J. Broughton. 1994. Biological activity of *Rhizobium* sp. NGR234 Nod-factors on *Macroptilium atropurpureum*. Molec. Plant-Microbe Inter. 6:764-774.

Rodriguez-Barreuco, C. and F. Bermudez de Castro. 1973. Cytokinin-induced pseudonodules on *Alnus glutinosa*. Physiol. Plant. 29:277-280.

Scheres, B., McKhann, H.I., Zalensky, A., Löbler, M., Bisseling, T., and Hirsch, A.M. 1992. The *PsENOD12* gene is expressed at two different sites in Afghanistan pea pseudonodules induced by auxin transport inhibitors. Plant Physiol. 100:1649-1655.

Syono, K, and Torrey, J.G. 1976. Identification of cytokinins of root nodules of the garden pea, *Pisum sativum* L. Plant Physiol. 57:602-606.

Torrey, J.G. 1961. Kinetin as trigger for mitosis in mature endomitotic plant cells. Exp. Cell Res. 23:291-299.

Wu, C., R. Dickstein, A.J. Cary, and J.H. Norris. 1996. The auxin transport inhibitor N-(1-naphthyl)phthalamic acid elicits pseudonodules on nonnodulating mutants of white sweetclover. Plant Physiol. 110:501-510.

Nod-factor Attachment, Calcium-fluxes, and Lipid-transfer Proteins in Symbiotic Signal Transduction.

H.R. Irving[1], N.M. Boukli[2], D. Toomre[3], A. Krause[2], C.A. Gehring[4], S. Jabbouri[2], A.A. Kabbara[5], K. Seidel[2], R.W. Parish[6], J.-C. Kader[7], A.A. Varki[3], and W.J. Broughton[2].

[1]Department of Pharmaceutical Biology & Pharmacology, Victorian College of Pharmacy, Monash University, Parkville Campus, 381 Royal Parade, Melbourne, Victoria 3052, Australia; [2]LBMPS, Université de Genève, 1292 Chambésy/Genève, Switzerland; [3]UCSD Cancer Centre, CCM-East Bldg, Room 1065, Mail Code # 0687, 9500 Gilman Drive, University of California, San Diego, La Jolla, CA 92093-0687, USA; [4]School of Biological & Chemical Sciences, Deakin University, Geelong Campus, Victoria 3217, Australia; [5]School of Zoology, La Trobe University, Melbourne, Victoria 3052, Australia; [6]School of Botany, La Trobe University, Melbourne, Victoria 3052, Australia; [7]Physiologie Cellulaire et Moléculaire, Université Pierre et Marie Curie, 75252 Paris, France.

Abstract. Temporal and spatial observations on the attachment of Nod-factors were made by biotinylating the reducing terminus of the lipo-chito-oligosaccharides of the broad host-range *Rhizobium* sp. NGR234 with the fluorescent reagent 2-amino-(6-amidobiotinyl)pyridine. Complex formation between the biotinylated Nod-factors and fluorescent streptavidin allowed localisation of the binding-site. At concentrations of $> 10^{-7}$ M, these fluorescently tagged moleculs bound rapidly and asymmetrically ($\cong 1$ min) to nodulation competent root-hairs but they did not bind to root-hairs of the non-host *Arabidopsis thaliana*. Early cellular events within the root-hairs were studied by loading root-segments with the calcium indicators Fura-2 and Fluo-3. Fluorescence ratio imaging showed that addition of NodNGR factors provoked almost immediate, plateau-like increases in intra-cellular free calcium $\{[Ca^{2+}]_i\}$ in root-hairs and epidermal cells. Confocal laser scanning microscopy (CLSM) revealed that calcium accumulation was concentrated at the tips and the sides of the responsive root-hairs. A gene encoding a lipid transfer-like protein (LPT2) was isolated from a *Vigna unguiculata* root-hair cDNA bank. Levels of the LTP2-transcript increased in root-hairs 24 h after treatment with NGR234, or its Nod-factors. The LTP2-gene was cloned into the pMal™ expression vector and LTP2 purified by affinity chromatography. It was unable to transfer phospholipids between liposomes and mitochondria. Anti-sense analysis in which the LTP2 coding region was cloned between the

NATO ASI Series, Vol. G 39
Biological Fixation of Nitrogen
for Ecology and Sustainable Agriculture
Edited by A. Legocki, H. Bothe, A. Pühler
© Springer-Verlag Berlin Heidelberg 1997

35S promoter and terminator sequences reduced nodulation when transformed into *V. unguiculata*.

Introduction. A two-way molecular exchange between rhizobia and legumes largely controls symbiotic interactions. Rhizobia are attracted to the legume rhizoplane by small molecular weight products of the roots. Amongst these compounds are flavonoids which interact with the NodD class of transcriptional activators to derepress downstream <u>nod</u>ulation (*nod*) genes (Fellay *et al.*, 1995). Enzymes encoded by the *nod*-genes are responsible for the synthesis of a class of lipo-oligosaccharides called Nod-factors which provoke deformation (Had) and curling (Hac) of the root-hairs. Extreme contortions trap rhizobia within their folds and allow the bacteria to enter the root (Relic´ *et al.*, 1993, 1994). Nod-factors have other important effects: they prepare the way for the invading bacteria by stimulating the development of pre-infection threads and nodule primordia (Mylona *et al.*, 1995) and they stimulate flavonoid release (Schmidt *et al.*, 1994), thus completing an auto-regulatory circuit. Although a number of control points within this circuit have been defined, those within the plant remain largely unexplored. In this communication we have analysed several components of the signal transduction pathway within the legume *Vigna unguiculata*.

Materials and Methods. NodNGR factors were purified according to published procedures (Price *et al.*, 1992; Jabbouri *et al.*, 1995) and the reducing terminus tagged with 2-amino-(6-amidobiotinyl)pyridine (Rothenberg *et al.*, 1993). Root-segments were incubated in biotinyleted Nod-factors (10^{-7} to 10^{-5} M) for 1 to 16 min and their attachment to root-hairs followed by confocal laser scanning microscopy. Changes in $[Ca^{2+}]_i$ were followed by the methods of Gehring *et al.* (1996). LPT2 was isolated and characterised as described by Krause *et al.* (1994). LTP2 was cloned into pMal™ and the protein purified according to the manufacturers protocol (New England Biolabs, Beverly, MA, USA). Lipid transfer activity was assayed as described by Kader *et al.* (1984). Transformed hairy-roots of composite plants were produced following the techniques of Krause and Broughton (1996).

Results and Discussion. A 1 min exposure of either intact roots or excised root-segments of *V. unguiculata* to 10^{-9} - 10^{-7} M NodNGR factors results in bulbous deformation of the root-hairs. Longer exposures (> 4 min) provoked more commonly observed Had-responses (Gehring *et al.*, 1996; N.M. Boukli, unpublished). Rapid responses of root-hairs to Nod-factor treatment have also been observed in *Pisum sativum* (Heidstra *et al.*, 1994) and *Trifolium repens* (Dazzo *et al.*, 1996). CLSM studies showed that the 488 nm emission band of an argon-ion laser did not excite appreciable fluorescence in non-treated root-hairs. Yet, biotinylated Nod-factors attached to root-hairs in an asymmetric manner (N.M. Boukli, C.A. Gehring, H. Irving, and D. Toomre, unpublished). Fluorescence, which was dependent on both biotinylated Nod-factors and fluorescent streptavidin, was appreciable after 1 min exposure to Nod-factors. Attachment to the roots of the non-hosts *Medicago sativa* and *Trifolium subterraneum* was not observed nor did the biotinylated factors react with the roots of the non-legume *Arabidopsis thaliana*. Together, these data show that less that 60 sec is required for NodNGR factors to attach to *Vigna* root-hairs and to initiate Had.

CLSM was also used to study the changes in intra-cellular free-calcium occurring in root-hairs following application of NodNGR factors. Root segments of two legumes {*Vicia sativa* (a non-host of NGR234) as well as *V. unguiculata*} and three non-legumes (*A. thaliana*, *Petroselinum crispum* and *Zea mays*) were loaded with Fluo-3. Addition of 5×10^{-11} to 10^{-9} M Nod-factors resulted in a steady increase of $[Ca^{2+}]_i$ in the cytosol of *Vigna* root-hairs over a 15 min period (Fig. 1*a* to *f*). An increase in fluorescence intensity is evident at the very tip of the root-hair 7 min after Nod-factor application (arrow, Fig. 1 *c*). Addition of 10^{-7} M NodNGR factors to *Vigna* root parenchyma cell preparations did not influence the levels of $[Ca^{2+}]_i$ (Fig. 1 *g,h*). Furthermore, similar concentrations of Nod-factors had no effect on $[Ca^{2+}]_i$ in root-hairs of maize (Fig, 1 *i,j*), or *Arabidopsis* and parsley (data not shown). Conformation was attained using a quantitative technique in which root-segments were loaded with Fura-2 and the effects of Nod-factors followed by fluorescence ratio imaging. Application 10^{-9} M NodNGR factors resulted, within seconds, in plateau-like increases in $[Ca^{2+}]_i$ in root-hairs and root-epidermal cells (Gehring *et al.*, 1996). Although Erhardt *et al.* (1996) observed spikes in $[Ca^{2+}]_i$ rather than plateaux when root-hairs of *M. saliva* were treated with *R. meliloti* Nod-factors, these data suggest that increases in intra-cellular free calcium are associated with Nod-factor induced root-hair deformation.

Nod-factors also induce the expression of a lipid transfer-like protein (LPT2) which is specifically expressed in root-hairs (Krause *et al.*, 1994). To test whether this protein is important early in the signal transduction pathway, it was cloned into an expression vector and purified by affinity chromatography.

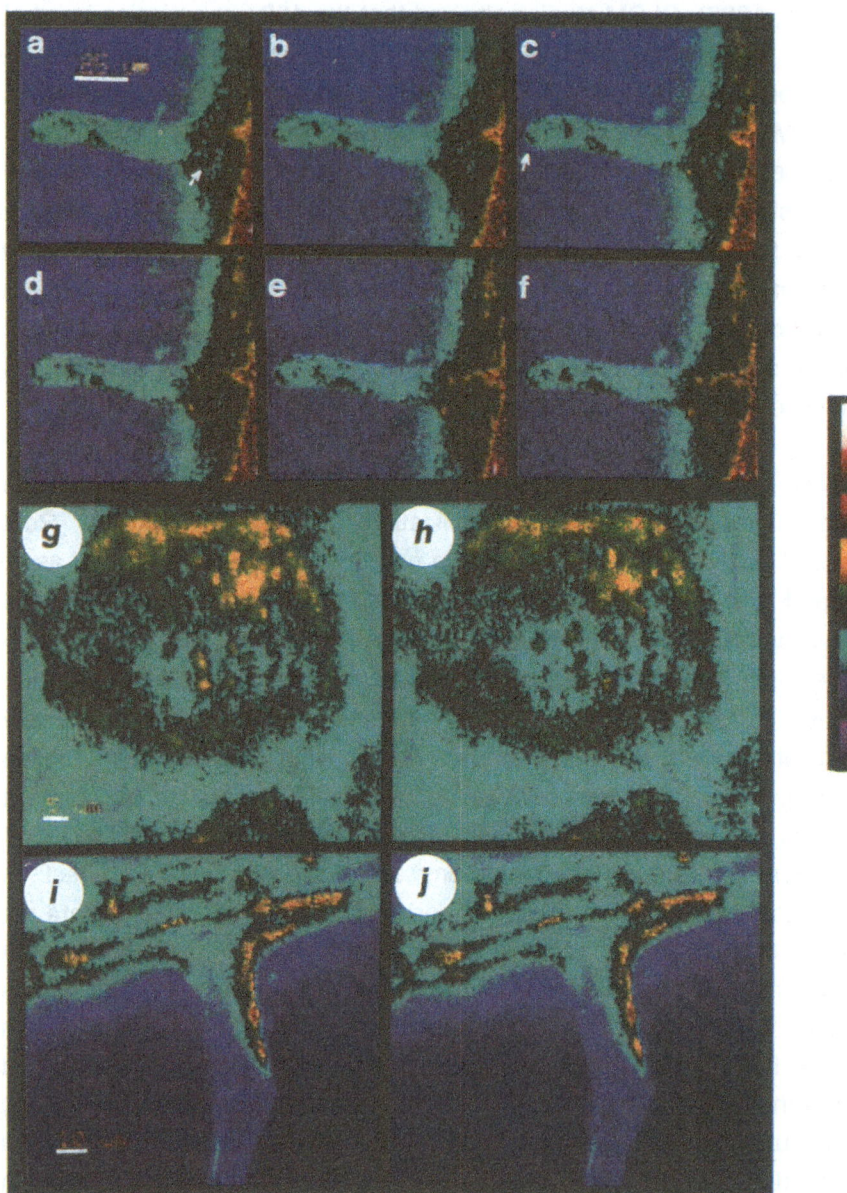

Fig. 1. $[Ca^{2+}]_i$-dependent cytosolic fluorescence in host parenchyma and non-host (maize) root-hair cells treated with NodNGR[S] factors. The colour bar represents a range of $[Ca^{2+}]_i$ from $\leq 0.01\mu M$ (purple/bottom) to ≥ 10 μM (white/top). Changes in $[Ca^{2+}]_i$-dependent cytosolic fluorescence of a *V. unguiculata* root-hair in response to 5×10^{-11} M NodNGR factors. The images were taken (a) 3 min, (b) 5 min, (c) 7 min, (d) 9 min, (e) 13 min, and (f) 15 min after Nod-factor addition in incubation buffer. The arrows point to the region of the

root-hair where the most marked changes occur. Fluorescent images of *V. unguiculata* root parenchyma cells scanned 3 min (g) and 7 min (h) after addition of 10^{-7} M NodNGR factors (Bar = 5 μm). Fluorescent images of a maize root-hair and adjacent epidermal cells scanned 4 min (I) and 10 min (j) after the addition of 10^{-7} M NodNGR factors (Bar = 10 μm).

There was no detectable transfer of phospholipids between liposomes and mitochondria (K. Seidel, J.-C. Kader, unpublished), but this does not exclude the possibility that LPT2 can transfer Nod-factors within plant cells. We are currently testing this possibility. As another way of evaluating the role of lipid-transfer, anti-sense LTP2 constructs (in pBIN19) were introduced into the hairy-roots of composite *V. unguiculata* plants. In this system, nodulation was reduced by 60% suggesting that LTP2 is an important intermediate in the symbiotic signal transduction pathway.

Acknowledgements. We are extremely grateful to D. Gerber and S. Relic´ for their help with many aspects of this work. Financial assistance was provided by the Swiss National Science Foundation, the Université de Genève, Leica Lasertechnik GmbH, Heidelberg, Germany, the Australian Research Council, and La Trobe University.

References.
Dazzo, F.B., Orgambide, G.G., Philip-Hollingsworth, S., Ninke, K.O., and Salzwedel, J.L. (1996) *J. Bacteriol.* **178**: 3621 - 3627.
Engelhardt, D.W., Wais, R., and Long, S.R. (1996) *Cell* **85**: 673 - 681.
Fellay, R., Rochepeau, P., Relic´, B., and Broughton, W.J. (1995) in *Pathogenesis and Host Specificity in Plant Diseases. Histopathological, Biochemical, Genetic and Molecular Bases.* Vol. 1 *Prokaryotes* Singh, R.P. Singh, and K. Khomoto, Pergamon/Elsevier Science Ltd., Oxford, p. 199-220.
Heidstra, R. Geurts, R., Franssen, H., Spaink, H.P., van Kammen, A, and Bisseling, T. (1994) *Plant Physiol.* **105**: 787 - 797.
Gehring, C.A., Irving, H.R., Karrara, A., Parish, R.W., Boukli, N.M., and Broughton, W.J. (1996) *Plant Journal*, submitted.
Jabbouri, S., Fellay, R., Talmont, F., Kamalaprija, P., Burger, U., Relic´, B., Promé, J.-C., and Broughton, W.J. (1995) *J. biol. Chem.* **270**: 22968 - 22973.
Kader, J.-C., Julienne, M., and Vergnolle, J. (1984) *European J. Biochem.* **139**: 411 - 416.
Krause, A., and Broughton, W.J. (1996) *Plant Cell Rep.*, in preparation.
Krause, A., Sigrist, C.J.A., Dehning, I., Sommer, H., and Broughton, W.J. (1994) *Mol. Plant-Microbe Interact.* **7**: 411-418.

Mylona, P., Pawlowski, K., and Bisseling, T. (1995) *Plant Cell* 7; 869-885.

Price, N.P.J., Relic´, B., Talmont, F., Lewin, A., Promé, D., Pueppke, S.G., Maillet, F., Dénarié, J., Promé, J.-C, and Broughton, W.J. (1992) *Mol. Microbiol.* **6**: 3575 -3584.

Relic´, B., Talmont, F., Kopcinska, J., Golinowski, W., Promé, J.-C., and Broughton, W.J. (1993)). *Mol. Plant-Microbe Interact.* 6: 764 - 774.

Relic´, B., Perret, X., Estrada-Garcia, M.T., Kopcinska, J., Golinowski, W., Krishnan, H.B., Pueppke, S.G., and Broughton, W.J. (1994) *Mol. Microbiol.* **13**: 171 - 178.

Rothenberg, B.E., Hayes, B.K., Toomre, D., Manzi, A.E., and Varki, A. (1993) *Proc. Natl. Acad. Sci.* **90**: 11939 - 11943.

Schmidt, P.E., Broughton, W.J., and Werner, D. (1994). *Mol. Plant-Microbe Interact.* **7**: 384-390.

Degradation of Nodulation Signals from *Rhizobium meliloti* by its Host Plants

Christian Staehelin[1], Maud Vanney[1], Fabrice Foucher[1], Eva Kondorosi[1], Michael Schultze[1], and Adam Kondorosi[1,2]

[1] Institut des Sciences Végétales, CNRS, F-91198 Gif-sur-Yvette, France
[2] Institute of Genetics, Biological Research Center, Hungarian Academy of Sciences, P.O. Box 521, H-6701 Szeged, Hungary

Introduction. *Rhizobium meliloti* has a relatively narrow host range which includes only the species of the genera *Medicago, Melilotus* and *Trigonella*. The lipo-chitooligosaccharide signals (Nod factors) of *R. meliloti* have a tetrameric or pentameric chitooligosaccharide backbone which is substituted by an *N*-linked acyl moiety, mainly by a C16:2 fatty acid. The Nod factors are structurally modified by a sulfate group on the reducing end and partly *O*-acetylated on the non-reducing end (Dénarié et al., 1996; Schultze and Kondorosi, 1996). These structural parameters influence the signal activity on the host plants of *R. meliloti*. The sulfate group has been shown to be necessary to induce specific responses of the host plant *Medicago* (Roche et al., 1991), while the *O*-acetyl group increases the activity of the Nod factors in long-term assays such as root hair deformation and cortical cell division (Truchet et al., 1991; Ardourel et al., 1994). Inoculation with rhizobial mutants producing non-*O*-acetylated Nod factors resulted in delayed nodulation on certain *Medicago* species including *M. sativa* (Ardourel et al., 1995). However, the capacity of *O*-acetylated and non-*O*-acetylated Nod factors was equal in inducing very early responses of *M. sativa* such as rapid plasma-membrane depolarization and intracellular alkalinization in root hair cells (Felle et al., 1995; Felle et al., 1996).

Nod factors are substrates for plant chitinases. Structural parameters of the Nod factors have been shown to influence their stability in the rhizosphere and protected the molecules against degradation by chitinases *in vitro* (Staehelin et al., 1994a,b). The degradation products, the acylated dimers and trimers, lack the capacity to induce root hair deformation on the host plant indicating an inactivation of the signal molecules by the hydrolytic reaction (Heidstra et al., 1994; Staehelin et al., 1994a,b). Using a set of tetrameric and pentameric Nod factors as substrate, six distinct hydrolytic activities were demonstrated in *M. sativa*. The activities of the enzymes were distinguished on the basis of their substrate specificities and the analysis of the cleavage sites (Staehelin et al., 1996). Most of these hydrolases are probably chitinases/lysozymes known to play a role in plant-pathogen interactions. Three of the isolated *Medicago* enzymes were characterized by their substrate specificity which was different from the specificity obtained for chitinases/lysozymes of a number of legumes and non-legumes.

NATO ASI Series, Vol. G 39
Biological Fixation of Nitrogen
for Ecology and Sustainable Agriculture
Edited by A. Legocki, H. Bothe, A. Pühler
© Springer-Verlag Berlin Heidelberg 1997

The most abundant and most active lipo-chitooligosaccharide of *R. meliloti* is NodRm-IV(C16:2,Ac,S). In the rhizosphere of *M. sativa*, this short and highly modified Nod factor is only degraded by a specific enzyme which exclusively releases lipo-disaccharides from tetrameric or pentameric Nod factors. Interestingly, the activity of this "dimer-forming" enzyme was stimulated up to 6-fold when roots were pretreated with *R. meliloti* Nod factors, indicating a rapid feedback inactivation of the Nod factors after their perception (Staehelin et al., 1995). The inducibility of this activity by the active Nod factors suggests a function of the "dimer-forming" enzyme in the nodule symbiosis. Therefore, we purified the protein from *M. sativa*, characterized its substrate specificity and determined its occurrence in different parts of the plant. Furthermore, we searched for related activities in host and non-host plants of *R. meliloti*.

The "dimer-forming" lipo-chitooligosaccharidase of *Medicago*. We purified the "dimer-forming" enzyme from extracts of Nod factor-treated young *M. sativa* roots. The hydrolase was found to bind strongly to concanavalin A (ConA), indicating that it is a glycoprotein containing *N*-linked high mannose-type oligosaccharides. As determined by SDS-PAGE, the molecular weight of the purified protein was about 65 kDa. Based on the determination of the activity of fractions after gel filtration chromatography, the native protein has a molecular weight of about 60 kDa. The hydrolase was characterized by Michaelis-Menten constants for Nod factors in the micromolar concentration range. The enzyme forms the β-anomer of the acylated disaccharide, indicating a retaining reaction mechanism. Chitotetraose, added to the assay mixture at 500-fold higher concentration than the Nod factor substrates, did not show any inhibitory effect on Nod factor degradation. When chitotetraose alone was tested as substrate, only a weak degradation activity was observed indicating a higher affinity of the enzyme for acylated chitooligosaccharides. Moreover, colloidal chitin was resistant against degradation by this enzyme. The substrate specificity and the amino acid sequences of peptides obtained from the purified protein indicate that it represents a new type of plant glycosylhydrolase which is distinct from known chitinases/lysozymes. Therefore, the enzyme has to be considered as a Nod factor hydrolase or lipo-chitooligosaccharidase.

The occurrence of this enzyme in different parts of the plant was tested using its property to bind to ConA and on its cleavage specificity to release acylated dimers. The lipo-chitooligosaccharidase of *M. sativa* was detected on the surface of intact roots, in root exudates and in extracts of roots and nodules as well. However, the enzyme could not be identified in seeds and in the aerial part of the plant. This tissue-specific localization gives further evidence that the enzyme is related to the *Rhizobium*-plant symbiosis.

The "dimer-forming" hydrolase was also identified in roots of several other *Medicago* species. However, the substrate specificity was found to be different with respect to the *O*-acetyl group of the Nod factors. While this structural modification of NodRm-IV(C16:2,Ac,S) partly protected the molecule against hydrolysis by the lipo-chitooligosaccharidase of *M. sativa* (Staehelin et al., 1995), the *O*-acetyl group did not protect the tetrameric Nod factor against degradation by the corresponding enzyme of *M. truncatula.* These data indicate a variation in the substrate specificity of lipo-chitooligosaccharidases from different *Medicago* species. Moreover, it is tempting to speculate that the requirement of the *O*-acetyl group for nodulation of certain *Medicago* species correlates with the requirement of the *O*-acetyl group for stability of the Nod factors. The *O*-acetyl group would facilitate nodule formation mainly in those *Medicago* species which inactivate the non-*O*-acetylated Nod factors more efficiently than the *O*-acetylated molecules.

Lipo-chitooligosaccharidases of *Melilotus* and *Trigonella*. Intact roots of *Melilotus alba,* another host plant of *R. meliloti,* had also the capacity to cleave NodRm-IV(C16:2,Ac,S). Similarly to the *Medicago* species, the degradation activity was stimulated by a pretreatment of the roots with low concentrations of Nod factors. However, in addition to the increased formation of the lipo-disaccharide NodRm-II(C16:2), we observed on this species an increase of the lipo-trisaccharide NodRm-III(C16:2) when the pentameric Nod factor NodRm-V(C16:2,S) was used as substrate. The ConA-binding fraction obtained from root extracts of *Melilotus* exhibited a similar cleavage specificity. The enzyme released from NodRm-V(C16:2,S) both, NodRm-II(C16:2) and NodRm-III(16:2). Excess amounts of chitotetraose did not inhibit the cleavage of the Nod factors by the *Melilotus* enzyme indicating a lipo-chitooligosaccharidase activity as it was found for the "dimer-forming" enzyme of the *Medicago* species. Similarly to *Melilotus,* another host plant of *R. meliloti, Trigonella coerulea,* had a Nod factor-inducible ConA-binding activity which hydrolyzed NodRm-V(C16:2,S) to the lipo-disaccharide and lipo-tri-saccharide.

Nod factor degradation by non-host plants. An extracellular activity able to cleave NodRm-IV(C16:2,Ac,S) has not been identified so far in the rhizosphere of non-host plants of *R. meliloti,* although such an activity was measured in the ConA-binding fraction obtained from total root extracts of several plants including non-legumes, indicating that these enzymes are intracellularly located. Most of these activities, however, seem to be different from the extracellular lipo-chitooligosaccharidases identified for the host plants of *R. meliloti.* The Nod factor-cleaving activity of soybean, for example, was nearly completely inhibited by high amounts of chitotetraose, indicating the absence of a specific lipo-chitooligo-saccharidase in the ConA-binding fraction.

Conclusion. NodRm-IV(C16:2,Ac,S) is resistant against degradation by known plant chitinases which have a substrate preference for longer and less modified molecules. However, all tested host plants of *R. meliloti* possess a lipo-chitooligosaccharidase inactivating the NodRm factors. We suggest that these enzymes may play a role in the symbiotic interaction after perception of the Nod factors. We postulate that the lipo-chitooligosaccharidases of *Medicago, Melilotus* and *Trigonella* have evolved for Nod factor inactivation that is necessary to suppress continuous stimulation by active Nod factors. We can speculate that only those legumes which are able to cleave NodRm-IV(C16:2,Ac,S) were able to develop a receptor for this molecule. This implicates that Nod factor receptors and Nod factor hydrolases might have common features. We hypothesize that lipo-chitooligosaccharidases are a hydrolytic variation of a Nod factor receptor theme.

Acknowledgments. This work was supported by the Swiss National Science Foundation (grant 83EU-044164 to C.S.) and by the European Economic Community (Human Capital and Mobility program CHRX-CT94-0656).

References

Ardourel, M., Demont, N., Debellé, F.D., Maillet, F., de Billy, F., Promé, J.C., Dénarié, J. and Truchet, G. 1994. *Plant Cell* 6: 1357-1374.

Ardourel, M., Lortet, G., Maillet, F., Roche, P., Truchet, G., Promé, J.C. and Rosenberg, C. 1995. *Mol. Microbiol.* 17: 687-699.

Dénarié, J., Debellé, F. and Promé, J.C. 1996. *Annu. Rev. Biochem.* 65: 503-535.

Felle, H.H., Kondorosi, É., Kondorosi, Á. and Schultze, M. 1995. *Plant J.* 7: 939-947.

Felle, H.H., Kondorosi, É., Kondorosi, Á. and Schultze, M. 1996. Rapid alkalinization in alfalfa root hairs in response to rhizobial lipochitooligosaccharide signals. *Plant J.* 10: 101-107.

Heidstra, R., Geurts, R., Franssen, H., Spaink, H.P., Van Kammen, A. and Bisseling, T. 1994. *Plant Physiol.* 105: 787-797.

Roche, P., Debellé, F., Maillet, F., Lerouge, P., Faucher, C., Truchet, G., Dénarie, J. and Promé, J.C. 1991. *Cell* 67: 1131-1143.

Schultze, M. and Kondorosi, Á. 1996. *Curr. Op. Genet. Dev.* 6: in press.

Staehelin, C., Granado, J., Müller, J., Wiemken, A., Mellor, R.B., Felix, G., Regenass, M., Broughton, W.J. and Boller, T. 1994a. *Proc. Natl. Acad. Sci. USA* 91: 2196-2200.

Staehelin, C., Schultze, M., Kondorosi, É., Mellor, R.B., Boller, T. and Kondorosi, Á. 1994b. *Plant J.* 5: 319-330.

Staehelin, C., Schultze, M., Kondorosi, É. and Kondorosi, Á. 1995. *Plant Physiol.* 108: 1607-1614.

Staehelin, C., Foucher, F., Kondorosi, É., Kondorosi, Á. and Schultze, M. 1996. *Chitin Enzymology, Vol. 2*, Muzarelli R.A.A., (ed.), Atec Edizioni, Italy, 371-378.

Truchet, G., Roche, P., Lerouge, P., Vasse, J., Camut, S., de Billy, F., Promé, J.C. and Dénarié, J. 1991. *Nature* 351: 670-673.

Induction of root cortical cell divisions by heterologous nodulation factors

Clara L. Diaz, Kees J.M. Boot, Helmi R.M. Schlaman, Christof Sautter*,
Ton A.N. van Brussel, Herman P. Spaink, Jan W. Kijne

Institute of Molecular Plant Sciences, Leiden University, Leiden, The Netherlands
*Institute of Plant Sciences, E.T.H., Zürich, Switzerland

Keywords: Lipochitin oligosaccharides, lectin, uridine

1 Lipochitin oligosaccharides

The symbiotic interaction between rhizobia and legume plants, resulting in formation of nitrogen-fixing root nodules, is host-plant-specific. For example, *Rhizobium leguminosarum* biovar viciae nodulates pea and vetch, but not clover, soybean or alfalfa, whereas *R. leguminosarum* biovar trifolii preferably nodulates clover. Key factors in mutual recognition of the symbiotic partners are lipochitin oligosaccharides (LCOs), produced by the rhizobia. Expression of host-plant-specificity is a two-step process: host plant-derived flavonoids can specifically induce production of LCOs by rhizobia, and LCOs can specifically induce root nodule formation in host plant roots.

The latter activity is dependent on the chemical structure of the LCOs. Specific LCOs produced by *R. leguminosarum* biovar viciae consist of an oligosaccharide backbone of ß-1,4-linked *N*-acetyl-D-glucosamine, in which the nonreducing sugar residue is substituted with an 18:4 unsaturated fatty acid and a 6-O-acetylgroup. These LCOs induce a variety of nodulation-related phenomena in host plants of the producing bacteria, including formation of nodule primordia (for a recent review, see Spaink, 1996). In contrast, these LCOs do not induce nodule primordia in clover roots. However, substitution of the 18:4 acyl moiety for C20:3 or C20:4 acyl moieties changes these LCOs into mitogenic signals for clover roots (Spaink et al., 1995).

2 Lectins

Host plant lectins have been proposed to also play a role in host-plant-specificity of rhizobia-legume interactions. Lectins are proteins with only one property in common: presence of at least one noncatalytic domain that can bind reversibly to a specific saccharide (Peumans and Van Damme, 1995). In earlier work (Diaz et al., 1989, 1995a), we showed that white clover hairy roots transformed with the pea lectin gene have acquired the ability to be nodulated by the pea symbiont *R. leguminosarum* biovar viciae. In these roots, pea lectin appeared to be present at root surface sites similar to those on pea roots (Diaz et al., 1995b). White clover hairy roots transformed with a pea lectin mutant containing defective sugar-binding sites only responded at background level (Van Eijsden et al., 1995). These results suggest that the presence of pea lectin facilitates recognition and/or response of white clover roots to LCOs produced by *R. leguminosarum* biovar viciae, and that the sugar-binding domain of pea lectin plays a role in this phenomenon. Part of this effect may be caused by enhanced binding of the pea symbionts to the transgenic white clover root hairs. In 1973, Hamblin and Kent suggested that legume lectins play a role in the *Rhizobium*-legume symbiosis by binding the symbiotic partner to host root hairs. One year later, Bohlool and Schmidt (1974) added the element of host-specificity to this suggestion by showing that soybean seed lectin preferably binds to *Bradyrhizobium japonicum* bacteria. Although in later years several studies have shown that less specific mechanisms contribute to rhizobial attachment, especially at high inoculum densities (for a review, see Smit et al. 1992), the hypothesis of lectin-enhanced attachment of rhizobia to host root hairs under conditions favourable for nodulation is still valid (Dazzo et al., 1976; Kijne et al., 1988).

3 Transgenic red clover roots

In order to separate LCO signaling from rhizobial attachment, we tested susceptibility of clover roots transformed with the pea lectin gene to heterologous LCOs. As a test plant, we selected red clover. In contrast to white clover, roots of this plant do not visibly respond to LCOs from *R. leguminosarum* biovar viciae. After application of these LCOs, loci of cortical cell divisions resembling root nodule primordia were observed in 60% of the transgenic red clover plants. Control roots transformed with an empty vector did not respond. Obviously, this result can not be explained by improved binding of the heterologous rhizobia. Pea lectin may function as a binding factor of LCOs produced by pea symbionts. However, it should be noted that the sugar-binding domain of pea lectin has been characterized into detail, and that it can not accomodate these LCOs in the way established for a specific ligand such as mannose (discussed by Kijne et al., 1994). Presently, we study the possibility that this pea lectin domain binds LCOs in a nonspecific way. The latter possibility was accentuated by our finding that, in contrast to control roots, red clover roots transformed with the pea lectin gene also responded to application of LCOs from *R. meliloti*, symbiont of alfalfa, and *R. loti*, the symbiotic partner of lotus. These LCOs differ from those of pea symbionts by carrying different fatty acids and different oligochitin modifications. The observed lack of specificity is consistent with our observation that presence of pea lectin in clover roots also stimulates nodulation by the homologous symbiont *R. leguminosarum* biovar trifolii (Diaz et al., 1995a).

4 Response to chitin oligomers

Irrespective of the mechanism by which pea lectin enables a response to heterologous

LCOs in red clover roots, our data show that legume roots essentially are able to respond to heterologous LCOs. Apparently, the presence of specific substituents on LCOs is irrelevant for this phenomenon. The latter notion is also illustrated by results of Schlaman et al. (reported by Spaink et al.,1996) showing that ballistic introduction of chitin oligomers into vetch or lotus roots induced cortical cell division activity. In order to obtain this response, co-targeting of uridine was required. Uridine has been identified by Smit et al. (1995) as a cell division factor in pea roots. Recent work by Boot et al. (manuscript in preparation) strongly suggests that uridine acts as an enhancer of auxin activity. Possibly, addition of uridine has to compensate for targeting of the chitin oligomers to less responsive cortical cells. Also for soybean, Minami et al. (1996) have demonstrated that application of a chitin oligomer induces a nodulation-related event, namely expression of the early nodulin ENOD40. This response appeared to be transient, which suggests that specific decorations of chitin oligomers are required for a sustained host plant response rather than for initial elicitation.

5 Role of pea lectin

Pea lectin enhances accumulation of capsulated *R. leguminosarum* biovar viciae bacteria on pea root hair tips (Kijne et al., 1988). Rhizobial nodulation genes involved in synthesis of LCOs do not seem to play a role in this process. Therefore, eventual host specificity in pea lectin-mediated attachment is dependent on additional nodulation genes of *Rhizobium*, such as genes required for synthesis of specific extracellular and/or capsular polysaccharides. In view of its location, one may expect that pea lectin can enhance attachment of pea symbionts to transgenic clover roots. This will result in a more concentrated delivery of pea-specific LCOs at the clover root hair tips. Then, in a yet unknown way, pea lectin enables elicitation of cortical cell division activity by these heterologous signal molecules. Pea lectin may facilitate binding and transport of LCOs, by its interaction with the plasmamembrane at the root hair tip (Booij et al., 1996, Diaz et al., 1995b). As an alternative, pea lectin may enable the cell division response by activating the inner cortical root cells. *Medicago* lectin genes are expressed early during nodulation at sites of cortical cell division (Bauchrowitz et al., 1996).

In short, our new results open the possibility that pea lectin is involved in signal transduction of LCOs or chitin oligomers.

References

Bauchrowitz MA et al (1996) Plant J 9:31-43
Bohlool BB, Schmidt E (1974) Science 185:269-271
Booij P et al (1996) Plant Mol Biol 31:169-173
Dazzo FB et al (1976) Appl Environ Microbiol 32:166-171
Diaz CL et al (1989) Nature 338: 579-581
Diaz CL et al (1995a) Mol Plant-Microbe Interact 8:348-356
Diaz CL et al (1995b) Plant Physiol 109:1167-1177

Hamblin J, Kent SP (1973) Nature New Biol 245:28-29
Kijne JW et al (1988) J Bacteriol 170:2994-3000
Kijne JW et al (1994) In: Proceedings of the 1st European Nitrogen Fixation
Conference. Kiss GB, Ende G, eds. Officina Press, Szeged, pp. 106-110
Minami E et al (1996) Plant J 10:23-32
Peumans WJ, Van Damme EJM (1995) Plant Physiol 109:347-352
Smit G et al (1992) Mol Microbiol 6:2897-2903
Smit G et al (1995) Plant Mol Biol 29:869-873
Spaink HP et al (1995) Mol Plant-Microbe Interact 8:155-164
Spaink HP et al (1996) The molecular basis of host specificity of rhizobia. In:
Proceedings 8th Int Congress Mol Plant-Microbe Interact, Knoxville, TN (in press)
Spaink HP (1996) Regulation of plant morphogenesis by lipo-chitin oligosaccharides.
Crit Rev Plant Sci (in press)
Van Eijsden RR et al (1995) Plant Mol Biol 29:431-439

ENOD40 expression precedes cell division and affects phytohormone perception at the onset of nodulation

Wei Cai Yang[2], Karin van de Sande[2], Katharina Pawlowski[2], Jürgen Schmidt[1], Richard Walden[1], Martha Matvienko[2], Henk Franssen[2] and Ton Bisseling[2]

[1] Max-Planck-Institut für Züchtungsforschung, Carl-von-Linné-Weg 10, 50829 Köln, Germany
[2] Department of Molecular Biology, Agricultural University, 6703 HA Wageningen, The Netherlands

Keywords.   Symbiosis, *Rhizobium*, legumes, pericycle

1. Introduction

Legume nodule organogenesis is initiated by local dedifferentiation of the root cortex. Nod factors initiate this mitotic reactivation in a spatially controlled manner (Van Brussel et al., 1992; Vijn et al., 1993). In most temperature legumes like pea and alfalfa, cell division is induced in the inner cortex opposite proto-xylem points (Libbenga and Harkes, 1973; Newcomb et al., 1979). Although the mode of action of Nod factors remains unresolved several studies suggest that these compounds probably establish a local change of the auxin/cytokinin ratio resulting in induction of cell division in the root cortex (Allen et al., 1953; Hirsch et al., 1989; Cooper and Long, 1994). In these dividing cortical cells, *ENOD40* is induced by Nod factors (Vijn et al., 1993; Yang et al., 1993). We were interested whether the expression of *ENOD40* is involved in the initiation of cortical cell division and investigated whether it is involved in establishing a change in phytohormone concentration of sensitivity as an early step in nodule formation.

2. Timing of *ENOD40* expression

ENOD40 has been identified in several legumes and in all these species the gene is induced in the nodule primordium as well as in the region of the root pericycle neighbouring the primordium (Vijn et al., 1993; Yang et al., 1993; Kouchi and Hata, 1993; Matvienko et al., in press). The timing of *ENOD40* expression in dividing cortical cell as well as in the root pericycle prior to cytologically visible events can be precisely followed using a spot inoculation procedure. Alfalfa roots were spot-inoculated with *R. meliloti* (Dudley et al., 1987) and root segments harvested 15 and 48 hrs after inoculation. After 15 hrs, cell divisions have not taken place yet, while by 48 hrs following inoculation a nodule primordium has been formed. *In situ* hybridization shows that at 48 hrs *ENOD40* RNA is present in the

NATO ASI Series, Vol. G 39
Biological Fixation of Nitrogen
for Ecology and Sustainable Agriculture
Edited by A. Legocki, H. Bothe, A. Pühler
© Springer-Verlag Berlin Heidelberg 1997

primordium cells as well as in the root pericycle, while at 15 hrs *ENOD40* is only expressed in the root pericycle. Thus *ENOD40* expression in the pericycle precedes the induction of cell division in the root cortex. This suggest that *Rhizobium* induced expression of *ENOD40* in the root pericycle might be part of the mechanism triggering a local change in auxin/cytokinin balance which ultimately leads to cell divisions in the root cortex. Hence we were interested in studying whether *ENOD40* expression changes the response of the host cell to phytohormones.

3. *ENOD40* confers tolerance to high auxin

Tobacco cells provide a model system to assay the biological effects of phytohormones. Protoplast division under defined culture conditions provides a powerful means of studying the effects of the expression of genes either directly modifying levels of active phytohormones (Brzobohaty et al., 1993) or their signal transduction (Hayashi et al., 1992; Walden et al., 1993). Wild-type tobacco mesophyll protopasts display optimal division (about 60%) at 5.5 µM NAA while they divide poorly (about 20%) at high NAA concentrations (13.8 µM NAA, Hayashi et al., 1992; Walden et al., 1993). To study whether *ENOD40* influences the auxin response of plant cells we compared auxin dependent division of protoplasts isolated from leaf tissue of homozygous transformed plants (11S-F3) and from untransformed plants (Van de Sande et al., 1996). The frequency of cell division of 11S-F3 protoplasts was the same as for wild-type protoplasts at NAA concentrations up to 5.5 µM. However, while the wild-type protoplasts showed a markedly reduced division frequency at NAA concentrations above 5.5 µM, protoplasts from 11S-F3 were able to divide at these NAA concentrations with reduced frequency. Hence, *ENOD40* expression removes the sensitivity of dividing protoplasts to high levels of auxin (Van de Sande et al., 1996).

4. *ENOD40* encodes a peptide

A tobacco *ENOD40* homolog was isolated and sequence comparison of the tobacco and legume *ENOD40* clones revealed two conserved areas. The area at the 5' end of all cDNAs (region 1) contained a highly conserved small open reading frame (ORF), starting with the first ATG available, encoding a peptide of 10 (tobacco), 12 (soybean), or 13 (pea, alfalfa, and vetch) amino acids. The second conserved sequence (region 2), located in the central part of *ENOD40*, lacked a conserved ORF (Van de Sande et al., 1996).

We showed that the small ORF of region 1 was actually translated in protoplasts by making a translational fusion between the ORF of *GmENOD40* and green fluorescent protein (GFP) (Sheen et al., 1995). Furthermore, we demonstrated that the peptide was made by the protoplasts using an antibody directed against the peptide (Van de Sande et al., 1996). Because *ENOD40* is expressed in dividing cortical cells during legume nodule formation, and the induction of *ENOD40* expression in the pericycle precedes the mitotic activation of cortical cells, it is probable that ENOD40 peptide plays a role in the start of nodule organogenesis.

References

Van Brussel AAN et al. (1992) Nature 344: 781.

Vijn I, Das Neves L, Van Kammen A, Franssen H, Bisseling T (1993) Science 260: 1764.

Libbenga KR, Harkes PAA (1973) Planta 114: 17.

Newcomb W, Sippel D, Peterson RL (1979) Can J Bot 54: 2163.

Allen EK, Allen ON, Newman AS (1953) Am J Bot 40: 429.

Hirsch AM, Bhuvaneswari TV, Torrey JG, Bisseling T (1989) Proc Nat Acad Sci USA 86: 1244.

Cooper JB, Long SR (1994) Plant Cell 6: 215.

Yang WC et al. (1993) Plant J 3: 573.

Kouchi H, Hata S (1993) Mol Gen Genet 238: 106

Matvienko M et al. (in press) Plant Mol Biol

Dudley ME, Jacobs TW, Long SR (1987) Planta 171: 2899.

Brzobohaty B et al. (1993) Science 262: 1051.

Hayashi H, Czaja I, Lubenow H, Schell J, Walden R (1992) Science 258: 1350.

Van de Sande et al. (1996) Science 273: 370.

Walden R, Czaja I, Schmülling T, Schell J (1993) Plant Cell Rep 12: 551.

Sheen J, Hwang S, Niwa Y, Kobazashi H, Galbraith DW (1995) Plant J 8: 777.

Enod40 expression and phytohormonal imbalances in nodule organogenesis

Crespi[1] M., Johansson[1] C., Charon[1] C., Frugier[1] F., Poirier[1] S. and Kondorosi[1,2] A.

[1]Institut des Sciences Végétales-CNRS, 91198 Gif sur Yvette, France and [2] Institute of Genetics, Biological Research Center, Szeged, Hungary

We are interested in the molecular mechanisms involved in nodule initiation in the *R. meliloti-Medicago* symbiosis. To that end, early nodulin genes expressed in the initially dividing cortical cells of *Medicago* have been identified: *Msenod40, Mscal and Msenod12A*. Using these molecular markers, evidence indicating that Nod factors and cytokinins may share certain common signalling elements will be presented. In addition, we propose that the induction of *enod40* in the cortex may be involved in the elicitation of cortical cell divisions.

1 Expression of early nodulin genes during nodule initiation

Molecular markers for nodule initiation have been identified in alfalfa in our laboratory using different approaches (Bauer et al, 1994; Crespi et al. 1994; Crespi et al. 1994b). *Msenod12A* codes for a proline rich protein homologous to *Psenod12* (Scheres et al. 1990) that is expressed early during alfalfa nodule development. Transgenic plants carrying a promoter *Msenod12A*-GUS fusion were used to show that the gene is expressed in the dividing cortical cells of the nodule primordium. Later GUS activity was found in the invasion zone of the mature nodule. By treating the roots of these transgenic plants with various growth regulators, cytokinins (BAP or kinetin at 1 mM) were found to induce cortical cell division similarly to Nod factors. Moreover, this induction occurred in the region containing growing root hairs and was blocked by the addition of combined nitrogen, revealing a similar control for cytokinin action and nodule initiation in alfalfa roots (Bauer et al., 1996). In addition, we have recently identified two genes expressed in the nodule primordium and in spontaneous nodules: *Mscal* and *Msenod40* (Crespi et al., 1994a). These genes were both induced by cytokinins in roots under similar conditions although *Msenod40* was induced earlier. These changes in gene expression correlated with the detection of dividing cortical cells and amyloplast deposition induced by both cytokinin and Nod factor treatment of roots. Moreover, auxin treatment, which commits the dividing cells towards lateral root formation, did not induce any of these genes (or did so after a much longer period of time).

Thus, exogenous application of purified Nod factors and cytokinins had similar effects with respect to the induction of cortical cell division, early nodulin expression and amyloplast deposition. In contrast, cytokinins could not elicit any response of the epidermal root hairs whereas Nod factors induced typical root hair curling. These results suggest that lipochitooligosaccharides and cytokinins may share certain signalling elements in the activation of the root cortical cells, further strengthening the role of phytohormonal imbalances in the elicitation of nodule initiation (Hirsch and Fang, 1994; Cooper and Long, 1994). It is possible, that at least certain common elements might be induced in the primordium. Interestingly, upon treatment with Nod factors or cytokinins, the *enod40* gene was induced earlier than the other molecular markers. Already after one day, significantly higher levels of *Msenod40* transcripts were detected in alfalfa roots. Therefore, we decided to study whether *Msenod40* may play a role in nodule initiation and act early upstream in a signal transduction pathway leading to early nodulin expression.

2. *Msenod40* expression induces cortical cell division in alfalfa

The *enod40* genes coding for 700 nucleotide long RNAs without any long ORF, have been identified in several legume species. Comparison of several *enod40* genes showed that only a small ORF (corresponding to 12 or 13 amino acids) is common among them (Vijn et al., 1995), despite a strong conservation of the nucleotide sequence. Moreover, the *enod40* sequences have the tendency to form particularly stable secondary structures, a property characteristic of biologically active RNAs. We proposed that these genes might act as "riboregulators"(Crespi et al., 1994), a class of RNAs involved in the control of cell division and differentiation. Their unusual nucleotide sequence as well as their temporal expression pattern during nodulation has led us to study the effects of *enod40* expression in alfalfa root cortical cells. Hence, we developed a transient assay based on particle gun bombardment of alfalfa roots. We used the transgenic alfalfa plants carrying the *Msenod12A*-GUS fusion since in this plant nodule initiation events can be easily monitored after GUS staining. Roots (37 independent roots corresponding to 10 plants) grown for two weeks in a medium without nitrogen were shot with a 35S-GUS construct conferring constitutive GUS expression. Two days after bombardment, several transformed cells (on average 5-6 spots per root) were detected in the epidermis and the first cortical cell layer showing a clear and distinct blue colour. Occasionally, more internal cortical cell layers were transformed. These cells were transiently expressing the GUS reporter gene since controls without plasmid did not give any blue coloration. This also demonstrated that shooting itself (or wounding) did not induce the *Msenod12A* promoter. Thus, these transgenic plants could be used to assay for genes whose expression might be capable of eliciting *Msenod12A* expression.

Once conditions for the transient expression assay in alfalfa roots of the transgenic plant carrying the *Msenod12A*-GUS fusion were optimised, the effects of overexpression of *enod40* were tested in this assay. When constructs carrying a 35S-*enod40* gene fusion were used, blue spots could be detected in the shot roots (on average one spot per plant in 28 plants with 3-6 roots each). Microscopical examination showed that these sites corresponded to dividing cortical cells expressing GUS. Although we could also detect cell divisions in the pericycle, these sites did not normally express *Msenod12A* and were surrounded by a typical fluorescence from the endodermis (in our fixation conditions), indicating that they probably correspond to an early lateral root primordium. Moreover, the presence of dividing cortical cells was finally confirmed by sectioning and histological analysis of the samples. In addition, the *Msenod12A*-expressing cells showed amyloplast accumulation. Roots of 40 transgenic plants shot with control vectors without insert or overexpressing other reporter genes did not induce any cortical cell division or the *Msenod12A* promoter-GUS fusion (blue spots). Nevertheless, divisions in the pericycle were also seen in these roots.

The relation of the induction of cortical cell divisions to nodulation was further supported, since this *enod40*-mediated response was blocked by the presence of combined nitrogen in the culture medium. Moreover, transgenic *Medicago* plants overexpressing *enod40*, prepared in our laboratory, showed extensive cortical cell divisions under nitrogen starvation. These results are in agreement with our findings in the transient assay and indicate that *enod40* overexpression, either stably or transiently, is able to activate certain cortical cells by inducing their division in alfalfa roots under nitrogen-limiting conditions. These dividing cells expressed the early nodulin *Msenod12A* and accumulated large amounts of amyloplasts, suggesting that *enod40* action might be related to the elicitation of phytohormonal imbalances in these inner root cortical cells required for the generation of the nodule primordium.

3. What is the active *enod40* gene product?

It has recently been reported that *enod40* expression rendered tobacco protoplasts tolerant to auxin (van de Sande et al., 1996). Both the region spaning the small peptide as well as the 3'UTR alone were able to modify this phytohormonal response. These results indicate that whereas the small peptide is the *enod40* gene product, the 3'UTR region may play an important regulatory role, probably controlling its production.

In our transient assay in alfalfa roots, we have tested *enod40* derivatives carrying deletions corresponding to regions spanning the small peptide, as well as the 3'UTR of the *Msenod40* gene. Two days after bombardment, both constructs were capable of eliciting cortical cell division and *Msenod12A* expression, in agreement with the tobacco experiments. Moreover, direct shooting of the *in vitro*-synthesized RNA also gave a positive response, whereas control RNAs or degraded *enod40* RNA did not. The *in vitro*-synthesized small peptide was added directly to roots which were subsequently shot with naked gold particles since the peptide did not bind to these. Preliminary experiments indicate that, after two days, the *Medicago* small peptide was able to elicit cortical cell division at 10^{-7}M. These experiments also suggested that the *enod40* small peptide and its 3'UTR are connected through a regulatory circuit controlling their action.

Recently, the 3'UTR of several genes has been implicated in the control of the spatial localization of morphogens in *Drosophila* and also in the induction of certain differentiation processes in mammalian cells (Cook, 1995). In the latter case, the 3'UTR of the tropomyosin gene was able to act *in trans* on the elicitation of tumour suppression (Rastinejad et al., 1993). Hence, the 3'UTRs might have novel functions in the cell concerning intracellular trafficking and organisation of the cytoplasmic compartments, possibly mediating the action of morphogens or hormones. Thus, analysis of the molecular mechanism of *enod40* action, possibly being the first riboregulator in plants (Crespi et al., 1994a), may unravel a novel way for translational control in plants. Furthermore, *enod40* function is likely involved in the modulation of hormonal responses in the root cortical cells which are required for the initiation of nodule organogenesis.

Acknowledgements: The authors wish to thank Petra Bauer for preparing the RNA from hormone-treated roots. F.F. and C.C. were the recipients of fellowships from the Ministere Francais de l'Enseignement Superieur et de la Recherche. C.J. was the recipient of a fellowship from the Swedish Council for Forestry and Agricultural Research.

References

Bauer, P., Crespi, M.D., Szécsi, J., Allison, L., Schultze, M., Ratet, P., Kondorosi, E. and Kondorosi A. (1994) Plant Physiol. 105, 585.

Bauer, P., Ratet, P., Crespi, M.D., Schultze M. and Kondorosi, A. (1996) Plant J. 10(1), 91.

Cook,D. (1995) BioEssays 17(3), 191.

Cooper, J.B., and Long, S.R. (1994) Plant Cell 6, 215.

Crespi, M., Jurkevitch, E., Poiret, M., D'Aubenton-Carafa, Y., Petrovics, G., Kondorosi, E., and Kondorosi, A. (1994a) EMBO J. 13: 5099.

Crespi, M., Jurkevitch, E., Brown, S., Coba, T., Frugier, F., Magyar, Z., Poiret, M., Savouré, A., Schultze, M., Dudits, D., Kondorosi, A. and Kondorosi, E. (1994b) In Proc. of the 1st European Congress of Nitrogen Fixation in Szeged, Hungary, eds. G. Kiss and G. Endre, Officina Press, 165-168.

Hirsch, A.M. and Fang, Y. (1994) Plant. Mol. Biol. 26, 5.

Rastinejad, F., Conboy, M.J., Rando, T.A. and Blau, H.M. (1993) Cell 75, 1107.

Scheres, B., van de Wiel, C., Zalensky, A., Horvath, B., Spaink, H., van Eck, H., Zwartkruis, F., Wolters, A-M., Gloudemans, T., van Kammen, A. and Bisseling, T. (1990b) Cell 60, 281.

Van de Sande, K., Pawlowski, K., Czaja, I., Wieneke, U., Schell, J., Schmidt, J., Walden, R., Matvienko, M., Wellink, J., Van Kammen, A., Franssen, H. and Bisseling, T. (1996) Science, 273, 370.

Vijn I., Yang W., Pallisgard N., Ostergaard E., van Kamen A. and Bisseling T. (1995). Plant Mol. Biol. 28, 1111.

N-Acyl Galactosamine Inhibition of Lipo-Chitooligosaccharide Action

Horst Röhrig, Jürgen Schmidt, Ursula Wieneke, Richard Walden, Jeff Schell and Michael John

Max-Planck-Institut für Züchtungsforschung, Carl-von-Linné-Weg 10, 50829 Köln, Germany

Keywords. Glycolipids, lipo-chitooligosaccharide synthesis, inhibition of nodulation

1. Introduction

In the rhizosphere, soil bacteria of the genus *Rhizobium* are induced to synthesize lipo-chitooligosaccharide (LCO) signal molecules that in turn trigger the formation of nodules on the roots of leguminous plants (Dénarié and Cullimore, 1993). The important role of LCOs as plant growth regulators has stimulated considerable interest in their synthesis. Therefore, we developed simplified procedures for the synthesis of LCOs which employ both an enzymatic and a chemical step (Röhrig et al., 1995 and 1996). We demonstrated that these synthetic LCO signals, which trigger nodule organogenesis in legumes, also promote cell division of tobacco protoplasts at very low concentrations in the absence of auxin and cytokinin. Furthermore, LCOs induce the expression of *AXI1* (Röhrig et al., 1995), a gene implicated in auxin signaling (Hayashi et al., 1992). More recently, using the protoplast division assay and transient expression assays with a construct in which the *AXI1* promoter was linked to the *GUS* reporter gene, we showed that the N-octadecenoylated monosaccharide glucosamine has all structural requirements for a biological active glycolipid. The assays also revealed that the N-acylated galactosamine epimer was inactive and specifically inhibited LCO action (Röhrig et al., 1996).

2. Synthesis of the Inhibitor N-Acyl GalN (ß 1-4 GlcNAc)₃

It was of interest to test whether the results obtained with cultured protoplasts from a nonleguminous plant also apply for the *Rhizobium*-legume interaction. Therefore, we analyzed whether the specific LCO inhibitor N-octadecenoyl GalN can also block LCO signaling in leguminous plants. We found that in plant nodulation experiments with *Rhizobium leguminosarum*, addition of this compound to the medium significantly reduced the number of nodules on the roots of *Vicia sativa* plants (data not shown). This observation suggests that the introduction of a minor structural variation at the nonreducing end of the glycolipid signal molecule can affect activity. To test this further, we synthesized a glycolipid molecule which differs from a biological active LCO only in the orientation of the hydroxyl group at the C-4 of the nonreducing sugar residue of the oligosaccharide backbone.

NATO ASI Series, Vol. G 39
Biological Fixation of Nitrogen
for Ecology and Sustainable Agriculture
Edited by A. Legocki, H. Bothe, A. Pühler
© Springer-Verlag Berlin Heidelberg 1997

The synthesis of the LCO analogue N-acyl $_{C18:1(9E)}$ GalN (ß 1-4 GlcNAc)$_3$ involves two enzymatic steps for the formation of the oligosaccharide backbone, followed by the chemical attachment of the fatty acyl chain (Fig. 1). To attach galactosamine to tri-N-acetyl-chitotriose we synthesized UDP-GalN from UDP-Glc and GalN-1-phosphate by an exchange reaction using the enzyme Gal-1-phosphate uridyl transferase (Maley, 1968). The nucleotide sugar donor UDP-GalN was subsequently used to synthesize the tetrasaccharide GalN (ß 1-4 GlcNAc)$_3$ by a transfer reaction catalyzed by galactosyltransferase (Palcic, 1994). Finally, the monounsaturated fatty acid *trans*-9-octadecenoic acid was attached to the free amino group of the GalN residue through the action of 2-chloro-1-methylpyridinium iodide as described (Röhrig et al., 1996). The acylated tetrasaccharide was obtained by HPLC purification on a preparative C18 column (Röhrig et al., 1994). Transient GUS expression assays in transfected tobacco protoplasts revealed that this compound was not able to induce *AXI1* expression.

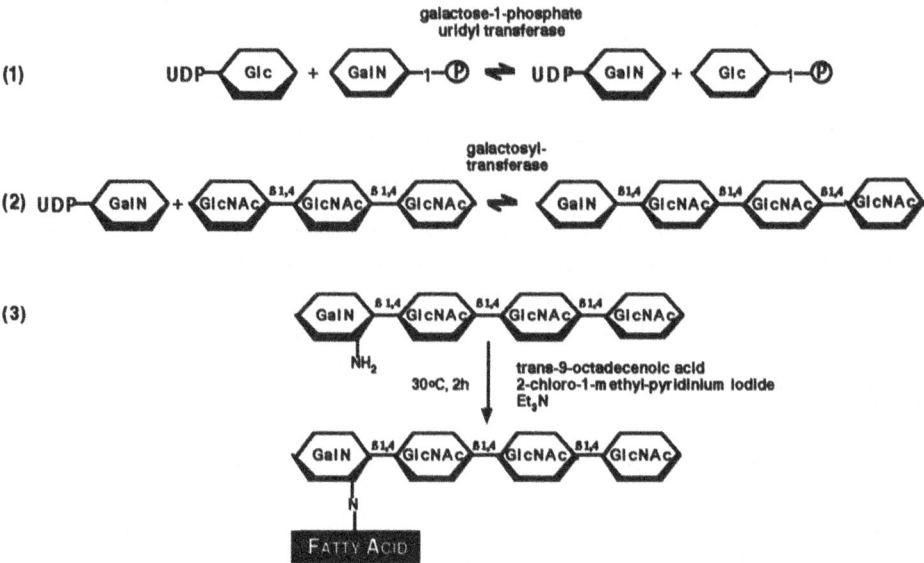

Fig. 1. Steps involved in the synthesis of the inhibitor N-acyl $_{C18:1(9E)}$ GalN (ß1-4 GlcNAc)$_3$

3. Effect of the Inhibitor on Nodule Formation

When *R. leguminosarum bv. viciae* was inoculated onto its host plant *Vicia sativa* together with the LCO analogue N-acyl GalN (ß 1-4 GlcNAc)$_3$, nodule formation was reduced to more than 50% in comparison with control experiments which contained a biological active LCO, or no glycolipid (Fig. 2).

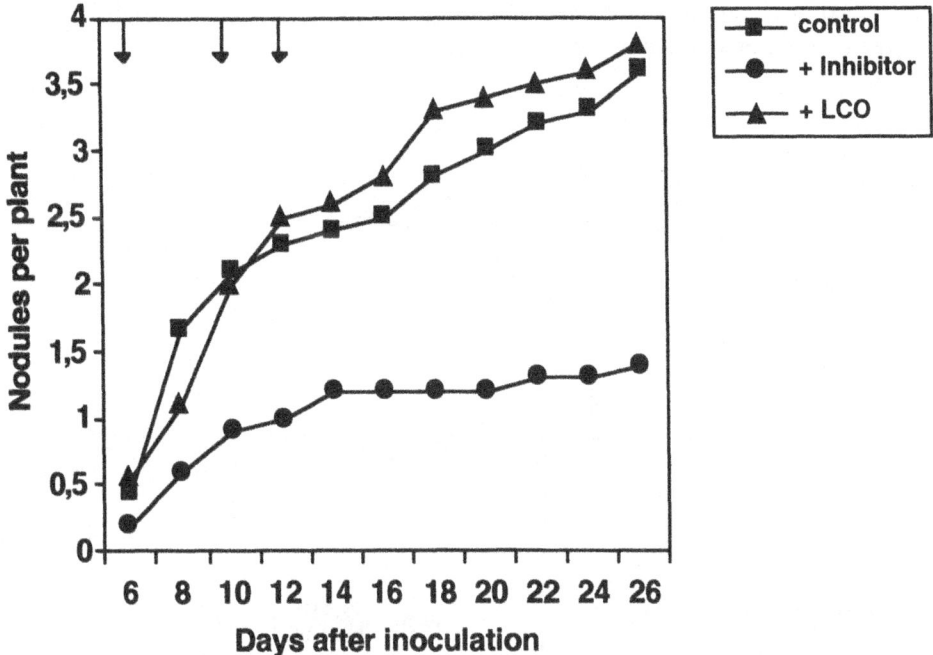

Fig. 2. Inhibition of nodulation by *N*-octadecenoyl Gal*N* (ß 1-4 GlcNAc)₃. Nodules were induced by *Rhizobium leguminosarum bv. viviae* on vetch. Ten plants were used for each experiment. *Vicia sativa* seedlings were grown in tubes containing 20 ml of sterile agar medium supplemented with, or without 10^{-7} M synthetic glycolipids. At the intervals indicated (arrows), fresh glycolipids were added. *R. leguminosarum* without glycolipids (■, control); *R. leguminosarum* with the LCO *N*-acyl $_{C18:1(9E)}$ Glc*N* (ß 1-4 GlcNAc)₃ (▲); *R. leguminosarum* with the inhibitor N-acyl $_{C18:1(9E)}$ Gal*N* (ß 1-4 GlcNAC)₃ (●). Data shown are representatives of three independent experiments with similar results.

Normal nodulation occured in the control experiments and there was no significant difference in the number of nodules formed in comparison with experiments in which the LCO was added. The LCO differs from the inhibitor molecule only in the monosaccharide present at the nonreducing end of the tetrasaccharide backbone (Fig. 3). In the Glc*N* moiety of the LCO all axial positions are occupied by H-atoms, whereas the bulkier -OH, -CH₂OH, and -NH₂ equatorial groups emerge at the periphery of the pyranose ring. In contrast, the GalN moiety of the LCO inhibitor carries a hydroxyl group at C-4 in an axial position which renders the corresponding glycolipid biologically inactive. This indicates, that a putative LCO binding protein can distinguish between this single point of difference. The observed reduction of nodulation after the addition of the Gal*N*-containing glycolipid suggests that this compound may compete with the LCO-signal secreted by the rhizobia for the recognition site of a putative receptor.

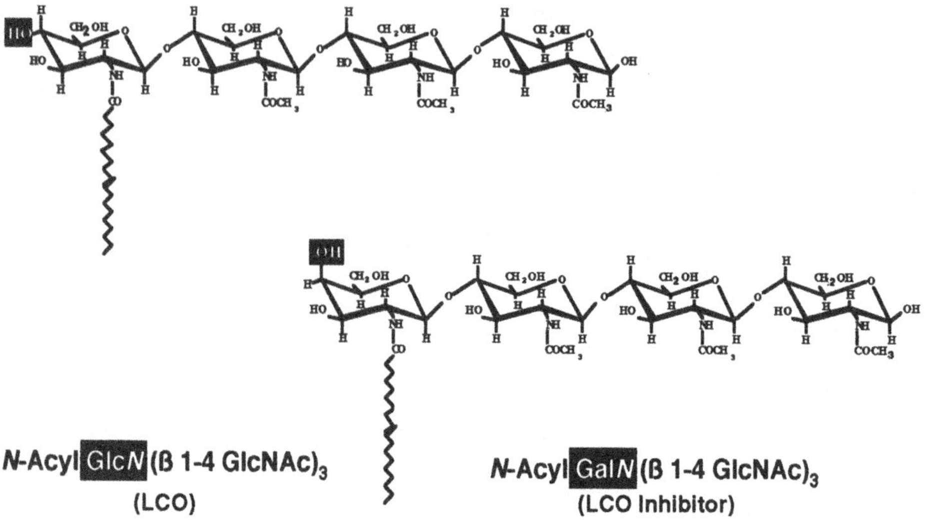

N-Acyl GlcN **(β 1-4 GlcNAc)₃**
(LCO)

N-Acyl GalN **(β 1-4 GlcNAc)₃**
(LCO Inhibitor)

Fig. 3. Comparison of the structures of a lipo-chitooligosaccharide with its inhibitory analogue.

4. Acknowledgements

This work was supported by the European Union (J. Schmidt).

5. References

Dénarié, J. & Cullimore, J. (1993) *Cell* **74**, 951-954.

Hayashi, H., Czaja, I., Lubenow, H., Schell, J. & Walden, R. (1992) *Science* **258**, 1350-1353.

Maley, F. (1968) *Meth. Enzymol.* **28**, 271-274.

Palcic, M., M. (1994) *Meth. Enzymol.* **230**, 300-316.

Röhrig, H., Schmidt, J., Wieneke, U., Kondorosi, E., Barlier, I., Schell, J. & John, M. (1994) *Proc. Natl. Acad. Sci. U.S.A.* **91**, 3122-3126.

Röhrig, H., Schmidt, J., Walden, R., Czaja, I., Miklasevics, E., Wieneke, U., Schell, J. & John, M. (1995) *Science* **269**, 841-843.

Röhrig, H., Schmidt, J., Walden, R., Czaja, I., Lubenow, H., Wieneke, U., Schell, J. & John, M. (1996) *Proc. Natl. Acad. Sci. U.S.A.*, in press.

Cell Cycle Regulation during Nodule Development

I. Meskiene, W.-C. Yang*, C. de Blank*, L. Bögre, K. Zwerger, M. Brandstötter, M. Mattauch, Ton Bisseling*, and Heribert Hirt

Institute of Microbiology and Genetics, Biocenter Vienna, Dr. Bohrg. 9, 1030 Vienna, Austria
* Dept. of Mol. Biology, Wageningen Agricult. Univ., Dreijenlaan 3, 6703HA Wageningen, The Netherlands

Keywords. cell cycle, cyclin, protein kinase, nodule, rhizobia

Cell proliferation in all eukaryotes requires the accurate coordination of the timing and the order of cell cycle events. Consequently, a number of biochemical pathways, called checkpoints [11], have evolved to ensure that the initiation of particular programmes is dependent upon the successful completion of others thereby coordinating cell cycle transitions. In mammalian cells the cell cycle transitions are controlled by extracellular and intracellular signals that may act in a positive or negative way. The signals are transmitted via networks of signalling pathways that ultimately change the activity of cyclin-dependent kinases (Cdks). Cdks are a family of serine/threonine protein kinases that regulate individual cell cycle transitions. Cdks require association with cyclins for activation, and the timing of activation is largely dependent upon cyclin expression [12]. Whereas in multicellular organisms, different members of the Cdk family act in different stages of the cell cycle [8], in unicellular eukaryotes, a single Cdk fulfills these functions [12]. The mechanisms of regulating Cdk activity consist of positive elements such as cyclin abundance, phosphorylation by kinases and dephosphorylation by phosphatases, and negative elements such as inhibitory phosphorylation by kinases [3] and abundance of Cdk inhibitors [4].

How do cells get started?

In the unicellular yeast *Saccharomyces cerevisiae*, cells decide to exit or continue the mitotic cycle in the G1 phase, at a point called START [12]. At this point, nutrient status and cell size are monitored and transition into S phase is allowed only if certain critical values are reached. However, once cells have passed START, cells become insensitive to these signals and undergo a complete cell cycle before becoming arrested in the next G1 phase. For the passage of START, expression of G1 cyclins (*CLN1*, *CLN2*, and *CLN3*) is required and their overexpression contracts G1 phase and decreases cell size. Cln3-Cdc28 is present throughout the cell cycle, and its kinase activity is necessary for activation of the *CLN1* and *CLN2* genes [17].
  Cells of multicellular organisms must respond to the same intracellular signals as the unicellular yeasts. However, they must also integrate the information of a multitude of extracellular signals, such as growth factors, mitogens, antimitogens, differentiation factors, spatial cues as well as signals from surrounding cells. In mammalian cells, these stimuli are monitored at a particular point of the G1 phase, called the restriction point [14]. Similar to yeast, mammalian cells become insensitive to these factors after passing the restriction point, and progress through the cell cycle until they reach the next G1 phase. G1 cyclins (D-and E-type cyclins) are key regulators of the transition through the G1 phase and their overexpression contracts G1 phase, reduces cell size and the dependency on mitogens [15]. The D-type cyclin associated Cdk4 and Cdk6 kinases phosphorylate the retinoblastoma protein (Rb), thereby releasing Rb-bound transcription factors of the E2F-DP1 family which in turn activate genes required for DNA replication [13]. D-type cyclins interact directly with Rb through a small domain, called the Rb-binding site. Recently, plant cells have also been found to contain homologs of D-type cyclins [2, 16]. The structural similarity as well as the presence of a conserved Rb-binding motif in all D-type plant cyclins suggest that the G1/S transition of plant cells is also based on an Rb-dependent mechanism.

Lateral Roots and Nodulation: two model systems for cell cycle programming

Although the major control point of mammalian cells appears to be the restriction point in the G1 phase, evidence for specific checkpoints in the G2 phase exists both in yeasts and animals. Plants contain a certain population of cells that are naturally arrested in the G2 phase and can be easily induced to reenter the cell cycle. These cells belong to the pericycle of the root and, after addition of auxin, undergo mitosis within a few hours [1]. Molecular studies on Arabidopsis have indicated that transcripts of *cdc2* but not of mitotic cyclin *cyc1* are normally present in the pericycle [5, 6, 9]. However, *cyc1* gene expression can be induced in this tissue after induction of lateral root formation by exogenous auxin [5].

NATO ASI Series, Vol. G 39
Biological Fixation of Nitrogen
for Ecology and Sustainable Agriculture
Edited by A. Legocki, H. Bothe, A. Pühler
© Springer-Verlag Berlin Heidelberg 1997

Although roots from many plant species respond to the addition of auxins by the formation of lateral roots, interaction with rhizobia or the addition of Nod factors induces nodule formation almost exclusively in legume roots. In contrast to lateral root formation, nodule primordia are formed from cortex cells that are stimulated to divide. The fact that only a few cortex cells start to proliferate and form nodule primordia led researchers to ponder on the underlying mechanism. Among the majority of root cortex cells with a 2C DNA content, a few cells are consistently found with a 4C DNA content. Because 2C and 4C DNA contents normally correspond to G1 and G2 cells, this observation was suggested to mean that only the small population of G2-arrested cells might be activatable by rhizobial mitogenic factors. The isolation of specific cell cycle marker genes offered the possibility to test this hypothesis by studying the reactivation of the cortex cells at the single cell level by in situ hybridization. Upon re-entry into the cell cycle from a G1 arrest point, DNA synthesis occurs before mitosis. Alternatively, re-entry from a G2 phase-arrest immediately triggers mitosis without DNA replication. It turned out that all the cortex cells investigated expressed an S phase-specific histone gene before a G2/M phase-specific cyclin gene [18]. These results unequivocally proved that cortex cells re-enter the cell cycle from a G1 arrest point (Fig.1).

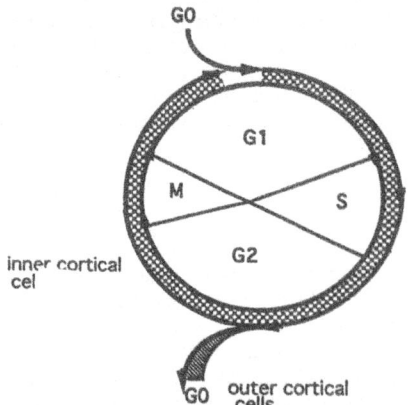

Fig. 1. Model of cell cycle events in root cortical cells after Nod factor stimulation. After mitogenic stimulation by Nod factor, inner and outer root cortex cells re-enter the cell cycle in the G1 phase. In contrast to the inner cortical cells that go through the entire cell cycle and proliferate, outer cortical cells transit through S phase but become arrested in G2.

Position-dependent cell cycle activation of root cortical cells during nodulation

Although auxin acts as a mitogen on root pericycle cells to reenter the cell cycle in the G2 phase giving rise to a lateral root primordium, cytokinin inhibits lateral root formation and mimicks Nod factors by activating root cortical cells to form nodule primordia [7]. Different hormones may not only have different effects on different tissues of the same organ, but may also stimulate cells of the same tissue to undergo different programs. This is shown by studying the expression of histone and cyclin genes during nodule formation in alfalfa roots. Although Nod factors induced expression of histone genes in inner and outer cortex cells, only inner cortical cells showed subsequent expression of mitotic cyclin genes [18]. These results suggested that the cell cycle is activated in both inner and outer cortex cells, and that both types of cells proceed through the S phase. Whereas inner cortical cells are allowed to progress through the entire cycle and proliferate, outer cortical cells proceed through START, transit S phase but get arrested before mitosis (Fig.1). Previous cytological studies suggested that the G2 arrest of the outer cortical cells is associated with the formation of particular structures, the so-called pre-infection threads. During activation of outer cortical cells, the nuclei move into a central position and radially aligned dense cytoplasmic strands form. The cytoplasmic strands are denoted as infection threads and bridge the outer cortical cells allowing the rhizobial bacteria to enter the plant in a later step. The similarity between the infection threads and phragmosomes led to the idea that the transient reentry of the outer cortical cells into the cell cycle merely serves the purpose to form cell cycle structural elements that can be efficiently used for the transport of the symbionts. Thus, besides other factors, the response of a cell to a particular mitogen also depends on the positional context of the cell in the tissue.

Re-entry into the cell cycle is a multistep programme that is associated with the expression of different cyclin genes

Mammalian D-type cyclins are induced as part of a delayed early response to mitogenic growth factors and peak near the G1/S transition, although expression persits as long as growth factors are present [15]. A very similar situation was observed when alfalfa leaves were mitogenically stimulated. The alfalfa D-type cyclin gene *cycMs4* was induced before onset of DNA synthesis. However, transcriptional induction of *cycMs3*, another cyclin gene, was observed hours before that of the *cycMs4* gene, indicating that several distinct steps occur during cell cycle activation of differentiated leaf cells (Fig. 2). After rhizobial activation, similar results were obtained as in mitogen-treated leaf pieces, indicating that root and leaf cells are activated by very similar mechanisms.

In mammalian cells, D-type cyclin expression is an early step during resumption of cell division, but immediate early events that control the re-entry into the cell cycle are apparent and depend on the MAP kinase-dependent expression of several transcripttion factors that subsequently activate expression of D-type cyclins and other genes. A multistep mechanism is also found in yeast cell cycle activation. In contrast to other cyclins, yeast cells have constitutive amounts of the Cln3 cyclin. It is assumed that activation of the Cln3-Cdc28 kinase triggers a positive feedback loop, resulting in the expression of the other G1 cyclins which will activate the Cdc28 kinase to sufficiently high levels to transit START [17]. The alfalfa cyclin CycMs3 might play a similar role in the plant cell cycle. Because *cycMs3* gene expression preceeds that of the D-type cyclin cycMs4, it is suggested that CycMs3 might be necessary to trigger the expression of later genes including other G1 cyclins [10, Fig. 2].

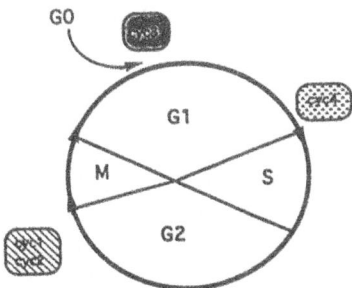

Fig. 2. Model of cell cycle events during re-entry of differentiated G0-arrested cells into the G1 phase of the proliferative cell cycle. Upon mitogenic stimulation of cells, *cycMs3* gene expression is induced, followed by induction of *cycMs4* before start of DNA replication. The expression of the *cycMs1* and *cycMs2* genes occurs when cells prepare for entry into mitosis.

Acknowledgments

The research is funded by grants from the Austrian Science Foundation and the European Human Capital and Mobility programme.

References

1.   Blakely LM, and Evans TA (1979) Plant Sci Lett 14: 79-83.
2.   Dahl M et al. (1995) Plant Cell 7: 1847-1857 (1995).
3.   Draetta GF (1993) Curr. Op. Cell Biol. 6: 842-846.
4.   Elledge SJ, and Harper JW (1994) Curr Op Cell Biol 6: 847-852.
5.   Ferreira PCG et al. (1994) Plant Cell 6: 1763-1774 (1994).
6.   Hemerly AS et al. (1993) Plant Cell 5: 1711-1723.
7.   Hirsch AM, and Fang Y (1994)
8.   Hirt H, and Heberle-Bors E (1994) Sem Dev Biol 5: 147-154.
9.   Martinez MC et al. (1992) Proc Natl Acad Sci USA 89: 7360-7364.
10. Meskiene I et al. (1995) Plant Cell 7: 759-771.
11. Murray A (1994) Curr Op Cell Biol 6: 872-876.
12. Nasmyth K (1993) Curr Op Cell Biol 2, 166-170.
13. Nevins JR (1992) Science 258: 424-429.
14. Pardee AB (1989) Science 246: 603-608.
15. Sherr CJ (1994) Cell 79: 551-555.
16. Soni R et al. (1995) Plant Cell 7, 85-103.
17. Tyers M et al. (1993) EMBO J 12, 1955-1968.
18. Yang W-C et al. (1994) Plant Cell 6: 1415-1426.

Part 3

Plant Genes
Involved in Nodulation

The Saga of the Nodulin Genes

Desh Pal S. Verma and Zonglie Hong

Department of Molecular Genetics and Plant Biotechnology Center, Ohio State University, Columbus, OH 43210, USA

The nodulins, nodule-specific host gene products, were first identified from soybean and nodulin-35 was detected to be the second most abundant protein in soybean root nodules[1]. Following this, a group of other nodulins were detected using various immunological approaches and 2-D gel analysis[2] and nodulin-35 was characterized to be a subunit of the nodule-specific uricase[3]. These early studies, coupled with DNA sequence hybridization[4] and differential screening of nodule cDNA libraries[5] established that a group of host genes encoding nodulins is induced following infection of the host plants by rhizobia. This group was later divided into early and late nodulins[6] and each group was generally considered to play a role in early root nodule morphogenesis and function, respectively. During the last 10 years almost 200 such genes, in addition to leghemoglobin genes, have been identified from different legume hosts.

Although a nomenclature for nodulins was suggested by van Kammen[7], it was not strictly followed, as the basis of this nomenclature was arbitrary. It has now become clear that infection, morphogenesis, and function of root nodule involves a variety of host gene products, most of which are also expressed in other tissues but at a much lower level. The expression of these genes is enhanced in root nodules. Although the paradime of nodulins has served a useful purpose, to understand the variety of factors that control expression of this diverse group of genes requires that we focus on each gene independently and study various factors that control the expression of these genes in different tissues. Even the leghemoglobin, considered to be strictly nodule-specific, has recently been identified in non-legumes and a non-symbiotic form of this gene is shown to be expressed in roots of legume plants[8]. The nodule-specific alleles of various genes have apparently evolved as a result of gene duplication in order to meet the specific demands of this highly specialized tissue. The infected cells of the symbiotic zone are hypoxic, hyper-osmotic and may have higher subcellular pH. Consequently, not only the structure of these genes but also mode of regulation may have to be altered. It is interesting to see

NATO ASI Series, Vol. G 39
Biological Fixation of Nitrogen
for Ecology and Sustainable Agriculture
Edited by A. Legocki, H. Bothe, A. Pühler
© Springer-Verlag Berlin Heidelberg 1997

that the entire developmental program may be under the control of Nod factors that act as morphogens and such factors may interact directly or *via* specific receptors with the novel transcription factors. In fact, nodule-specific transcription factors have recently been identified. They too, however, may have been modified from those expressed in flower[8].

The biological functions of nodulins are very diverse. We, therefore, suggest that each nodulin be considered at its own whether it is an allele of a gene expressed in other tissues or it encodes a novel gene product . Also, whether it represents a structural protein of the cell wall or membrane(s) or has an enzymatic activity, etc need to be determined. Thus, we have regrouped nodulin genes characterized from different legume plants according to the above criteria and have assigned a mnemonic as suggested by the Commission on Plant Gene Nomenclature[9]. Those that have isoforms in other tissues but a clear nodule-specific form exists are marked by subscript "n", such as glutamine synthetase, $Gln1_n$. There are few nodulins such as nodulin-24 and nodulin-23 in soybean which do not seem to have isoforms expressed in other tissues[9a].

Table 1 represents an example of suggested mnemonics for nodulins from soybean, pea and alfalfa. Mnemonics for nodulins from other legumes will be accessible on the CPGN database. As shown in Table 1, the function of many nodulins remains to be determined. As the scope of nodulin genes varies, the regulation of these genes is also very diverse, ranging from developmental to metabolic control. We have demonstrated that $Gln1_n$ is induced directly by ammonia[9b] and the first gene of the purine biosynthesis pathway is induced by glutamine[9c].

Although over 100 nodulation mutations in different legume species have been identified, none of the corresponding genes have yet been cloned. It may be necessary to use chromosome mapping or tagging approach to identify these genes and determine their function in root nodule symbiosis. This has been hampered by the large genome and difficulties in tagging the genome using specific transposons. A model legume like *Lotus* may prove useful in this context. Recently, it has been demonstrated that some of the nodulation mutants may also affect infection by mycorrhiza indicating a common mechanism is involved in the infection process. Since mycorrhizal infection is more prevelant in plants and not restricted to legumes, understanding the function of these genes may allow broadning of the rhizobial host range.

Table 1. Suggested classification and nomanclature of sequenced nodulin genes for symbiotic nitrogen fixation from soybean, pea and alfalfa

Class	Gene product	Gene Family	Gene Synonym	Gene of a Plant Species	Ref.
1.Structural proteins					
- Membrane proteins					
	Ion channel	$Icl1$	nodulin-26	GLYmax;Cnl1	10, 11
	Sulfate transporter	$Str1$	GmN70	GLYmax;Str1	12, 13
	Peribacteroid membrane protein	$Pbp1$	nodulin-24	GLYmax;Pbp1	15
	Peribacteroid membrane protein	$Pbp2$	nodulin-16	GLYmax;Pbp2	16
- Cell wall proteins					
	Prolin-rich protein	$Prp1;1$	nodulin-75A; GmENOD2A	GLYmax;Prp1;1	17, 18
	Prolin-rich protein	$Prp1;2$	nodulin-75B GmENOD2B	GLYmax;Prp1;2	17, 18
	Prolin-rich protein	$Prp1$	PsENOD2	PISsat;Prp1	19
	Prolin-rich protein	$Prp1$	MsENOD2	MEDsat;Prp1	20
	Prolin-rich protein	$Prp2$	PsENOD5	PISsat;Prp2	21
	Prolin-rich protein	$Prp3;1$	PsENOD12A	PISsat;Prp3;1	22
	Prolin-rich protein	$Prp3;2$	PsENOD12B	PISsat;Prp3;2	22
	Prolin-rich protein	$Prp3;1$	MsENOD12A	MEDsat;Prp3;1	23
	Prolin-rich protein	$Prp3;2$	MsENOD12B	MEDsat;Prp3;2	23
	Cell wall bound protein	$Cwb1$	MsENOD8	MEDsat;Cwb1	24
2.Enzymes					
- Nitrogen assimilation					
	Glutamine synthetase	$Gln1_n$	GSGmE	GLYmax;Gln1$_n$	25
	Glutamine synthetase	$Gln1_n$	PsGSR3	PISsat;Gln1$_n$	26
	Glutamine synthetase	$Gln1_n$	PGS13	MEDsat;Gln1$_n$	27
	Uricase	$Urc1_n$	nodulin-35	GLYmax;Urc1$_n$	28
- Carbon metabolism					
	Sucrose synthase	$Sus1_n$	nodulin-100	GLYmax;Sus1$_n$	29
	isopropylmalate synthase	$Ims1_n$	GmN56	GLYmax;Ims1$_n$	30

(continued)

Class	Gene product	Suggested mnemonics	Trivial name	Gene of a Plant Species	Ref.
- Oxygen delivery/protection					
	Leghemoglobin	$Lhg1;1_n$	Lba	GLYmax;Lhg1;1	31
	Leghemoglobin	$Lhg1;2_n$	Lbb	GLYmax;Lhg1;2	31
	Leghemoglobin	$Lhg1;3_n$	Lbc1	GLYmax;Lhg1;3	31
	Leghemoglobin	$Lhg1;4_n$	Lbc2	GLYmax;Lhg1;4	31
	Leghemoglobin	$Lhg1;5_n$	Lbc3	GLYmax;Lhg1;5	31
	Leghemoglobin	$Lhg1;1_n$	Ms460	MEDsat;Lhg1;1	32
	Leghemoglobin	$Lhg1;2_n$	Ms560	MEDsat;Lhg1;2	32
	Leghemoglobin	$Lhg1;3_n$	Ms561	MEDsat;Lhg1;3	32
	Leghemoglobin	$Lhg1;4_n$	Ms11	MEDsat;Lhg1;4	32
	Leghemoglobin	$Lhg1;5_n$	MsIII	MEDsat;Lhg1;5	32
	Protochlorophyllide reductase	$Por1_n$	MtNOD712	MEDsat;Pcr1	33
- Stress/pathogen-induced enzymes					
	Chitinase	Chi1;1	2-8-X	VICfab;Cht1A	34
	Chitinase	Chi1;2	176-X	VICfab;Cht1B	34
- Extracellular proteins					
	Protease inhibitor	$Pi1_n$	Nodulin-21	PSOtet;Pi1n	35
	Nodule-specific lectin	Ns11	PsNlec1	PISsat;Ns11	36
	a-mannosidase	Msd1			37
	Peribacteroid fluid proteins	To be characterized			38
3. Signal transduction pathway proteins					
- Transcription factor					
	MADS-box-containing protein	Mbc1	NMH7	MEDsat;Mbc1	10
- Peptide hormones					
	Auxin modulator	$Amd1;1_n$	GmENOD40-1	GLYmax;Amd1;1n	39, 40
	Auxin modulator	$Amd1;2_n$	GmENOD40-2	GLYmax;Amd1;2n	39, 40
	Auxin modulator	$Amd1_n$	PsENOD40	PISsat;Amd1n	39, 40
	Auxin modulator	$Amd1_n$	MsENOD40	MEDsat;Amd1n	39, 40
- Kinases/phosphatases					
	Several genes are being characterized				

(continued)

Class	Gene product	Suggested mnemonics	Trivial name	Gene of a Plant Species	Ref.
4. Uncharacterized proteins/genes					
	putative metal-binding protein	Mbp1	nodulin-20	GLYmax;Mbp1	41
	putative metal-binding protein	Mbp2	nodulin-22	GLYmax;Mbp2	41
	putative metal-binding protein	Mbp3	nodulin-23	GLYmax;Mbp3	42, 43
	putative metal-binding protein	Mbp4	nodulin-26b	GLYmax;Mbp4	42, 43
	putative metal-binding protein	Mbp5	nodulin-27	GLYmax;Mbp5	42, 43
	putative metal-binding protein	Mbp6	nodulin-44	GLYmax;Mbp6	42, 43
	putative metal-binding protein	Mbp7	15-9-A	GLYmax;Mbp7	44
	putative metal-binding protein	Mbp8A	PsENOD3	PISsat;Mbp8A	21
	putative metal-binding protein	Mbp8B	PsENOD14	PISsat;Mbp8B	21
	putative metal-binding protein	Mbp8C	PsENOD6	PISsat;Mbp8C	45
	putative Copper-binding protein	Cbp1A	GmENOD55-1	GLYmax;Cbp1A	46
	putative Copper-binding protein	Cbp1B	GmENOD55-2	GLYmax;Cbp1B	46
	Methionine-rich protein	Mrp1	nodulin-21	GLYmax;Mrp1	47
	Unknown-function	----	GmN93	GLYmax;----	13
	Unknown-function	----	PsENOD7	PISsat;----	X93172
	Unknown-function	----	MsNOD25A	MEDsat;----	48, 49
	Unknown-function	----	MsNOD25	MEDsat;----	48, 49
	Unknown-function	----	MsNOD25R	MEDsat;----	48, 49

References

1. R Legocki & DPS Verma, 1979, Science, 205, 190-192
2. R Legocki & DPS Verma, 1980, Cell, 20, 153-163
3. H Bergmann et al., 1983, EMBO J. 2, 2333-2339
4. S Auger & DPS Verma, 1981, Biochem. 20, 1300-1306
5. F Fuller et al., 1983, Proc. Natl. Acad. Sci. USA 80, 2594-2598
6. DPS Verma et al., 1988, In: Molecular Genetics of Plant-Microbe Interactions, Eds., R. Palacios & D.P.S. Verma, APS Press, St. Paul, Minnesota, pp. 315-320
7. A van Kammen, 1984, Plant Mol. Biol. Reptr. 2, 43-45
8. C Anderson et al., 1996, Proc. Natl. Acad. Sci. USA 93, 5682-5687
9. CPGN. 1994. Plant Mol. Biol. Reptr. 12(2) Supplement: S1-S109
9a. DPS Verma et al., 1986 Plant Mol Biol. 7, 51-61
9b. G-H Miao et al., 1991 Plant Cell 3, 11-22
9c. J Kim et al.,. 1995 Plant J. 7, 77-86
10. J Heard & K Dunn, 1995, Proc. Natl. Acad. Sci. USA 92, 5273-5277
11. M Fortin et al., 1987, Nucleic Acids Res. 15, 813-824
12. J Lee et al., 1995, J. Biol. Chem. 270, 27051-27057
13. H Kochi & S Hata 1993, Mol. Gen Genet. 238, 106-119
14. F Smith et al., 1995, Proc. Natl. Acad. Sci. USA 92, 9373-9377
15. F Fuller et al., 1983, Proc. Natl. Acad. Sci. USA 80, 2594-2598
16. W Nirunsuksiri & C Sengupta-Gopalan, 1990, Plant Mol. Biol. 15, 835-849
17. H Franssen et al., 1987, Proc. Natl. Acad. Sci. USA 84, 4495-4499
18. H Franssen et al., 1990, Plant Mol. Biol., 14, 103-106
19. C van de Wiel et al., 1990, EMBO J., 9, 1-7
20. R Dickstein et al., 1988, Genes Dev. 2, 677-687
21. B Scheres et al., 1990, Plant Cell 2, 687-700
22. B Scheres et al., 1990, Cell 2, 60, 281-294
23. L Allison et al., 1993, Plant Mol. Biol. 21, 375-380
24. R Dickstein et al., 1993, Mol. Plant Microbe Interact. 6, 715-721
25. D Roche et al., 1993, Plant Mol. Biol. 22, 971-983
26. S Tingey et al., 1987, EMBO J. 6, 1-9
27. K Dunn et al., 1988, Mol. Plant Microbe Interact. 1, 66-74
28. T Nguyen et al., 1985, Proc. Natl. Acad. Sci. USA 82, 5040-5044
29. F Thummler & DPS Verma, 1987, J. Biol. Chem. 262, 14730-14736
30. H Kochi & S Hata 1995, Mol. Plant Microbe Interact. 8, 172-176
31. N Brisson & DPS Verma, 1982, Proc. Natl. Acad. Sci. USA 79, 4055-4059
32. E Davidowitz et al., 1991, Plant Mol. Biol. 16, 161-165
33. R Wilson & J Cooper, 1994, Plant Physiol. 104, 289-290
34. A Perlick et al., 1996, Plant Physiol. 110, 147-154

35. J Manen et al., 1991, Plant Cell, 3, 259-270
36. I Kardaisky et al., 1996, Plant Physiol. 111: 49-60
37. A Kinnback et al, 1987, J. Exp. Bot. 38, 1373-1377
38. M Fortin et al., 1985, EMBO J. 4, 3041-3-46
39. W Yang et al., 1993, Plant J., 3, 573-585
40. K van de Sande et al., 1996, Science 273, 370-373
41. N Sandal et al., 1987, Nucleic Acids Res. 15, 1507-1519
42. V Mauro et al., 1985, Nucleic Acids Res. 13, 239-249
43. C Sengupta-Gopalan et al., 1986, Mol. Gen. Genet. 203, 410-420
44. S Gottlob-McHugh & D Johnson, 1991, Can. J. Bot. 69, 2663-2669
45. I Kardailsky et al., 1993, Plant Mol. Biol. 23, 1029-1037
46. C de Blank et al., 1993, Plant Mol. Biol. 22, 1167-1171
47. A Delauney et al., 1990, Plant Mol. Biol. 14, 449-451
48. G Kiss et al., 1990, Plant Mol. Biol. 14, 467-475
49. Z Vegh et al., 1990, Plant Mol. Biol. 15, 295-306

Identification of New *Medicago truncatula* Nodulin Genes : Comparison of Two Molecular Approaches

Fernanda de Carvalho-Niebel, Nicole Lescure, Julie Cullimore and Pascal Gamas

Laboratoire de Biologie Moléculaire des Relations Plantes-Microorganismes,
INRA-CNRS, BP27, 31326 Castanet-Tolosan Cédex, France

1. Introduction

During nodulation many plant genes are induced or enhanced in expression (in comparison to roots) presumably in order to fulfil roles in the production and functioning of the new organ. These genes have been termed nodulins and have been further classified into late and early types depending on their timing of expression. Initially nodulin clones were isolated by differential screening of nodule cDNA banks with nodule and root probes. This technique is biased towards isolating clones for genes that are strongly induced such as those encoding leghaemoglobins and proline-rich proteins. Despite this bias, studies on these genes have provided an enormous amount of information on the structure and development of the nodule (see Franssen et al., 1992; Mylona et al., 1995).

We believe that studies on a larger number of nodulin genes would increase our knowledge and understanding of nodulation. We therefore decided to set up molecular approaches to isolate clones for such genes. These clones would serve two purposes. Firstly they would be used as new molecular markers to define the different stages of nodulation. Secondly, the nature of the encoded gene products may give some insight into the biological processes taking place during nodulation.

For this work we have chosen the model legume *M. truncatula* (Barker et al.,1990). Its small genome, homogeneous seed populations, indeterminate nodule structure and manipulability (with regard to growth and inoculation conditions) all aid the isolation of suitable material for the cloning of nodulin genes. We chose to concentrate on the isolation of new early nodulin genes and the two techniques we have used are well suited to the isolation of clones for transcripts that are not particularly abundant.

2. Subtractive Hybridisation Strategy

The basis of this strategy is to preferentially eliminate by hybridisation 'non-induced' cDNAs from a population containing 'induced' ones. There are many ways in which this can be done. Initially we decided to use a very direct approach starting with the preparation of single-stranded cDNA from mRNA isolated from roots harvested 6, 24 and 48 hours after inoculation with *R. meliloti* (the induced state). This cDNA was hybridised with biotinylated mRNA prepared from roots inoculated in the same way

NATO ASI Series, Vol. G 39
Biological Fixation of Nitrogen
for Ecology and Sustainable Agriculture
Edited by A. Legocki, H. Bothe, A. Pühler
© Springer-Verlag Berlin Heidelberg 1997

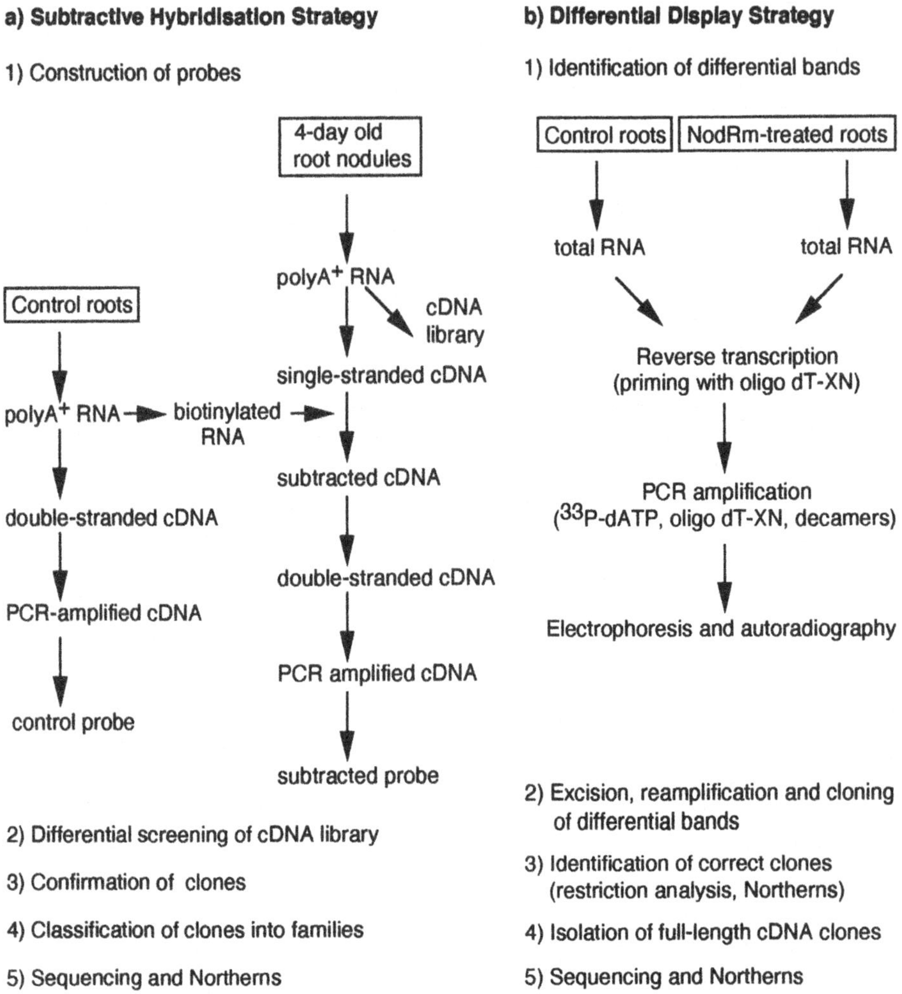

a) Subtractive Hybridisation Strategy

1) Construction of probes

b) Differential Display Strategy

1) Identification of differential bands

4-day old
root nodules

Control roots | NodRm-treated roots

polyA⁺ RNA

cDNA
library

Control roots

total RNA   total RNA

polyA⁺ RNA → biotinylated →
RNA

single-stranded cDNA

Reverse transcription
(priming with oligo dT-XN)

double-stranded cDNA

subtracted cDNA

PCR-amplified cDNA

double-stranded cDNA

PCR amplification
(^{33}P-dATP, oligo dT-XN, decamers)

control probe

PCR amplified cDNA

Electrophoresis and autoradiography

subtracted probe

2) Excision, reamplification and cloning
of differential bands

2) Differential screening of cDNA library

3) Confirmation of clones

3) Identification of correct clones
(restriction analysis, Northerns)

4) Classification of clones into families

4) Isolation of full-length cDNA clones

5) Sequencing and Northerns

5) Sequencing and Northerns

Fig. 1. Two strategies for the isolation of clones for new *M. truncatula* nodulins or Nod factor induced genes.

but with a Nod⁻ strain (the non-induced state). The biotinylated material, including the cDNA/mRNA hybrids, were then eliminated by streptavidin extraction. After two rounds of hybridisation the remaining cDNA was cloned. This so called subtracted bank was then either differentially screened with nodule and root probes or tested by analysing the expression patterns of individual clones. This strategy was largely unsuccessful as very few of the clones were proven to encode nodulins. Two problems were probably mainly responsible. Firstly the subtraction of non-induced species is never complete and the small differences between our 'induced' and 'non-induced' mRNA populations were probably not large enough to result in the 'induced

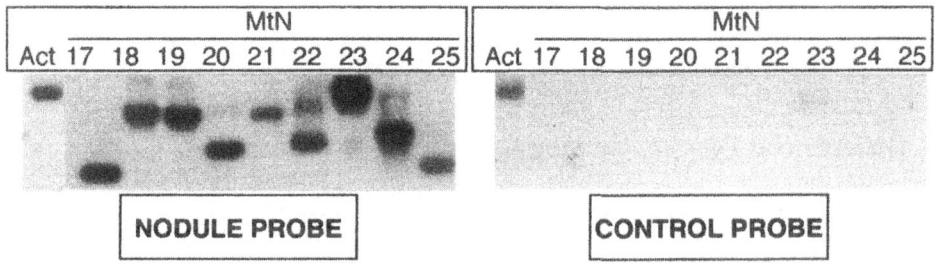

Fig. 2. Confirmation of potential nodulin clones by a 'South-Northern' approach. The PCR-amplified inserts of clones for a non-nodulin, actin (Act) and a selection of *M. truncatula* nodulins (MtN) were subjected to Southern analysis using as probes ^{32}P-labelled cDNA prepared from 4-day old nodules or control roots.

species' being the more abundant at the end of the subtraction. Secondly the small amount of cDNA remaining after two rounds of subtractive hybridisation was difficult to clone.

The strategy was thus modified to start with 4 days' old, non-nitrogen-fixing nodules as the 'induced' condition and the previously-described root material as the 'non-induced' (see Fig. 1a). A single round of subtractive hybridisation (performed as before) resulted in the removal of about 90% of the cDNA. The remaining cDNA was then amplified by PCR and used as a probe to differentially screen a (non-subtracted) 4 days' old nodule cDNA bank. The elimination of most of the 'non-induced' cDNAs from the nodule 'subtracted' probe should have increased the sensitivity of the screening by a factor of about 10 and thus allowed the detection of clones for lower abundance mRNAs. Using this approach, 473 clones were selected from a screen of 30,000 clones from the bank. Those corresponding to a variety of known nodulins were eliminated and the rest were re-tested by a 'South-Northern' approach: their inserts were amplified by PCR and screened on duplicate Southerns with root and nodule cDNA probes (see Fig. 2). Those giving a differential response were classified into families by a series of cross-hybridisations, resulting in the identification of 29 new nodulin genes (Gamas et al., 1996). The majority of these were represented by less than three clones whereas others were considerably more abundant.

Further studies by Northern analysis classified the genes according to the kinetics of expression during nodulation. About half of the early-expressed genes have homologues which are expressed in the spontaneous nodules (see Truchet et al., 1989) from non-infected *M. sativa* plants. Three of the early-expressed genes showed some induction by treatment of roots with purified Nod factors, but in each case the induction was poor in comparison to the Nod factor inducible gene, *Mtenod11*.

Partial sequencing of a clone from each class revealed that at least 10 of the new nodulin genes show significant homologies to sequences already described in the data bases. These include carbonic anhydrase, ascorbate oxidase, cycloartenol synthase, a metalloendoproteinase, a PR-like protein and two proline-rich proteins (Gamas et al., 1996).

3. Differential Display Strategy

The differential display technique is a PCR-based approach to identify cDNAs for genes that are differentially expressed (Liang and Pardee, 1992). Initially cDNA is prepared from 'induced' and 'non-induced' mRNA populations using oligo dTs containing an additional two nucleotides (primers oligo dT-XN) which should anchor at the start of the poly A tail of a subset of mRNAs. A subset of the cDNA is then amplified using combinations of the oligo dT-XN primers and random decamers and the resulting bands are visualised on polyacrylamide sequencing gels. By comparison between the tracks of the 'induced' and 'non-induced' samples, bands related to genes that are differentially expressed can be identified. We chose to use this approach to identify genes that are induced early in response to Nod factor treatment of roots (see Fig. 1b). Initially, mRNA was isolated from roots of plants grown in growth pouches and treated with plant growth medium with or without 10^{-7}M NodRm factors. Initially we tried this approach with plants grown for 4 days in pouches and treated with Nod factors for 2 and 24 hours. A total of 51 different oligo dT-XN/decamer pairs were used. As about 100 bands can be clearly visualised on a gel, a maximum of 5100 cDNAs were analysed in this way. Only two bands appeared to be differential and subsequent analysis revealed that they were unlikely to be related to nodulin genes. A subsequent experiment used mRNA from roots that had been grown for 10 days in pouches and treated for 48 hours with 10^{-7}M NodRm factors. From a combination of only 20 oligo dT-XN/decamer pairs, three bands were reproducibly found to be differential. These were excised from the gels, re-amplified and cloned. The clones were analysed by restriction digestion using '4-bp recognition' enzymes and as expected some of the bands were found to contain more than one cDNA. For each band the most abundant clone type was used to prepare probes for Northern analysis. By this method, one clone has now been confirmed to be related to a gene that is induced by Nod factors and full-length cDNA clones have now been isolated.

4. Discussion

Using both subtractive hybridisation and differential display techniques we have isolated clones for novel early nodulins of *M. truncatula*. The increased sensitivity of these techniques, in comparison to conventional differential screening, has allowed us to isolate clones for genes whose transcripts never attain a high abundance but are nonetheless induced in comparison to roots. We have found that the key to success for both techniques was the choice of the starting plant material: conditions must be found where the genes of interest are likely to be most induced. In addition rigorous testing of potential candidates is required to eliminate false positives. This might not be straightforward for poorly expressed genes and the use of techniques such as reverse transcriptase polymerase chain reaction (RT-PCR) may be necessary.

Both techniques have several advantages and disadvantages. The subtractive hybridisation technique allows a large number of cDNAs to be screened and may directly give full-sized cDNAs. However it requires relatively large amounts of starting material and only compares one induced condition with one non-induced. Moreover the classification of the clones into families, by a series of cross-hybridisations, is a time consuming step. The advantages of the differential display technique are that only small amounts of plant material are needed and that several different conditions can be used for the initial work. However, for this technique a large number of combinations of primers need to be tested in order to analyse a substantial proportion of the cDNAs from the mRNA populations (see above). Moreover the cloning of the differential bands may not be straightforward and it is then necessary to obtain full-length cDNA clones by other methods (eg screening of cDNA libraries or rapid amplification of cDNA ends (RACE)).

To date two other groups have, to our knowledge, published work using a subtractive hybridisation approach to isolate clones for nodulins. Kouchi and Hata (1993) isolated a number of soybean nodulin cDNAs using nodules as the starting material and Cook et al. (1995) have isolated clones for a peroxidase gene *of M. truncatula* using inoculated roots. Recently, Goomachtig et al. (1995) have used differential display to identify several novel nodulins from *Sesbania rostrata*. It is thus clear that these molecular techniques are well suited to the isolation of clones for new nodulins and it remains now to use these clones to further dissect nodule development.

4. References

Barker, D.G., Bianchi, S., Blondon, F., Datté, Y., Duc, G., Flament, P., Gallusci, P., Génier, P., Guy, P., Muel, X., Tourneur, J., Dénarié, J., and Huguet, T. 1990. Plant Mol. Biol. Rep. 8:40-49.

Cook, D., Dreyer, D., Bonnet, D., Howell, M., Nony, E., and VandenBosch, K. 1995. Plant Cell 7:43-55.

Franssen, H.J., Nap, J.-P., and Bisseling, T. 1992. Pages 598-624 in: Biological Nitrogen Fixation (G. Stacey, R.H. Burris and H.J. Evans, eds) New York: Chapman and Hall.

Gamas, P., de Carvalho-Niebel, F., Lescure, N., and Cullimore, J.V. 1996. Molec. Plant Microbe. Interacts. 9: 233-242.

Goomachtig, S., Valerio-Lepiniec, M., Szczyglowski, K., van Montagu, M., Holsters, M., de Bruijn, F.J. 1995. Molec. Plant Microbe. Interacts. 8: 816-824.

Kouchi, H., and Hata, S. 1993. Mol. Gen. Genet. 238:106-119.

Liang, P. and Pardee, A.B. 1992. Science 257 : 967-971.

Mylona, P., Pawlowski, K., and Bisseling, T. 1995. Plant Cell 7:869-885.

Truchet, G., Barker, D.G., Camut, S., de Billy, F., Vasse, J., and Huguet, T. 1989. Mol. Gen. Genet. 219,:65-68.

Plant Gene Expression during Stem Nodule development on *Sesbania rostrata*

Sofie Goormachtig, Marcio Alves Ferreira, Sam Lievens, Viviana Corich, Peter Mergaert, Wim D'Haeze, Manuel Fernández-López, Mengshen Gao, Marc Van Montagu, and Marcelle Holsters

Laboratorium voor Genetica, Department of Genetics, Flanders Interuniversity Institute for Biotechnology (VIB), Universiteit Gent, K.L. Ledeganckstraat 35, B-9000 Gent, Belgium

1 Introduction

In *Rhizobium*—legume interactions, nitrogen-fixing nodules typically develop on the roots of a specific host. On a few legumes, also stem nodules are formed. A particularly well studied example is *Sesbania rostrata*, an annual leguminous plant from West-Africa. As an adaptation to growth in wetland soils, *S. rostrata* carries dormant, adventitious root primordia in vertical rows all along its stem. These primordia develop into nodules upon infection with an appropriate microbial partner, such as *Azorhizobium caulinodans* (Dreyfus *et al.*, 1988). Root nodulation on *S. rostrata* takes place at lateral root bases (Ndoye *et al.*, 1995). Both stem and root nodulation occur by intercellular invasion and lead to the formation of mature determinate nodules.

Stem-nodule development on *S. rostrata* has previously been studied using different microscopic techniques (Tsien *et al.*, 1983; Duhoux 1984). An adventitious root primordium consists of a central vascular system, connected to the vascular system of the stem and surrounded by cortical tissues. At the tip, a dormant root meristem is present. The protrusion of the incipient root through the stem, creates a fissure wherein bacteria can proliferate. Densely populated intercellular infection pockets are formed opposite of which cortical cells dedifferentiate to form a nodule primordium. From the infection pockets, intercellular infection threads grow inside towards the nodule primordia. The first infected plant cells are proximally located, close to the vascular bundle of the root primordium. One week after inoculation, cell division activity ceases and further growth of the central tissue occurs via cell enlargement.

Upon simultaneous inoculation of the stem-located nodulation sites with a bacterial suspension, nodules start to develop simultaneously and synchronously. We took advantage of the abundance and accessibility of developing stem nodules to acquire a better insight into this organ development through the study of plant gene expression.

NATO ASI Series, Vol. G 39
Biological Fixation of Nitrogen
for Ecology and Sustainable Agriculture
Edited by A. Legocki, H. Bothe, A. Pühler
© Springer-Verlag Berlin Heidelberg 1997

2 Results and Discussion

2.1 Cell Cycle Gene Expression

cDNA clones corresponding to a cyclin-dependent kinase (CDK) gene homolog and a mitotic cyclin gene homolog of *S. rostrata* were isolated using an reverse transcriptase polymerase chain reaction approach and degenerate primers derived from amino acid sequence motifs of different CDK or cyclin proteins (Goormachtig *et al.*, 1996). CDKs are the catalytic subunits of protein complexes with different cyclins that control progression through the cell cycle. The deduced amino acid sequence of the *S. rostrata* cyclin cDNA (*CycB1;Sr*) showed 90% similarity with the mitotic soybean cyclins s13-6 and s13-7 (Hata *et al.*, 1991). The CDK homolog Cdc2-1Sr is 99% similar to Cdc-S5 from soybean (Miao *et al.*, 1991) and a Cdc2 kinase from *Vigna acutinifolia* (Hong *et al.*, 1993). An *H4-1Sr* clone was isolated after screening of a cDNA library with an H4 probe from pea.

Expression of CDKs has been related to competence to divide (Martinez *et al.*, 1992). Mitotic cyclins control the G2-M phase transition in the cell cycle and H4 is expressed in the DNA synthesis phase. The expression patterns of these genes was followed by *in situ* hybridizations. Antisense ^{35}S-labeled RNA probes were hybridized against sections of developing stem nodules. *Cdc2-1Sr* transcripts were found in all cells of uninfected as well as infected primordia, perhaps reflecting the totipotent character of these structures. In uninfected root primordia, few cells of the apical meristem contained *CycB1;Sr* transcripts. On the contrary, *H4-1Sr* transcripts were detected in many cells of the apical root meristem. The discrepancy between the abundance of expression of the H4 and cyclin genes could be explained by an arrest in late S phase or by endoreduplication. In either case, a double DNA content is expected in the cells expressing H4. However, flow cytometric measurements revealed that the meristem cells have a 2C content (Spencer-Barreto *et al.*, 1996). The abundant presence of H4 transcripts in the dormant root meristem perhaps reflects the readiness of this tissue to start division for root growth upon waterlogging stress.

Upon inoculation of the root primordia with *A. caulinodans*, a patchy pattern of *CycB1;Sr* and *H4-1Sr* expression appeared in the cortex of the root primordium and nodule primordia were found. With the onset of bacterial invasion of plant cells, a distal meristematic zone was delimited wherein *CycB1;Sr* and *H4* expression was concentrated. At this stage, different developmental zones could be recognized: a proximal infected zone, an infection zone with infection threads and a distal meristem. At one week post inoculation, meristem activity ceased, indicated by the loss of *CycB1;Sr* and *H4-1Sr* expression. In conclusion, stem nodules develop in proximal-distal direction and show during ontogeny many features of indeterminate nodules.

2.2 Expression Pattern of the Early Nodulin Gene *Enod2*

Expression of the early nodulin ENOD2 is a molecular marker for the nodule parenchyma. Two *Enod2* cDNAs and one gene copy were isolated from *Sesbania rostrata* (Strittmatter *et al.*, 1989; Dehio and de Bruijn 1992). Using the *Enod2* gene as a probe, a low basal level of transcription was detected in adventitious root primordia. *In situ* hybridizations revealed that *Enod2* transcripts were present in the outermost cortical cell layer of the adventitious root primordium and that this expression remained until about 7 days after inoculation, when the nodule parenchyma was completely differentiated. Expression in this exterior cortical cell layer could reflect a function for *Enod2* in constructing an oxygen barrier in the root primordium or in preventing desiccation. Using *in situ* hybridization the differentiation of the nodule parenchyma was studied. *Enod2* transcripts were first detected in proximal cell layers of the nodule primordium, near the stele and expression moved gradually in a distal direction. Finally, the nodule parenchyma surrounded the complete central tissue. This confirms the proximal-distal differentiation direction of the stem nodules.

2.3 Expression Patterns of Chitinase Genes

Two clones, identified via the differential display approach, were shown to be homologous to class III chitinase genes (Goormachtig *et al.*, 1995). RNA blot analysis revealed that both chitinase homologues have a different temporal expression pattern. Didi-24 is strongly induced one day after infection whereas didi-13 shows a slow induction, its expression peaks about five days post-inoculation.

In situ hybridizations revealed a different spatio-temporal expression pattern for both genes (Goormachtig *et al.*, in preparation). Didi-24 was weakly expressed in outer cortical cells of uninfected root primordia and strongly enhanced upon inoculation. In mature nodules the signal disappeared again. Didi-13 transcripts were not detected in uninfected root primordia. At one day after infection, low expression was observed in the cortex of the root primordium. Once the nodule primordia were forming, strong expression was detected in the nodule parenchyma, in the uninfected cells of the central tissue and around infection pockets and infection threads.

What could be the function of these chitinases during nodule development? What are the substrates or the targets of the enzymes? In order to answer the latter question, the genes will be expressed in *Escherichia coli*. Putative substrates are the Nod factors, with their chitooligosaccharide backbone. A Nod factor-degrading activity is induced during *S. rostrata* stem nodule development (S. Goormachtig, unpublished results). But also the bacteria could be targets, as many class III chitinases have lysozyme activity. The presence of *didi13*, for instance, in the uninfected cells could serve as a tool to prevent bacteria to enter these cells. In conclusion, the spatio-temporal expression pattern of the chitinase genes suggests a major role in this organ development, by controlling the action range of Nod factors or the viability of the bacteria.

Acknowledgements

The authors thank H. Fransen for providing the pea H4 probe, B. Dreyfus for *S. rostrata* seeds and M. De Cock for preparing the manuscript. The research was supported by grants from the

Belgian Programme on Interuniversity Poles of Attraction (Prime Minister's Office, Science Policy Programming, No. 38) and in part by the European Communities' BIOTECH Programme, as part of the Project of Technological Priority 1993-1996, and by the Human Capital and Mobility Programme (CHRX-CT94-0656). M.A.F., W.D., and M.F.L. are indebted to the Coordenação de Aperfeiçoamento de Pessoal de Nível Superior (CAPES 1328/94-4) for predoctoral fellowships, the Vlaams Instituut voor de Bevordering van het Wetenschappelijk-Technologisch Onderzoek in de Industrie, and to the Spanish Consejo Superior de Investigaciones Científicas and the European Union (ERBCHBICT-941082) for predoctoral and postdoctoral fellowships, respectively. S.L. and M.H. are Research Assistant and Research Director of the National Fund for Scientific Research (Belgium), respectively.

References

Dehio, C., and de Bruijn, F.J. (1992). The early nodulin gene *SrEnod2* from *Sesbania rostrata* is inducible by cytokinin. *Plant J.* 2, 117-128.

Dreyfus, B., Garcia, J.L., and Gillis, M. (1988). Characterization of *Azorhizobium caulinodans* gen. nov., sp. nov., a stem-nodulating nitrogen-fixing bacterium isolated from *Sesbania rostrata*. *Int. J. Syst. Bacteriol.* 38, 89-98.

Duhoux, E. (1984). Ontogénèse des nodules caulinaires du *Sesbania rostrata* (légumineuses). *Can. J. Bot.* 62, 982-994.

Goormachtig, S., Valerio-Lepiniec, M., Szczyglowski, K., Van Montagu, M., Holsters, M., and de Bruijn, F.J. (1995). Use of differential display to identify novel *Sesbania rostrata* genes enhanced by *Azorhizobium caulinodans* infection. *Mol. Plant-Microbe Interact.*, 8, 816-824.

Goormachtig, S., Alves-Ferreira, M., Van Montagu, M., Engler, G., and Holsters, M. (1996). Expression of cell cycle genes during *Sesbania rostrata* stem nodule development. *Mol. Plant-Microbe Interact.*, submitted.

Hata, S., Kouchi, H., Suzuka, I., and Ishii, T. (1991). Isolation and characterization of cDNA clones for plant cyclins. *EMBO J.* 10, 2681-2688.

Hong, Z., Miao, G.-H., and Verma, D.P.S. (1993). p34^{cdc2} protein kinase homolog from mothbean (*Vigna aconitifolia*). *Plant Physiol.* 101, 1399-1400.

Martinez, M.C., Jørgensen, J.-E., Lawton, M.A., Lamb, C.J., and Doerner, P.W. (1992). Spatial pattern of *cdc2* expression in relation to meristem activity and cell proliferation during plant development. *Proc. Natl. Acad. Sci. USA* 89, 7360-7364.

Miao, G.-H., Hirel, B., Marsolier, M.C., Ridge, R.W., and Verma, D.P.S. (1991). Ammonia-regulated expression of a soybean gene encoding cytosolic glutamine synthetase in transgenic *Lotus corniculatus*. *Plant Cell* 3, 11-22.

Ndoye, I., de Billy, F., Vasse, J., Dreyfus, B., and Truchet, G. (1994). Root nodulation of *Sesbania rostrata*. *J. Bacteriol.* 176, 1060-1068.

Spencer-Barreto, M.M., Cottignies, A., Chamel, A., and Duhoux, E. (1996). Caractéristiques cytophysiologiques des ébauches racinaires adventives de la tige de *Sesbania rostrata* Brem (Leguminosae). *Acta Bot.*, in press.

Strittmatter, G., Chia, T-F., Trinh, T.H., Katagiri, F., Kuhlemeier, C., and Chua, N.-H. (1989). Characterization of nodule-specific cDNA clones from Sesbania rostrata and expression of the corresponding genes during the initial stages of stem nodules and root nodules formation. *Mol. Plant-Microbe Interact.* 2, 122-127.

Tsien, H.C., Dreyfus, B.L., and Schmidt, E.L. (1983). Initial stages in the morphogenesis of nitrogen-fixing stem nodules of *Sesbania rostrata*. *J. Bacteriol.* 156, 888-897.

IDENTIFICATION OF *TRANS*-ACTING FACTORS REGULATING NODULIN GENE EXPRESSION

Erik Østergaard Jensen[1], Niels Pallisgaard[1], Henning Christiansen[1], Irma Vijn[2], Ton Bisseling[2], Mette Grønbæk[1], Kirsten Nielsen[1], Jan-Elo Jørgensen,[1] Knud Larsen,[1] Anette Chemnitz Hansen[1], Magdalena Mielczarek[1], Izabela Sniezko[1] and Kjeld A. Marcker[1]

[1]Laboratory of Gene Expression, Department of Molecular and Structural Biology, University of Aarhus, Denmark.
[2]Department of Molecular Biology, Agricultural University of Wageningen, The Netherlands

1 Introduction

Functional studies of nodulin gene promoters in transgenic legumes have identified a number of *cis*-regulatory elements important for nodule specific expression. One example is the soybean leghaemoglobin (lb) c3 promoter which were investigated in details in transgenic *Lotus corniculatus* plants. By analysing 5' promoter deletions and hybrid promoters fused to the GUS and CAT reporter genes the following elements were identified in the *lbc3* promoter; A strong positive element, a weak positive element, an organ specific element and a negative element (Figure 1, Stougaard et al. 1990).

Figure 1. Schematic representation of the *cis*-regulatory elements in the soybean lbc3 promoter. SPE: strong positive element; WPE: weak positive element; OSE: organ specific element; NE: negative element.

The minimal promoter sequences directing nodule specific expression are contained within a fragment from -230 to +1. However, a truncated promoter containing sequences only up to -139 can direct nodule-specific expression when a 35S enhancer is fused to the construct (Stougaard et al. 1990). By replacement mutagenesis it was shown that two conserved nodulin motifs (CTCTT;AAAGAT) within in the organ specific element were important for the function of this element (Ramlov et al. 1993). To study the function of the identified *cis*-elements, attempts were made to identify putative transcription factors interacting with these elements. This lead to the identification of a nodule-specific nuclear protein, NAT2, interacting with two AT-rich DNA sequences in the weak positive element (Jensen et al. 1988). It was furthermore shown that the NAT2 binding site when fused to a -139 *lbc3* promoter deletion could activate the otherwise silent construct (Laursen et al. 1994). However, due to the various

NATO ASI Series, Vol. G 39
Biological Fixation of Nitrogen
for Ecology and Sustainable Agriculture
Edited by A. Legocki, H. Bothe, A. Pühler
© Springer-Verlag Berlin Heidelberg 1997

properties of the NAT2 protein, it was not possible to purify the protein to homogeneity, which eventually could have lead to the identification of the corresponding gene.

Due to the technical problems, we decided to take a different approach and we screened a nodule λgt11 cDNA expression library using the identified *cis*-elements as probes.

2 South-Western screening of nodule cDNA library

Rather than searching for a single protein interacting with a specific sequence in the promoter e.g. the nodulin motifs, we decided to go for all proteins interacting with the proximal promoter region, since all the *cis*-regulatory elements required for nodule specific expression are located on this part.

We made a set of 10 double-stranded overlapping oligonucleotides covering the promoter region from -245 to +1 (figure 2). Each of the oligonucleotides were then concatenated before they were used as probes. We used this strategy beacuse DNA probe with multiple bindings sites for the protein of interest has a higher affinity for the protein than a probe containing only a single site.

Figure 2. Positions of the oligonucleotides used as probes for the South-Western screening

Using a cocktail of all ten oligonucleotides we screened 600.000 pfu from an un-amplified soybean nodule λgt11 cDNA library. The screening gave 58 positive signal and the corresponding phages were purified. The clones were hybridesed to each other and it turned out that 9 of clones were unique and 49 of the clones represented the same mRNA. The purified phages were then plated and hybridesed to the oligoncleotides one by one. About half of the phages encoded proteins that showed much stronger binding to some of the oligonucleotides than others. This kind of analysis is not very precise and to get a detailed information about the binding affinities, we subcloned the inserts into the pMAL-p2 expression vector. The corresponding proteins were expressed in the bacteria and purified on a maltose column. Filter-binding experiments were then performed. The expressed and purified proteins were applied to a nitrocellulose filter in parallel lines and the labelled oligonucleotides were hybridesed to filters in channels going perpendicular to the protein lines (figure 3). The actual recognitions sequences in the promoter were subsequently determined by the DNaseI - foot-printing (table I)

Finally, the cDNA inserts were subcloned into pGEMEX and the complete nucleotide sequences of the inserts were determined. Data base searches were performed to identify homologies to already identified genes (table I). To study the expression of the genes corresponding to the cDNA inserts, an RNase protection experiment was made and as shown in table I, several of the genes were induced in the nodules.

Figure 3. Filter-binding assay. The expressed proteins are applied in horizontal lines and the labelled oligonucleotides were added perpendicular to the protein lines. Oligo numbers corresponds to the numbers given in figure 2.

3 Conclusion

The South-Western screening of soybean nodule specific cDNA expression library identified 58 cDNA clones encoding 10 different DNA binding proteins. Five of the clones encodes proteins that bound to the leghaemoglobin promoter in a sequence-specific manner. One of these proteins, gmSN4, belongs to the homeodomain family. However, the homology to other homeodomain proteins is so low that it would not have been possible to isolate the cDNA by DNA hybridisation. The strong induction observed one day after the infection may argue against a possible role as a regulator of the leghaemoglobin genes. However, recently, we observed a low level expression of a soybean leghaemoglobin gene in transgenic *Vicia hirsuta* nodule primordia in addition to the high level expression in the central infected zone of the nodule. A second protein, gmSN14, appears to be the only of its kind identified so far. The DNA binding domain of this protein contain two cysteine domain with each ten cysteine residues. The high level expression of this protein late in nodule development suggest this protein to be a transcriptional factor involved in the high level expression of the leghaemoglobin genes.

GmSN49 is similar to a P-box binding protein from parsley. Since this protein is a putative activator of pathogenesis related proteins it might suggest that the induction of this protein in infected roots is a direct consequence of the bacterial infection. Whether gmN49 is involved in a defence reaction or in the nodule formation cannot be concluded from the present data.

In conclusion, the approach taken has resulted in the idenfication and isolation of novel DNA-binding protein that are putative regulators of the leghaemoglobin expression.

Name	Homologies	DNA recognition	Comments
gmN4	Homeodomain proteins	AT-sequences e.g. TTATTGTC	Strongly up-regulated 1 day after infection
gmN14	Not found	TGAAAA	Nodule specific - peaks day 12
gmN49	P-box binding protein from parsley (BPF-1)	TAAGT	Up-regulated 3-4 fold 2-3 days after infect.
gmN25	80% identical to HMG-box protein from Arabidopsis - HMG domain KDPNSPKR	CACC or similar C-A sequences	
gmN31	90% identical HMG/histone protein from soybean - Contains an AT-hook, RGRP	AT-sequences like TAAAT	
gmN15	Histone H1 from Arabidopsis	Not specific	Constitutively expressed
gmN23	Histone H1 from pea	Not specific	
gmN55	Histone H1 from pea	Not specific	
gmN7	Not found	Not specific	
gmN54	Not found	Not specific	multiple cDNAs

Table I. Presentation of the ten different cDNA clones isolated from a soybean λgt11cDNA expression library using the soybean *lb* promoter as a probe

References

Jensen, E. O., Marcker, K. A., Schell, J. and de-Bruijn, F. J. (1988). Embo J **7**: 1265-1271.

Laursen, N. B., Larsen, K., Knudsen, J. Y., Hoffmann, H. J., Poulsen, C., Marcker, K. A. and Jensen, E. O. (1994). Plant Cell **6**: 659-668.

Ramlov, K. B., Laursen, N. B., Stougaard, J. and Marcker, K. A. (1993). Plant J **4**: 577-80.

Stougaard, J., Jorgensen, J. E., Christensen, T., Kuhle, A. and Marcker, K. A. (1990). Mol Gen Genet **220**: 353-60.

Analysis of genes expressed in root nodules of broad bean (*Vicia faba* L.)

Andreas M. Perlick, Martin Frühling, Gerald Schröder, Ulrike Albus, S. Christian Frosch, Jörg Becker, Steffen Böhner, Hans-Joachim Quandt, Inge Broer, Helge Küster, and Alfred Pühler

Universität Bielefeld, Lehrstuhl für Genetik, Postfach 100131, 33501 Bielefeld, Germany

Soil bacteria from the genera *Rhizobium*, *Bradyrhizobium* and *Azorhizobium* induce the formation of symbiotic organs, designated root nodules, on the roots of legumes. Within mature root nodules, nitrogen fixation is carried out by the microsymbiont. In our group, we focus on the molecular analysis of the plant's part of the symbiotic interaction of broad bean (*Vicia faba* L.) with *R. leguminosarum* bv. *viciae*.

Identification and sequence analysis of 44 genes expressed in *V. faba* nodules

We identified transcript sequences representing 44 different genes active in broad bean root nodules (see table 1). By Northern and cDNA-cDNA "Southnorthern" hybridizations (Frühling *et al.* 1996), the expression of all 44 genes was analyzed. In addition to a number of transcripts occurring in most broad bean tissues, 19 transcripts could solely be detected in nodules and thus encode nodulins. Sequence analysis led to the identifications of 8 classes of broad bean genes expressed in root nodules: (1) leghemoglobins, (2) proline-rich proteins (PRPs), (3) glycine-rich proteins (GRPs), (4) cysteine-cluster proteins (CCPs), (5) other nodulins, (6) enzymes of the nodule carbon (Küster *et al.* 1993), nitrogen and sulfur metabolism, (7) other gene products, (8) sequences without significant homologies to other gene products of known function.

Three of the four leghemoglobin genes identified are expressed exclusively in nodules whereas VfLb29 is the only broad bean nodulin gene found to be expressed in mycorrhizal roots colonized by *Glomus fasciculatum* in the absence of rhizobia (Frühling *et al.* 1996). In addition to the well studied nodulins ENOD2, ENOD5 and ENOD12, we found a nodule extensin and another proline-rich protein displaying the characteristic Ser-Pro$_4$ and Pro$_3$X$_4$ motifs, respectively. These proline-rich motifs, which were also found in ENOD2 and ENOD12 proteins, are characteristic of cell wall proteins. This led to the hypothesis that these nodule proteins might have a cell wall modifying function in root nodules. The five different nodule-specific GRPs with a glycine-content of 15% to 32% (Küster *et al.* 1995a, Schröder *et al.* 1996) belong to a novel class of nodulins not detected in other legumes until now. Five nodule-specific transcripts specified small proteins characterized by the occurence of two conserved cysteine clusters, which, as the homologous ENOD3/14 and NOD6 nodulins from pea, might be involved in the binding of metal ions in the nodule tissue.

Several VfNOD28/32 transcripts encoding modular nodulins homologous to the nodulin-25 from *Medicago sativa* (Küster *et al.* 1994 and 1996) differed in the absence of exon sequences specifying individual modules. Possibly one VfNOD28/32 gene gives rise to a large number of transcripts by alternative splicing. VfENOD32 transcripts, which were identified in nodules and flowers but not in other tissues, encoded a broad bean homologue of the *V. narbonensis* α/β$_8$-barrel seed protein narbonin (Perlick *et al.* 1996a). Additionally, the VfENOD32 deduced amino acid sequence displayed local homologies to strongly conserved parts of the active centre of class III chitinases. Thus, the VfENOD32 proteins might be involved in the binding or modification of chitin-like oligosaccharides in root nodules. Finally, the

NATO ASI Series, Vol. G 39
Biological Fixation of Nitrogen
for Ecology and Sustainable Agriculture
Edited by A. Legocki, H. Bothe, A. Pühler
© Springer-Verlag Berlin Heidelberg 1997

clone-group	gene	expression in (nodule, root, leaf, seed, epicotyl, shoot, flower)	transcript size [kb]	gene product [kDa] (without signal peptide)	homology/ structural properties
VfLb49	VfLb49		0.75	15.8	leghemoglobin
VfNDS-B	VfLb-B		0.75	15.9	leghemoglobin
VfNDS-K	VfLb-K		0.8	15.9	leghemoglobin
VfNDS-X29	VfLb29		0.8	16.5	leghemoglobin
VfNDS-D	VfPRP1		1.1		proline-rich protein
VfNDS-E	VfExt1		1.2		extensin
VfNDS-H	VfENOD2		1.7	(53.4)	ENOD2 protein
VfNDS-X7	VfENOD12		0.6	10.2 (7.6)	ENOD12 protein
VfNDS-X11	VfENOD5		0.65	15.1 (12.3)	ENOD5 protein
VfNDS-F	VfNOD-GRP1		0.7	14.8	glycine-rich protein
VfNDS-G	VfENOD-GRP2		0.65	13.8 (11.5)	glycine-rich protein
VfNDS-X14	VfENOD-GRP3		0.8	19.0 (16.6)	glycine-rich protein
VfNDS-X20	VfNOD-GRP4		0.6	10.5	glycine-rich protein
VfNDS-X22	VfENOD-GRP5		0.75	(12.3)	glycine-rich protein
VfNDS-M	VfGRP6		1.0		glycine-rich protein
VfNDS-J	VfNOD-CCP1		0.45	7.4	cysteine-cluster protein
VfNDS-X12	VfNOD-CCP2		0.45	7.0 (4.1)	cysteine-cluster protein
VfNDS-X19	VfNOD-CCP3		0.45	7.6	cysteine-cluster protein
VfNDS-X25	VfNOD-CCP4		0.45	7.4 (4.7)	cysteine-cluster protein
VfNDS-X30	VfNOD-CCP5		0.45	6.7 (3.7)	cysteine-cluster protein
VfNDS-A	VfENOD32		1.2	32.5	narbonin
VfNDS-L	VfNOD28/32		1.3	29.5-31.8	MsNOD25
VfNDS-X6	VfENOD40		1.0	1.5	ENOD40
VfNDS-X9	VfENOD18		0.8	18.0	expressed sequence tags
VfNDS-C	VfSucS		2.8	92.5	sucrose synthase
VfNDS-X4	VfCS		1.5		cysteine synthase
VfNDS-X5	VfAS1		2.2	66.2	asparagine synthetase
VfNDS-I	VfEF-1α		1.6	49.2	EF-1α
VfNDS-X10	VfGAST1		0.7		GAST1
VfNDS-X13	VfLOX1		2.9	96.5	lipoxygenase
VfNDS-X15	VfIFR		1.2		isoflavone oxydoreductase
VfNDS-X16	Vf14-3-3		1.1	29.4	14-3-3 proteins
VfNDS-X17	VfPI-1		0.85	23.8 (20.9)	protease inhibitor
VfNDS-X18	VfDAHP		1.9		DAHP
VfNDS-X21	VfSAMDC		1.9	38.5	S-adenosyl methionine decarboxylase
VfNDS-X23	VfUbq		0.9	28.3 (8.5)	ubiquitin
VfNDS-X24			1.0	24.0	expressed sequence tags
VfNDS-X1			1.7		
VfNDS-X2			1.4		
VfNDS-X3			1.0		
VfNDS-X8					
VfNDS-X26			1.1		
VfNDS-X27			4.0		
VfNDS-X28			1.1		

Table 1: Overview on the expression and sequence properties of 44 genes avtive in *V. faba* root nodules. In the left column, the clone groups from the nodule-specific cDNA library are listed. In addition to the previously isolated VfLb49 clone group representing a leghemoglobin gene, these groups were designated VfNDS (*V. faba* nodule differential screening)-A to -M and -X1 to -X30 (Perlick and Pühler 1993). The expression of these genes in different tissues, as judged by Northern blotting, is indicated as follows: black, strong expression; dark grey, expression clearly detectable; light grey, low level of expression; white, no expression detectable. Homologies of the deduced gene products or characteristic sequence properties are indicated.

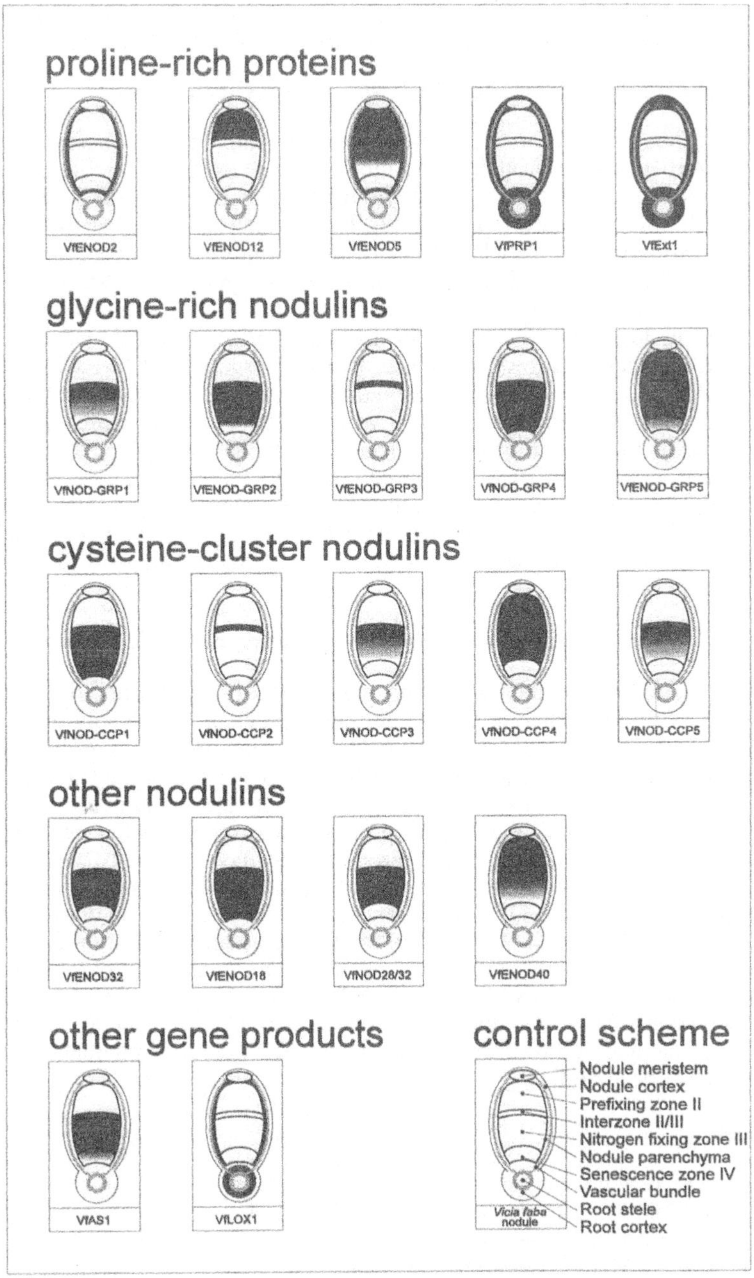

Figure 1: Overview on the expression of 21 genes in *V. faba* root nodules. Schematic representations of the results from tissue-print hybridizations are given using black colouring to indicate regions, where hybridizing transcripts significantly above the background level were detected. The names of the genes under investigation are defined in table 1. Details of the tissue-print technique are reported in Schröder *et al.* 1996.

lipoxygenase gene VfLOX1 is active in most broad bean tissues, but its expression is upregulated in nodules with respect to uninfected roots (Perlick *et al.* 1996b).

Localization of nodule transcripts by tissue-print hybridizations

As a further step towards the identification of potential functions for the nodule proteins identified, tissue-print hybridizations were performed (see figure 1). This way we could demonstrate that the VfENOD2 protein, the PRPs VfPRP1 and VfExt1 and the lipoxygenase VfLOX1 fulfill their function in the peripheral nodule tissues. Interestingly, the VfExt1 transcript was the only transcript examined, which was also identified in the nodule meristem. An expression in the peripheral nodule tissues could be indicative of a defence-related function for the encoded proteins. All other transcripts tested were detected in the central nodule tissues, where bacterial invasion and nitrogen fixation takes place. It was a striking fact that within the two groups of GRP and CCP genes the longitudinal transcript distribution of individual genes varied markedly. For both groups we found genes being expressed from the prefixing zone II to the nitrogen fixing zone III (VfENOD-GRP5, VfNOD-CCP4), whereas other genes were expressed predominantly in the interzone II/III region (VfENOD-GRP3, VfNOD-CCP2). These localization data imply a comparable, but possibly specific function for each GRP or CCP in nodule development. In addition, our transcript localizations identifed marker genes characteristic of different nodule tissues of broad bean.

Analysis of *V. faba* nodulin gene promoters in transgenic *Vicia hirsuta* nodules

To be able to investigate the expression regulation of *V. faba* nodulin genes, we established a system to generate transgenic root nodules on the closely related vetch *V. hirsuta* (Quandt *et al.* 1993). The activity of the promoters of the *V. faba* leghemoglobin gene Vflb3 and the VfENOD-GRP3 gene encoding a glycine-rich nodulin (see table 1) were assayed in transgenic roots and nodules using the *gus*A reporter gene. As expected, the Vfglb3 promoter was active in the nitrogen-fixing zone III of nodules. In addition, this promoter displayed a weak activity in the nodule meristem as well as in root tips. In accordance to the transcript localizations (see figure 1), the VfENOD-GRP3 promoter mediated a predominant GUS expression in the interzone II-III region of transgenic nodules. The minimal promoter fragment active in transgenic nodules was located from position -239 to +10 relative to the transcriptional start (Küster *et al.* 1995b). By analysing different GRP and CCP gene promoters, we intend to investigate the molecular mechanisms determinating the activation of these genes at specific points of time and in specific tissues of broad bean nodules.

References

- Frühling *et al.*, submitted (1996).
- Küster *et al.*, Mol Plant-Microbe Interact **6**: 507-514 (1993).
- Küster *et al.*, Plant Mol Biol **24**: 143-157 (1994).
- Küster *et al.*, Plant Mol Biol **28**: 405-421 (1995a).
- Küster *et al.*, Plant Mol Biol **29**: 759-772 (1995b).
- Küster *et al.*, Mol Gen Genet, in press (1996).
- Perlick and Pühler, Plant Mol Biol **22**: 957-970 (1993).
- Perlick *et al.*, Plant Physiol **110**: 147-154 (1996a).
- Perlick *et al.*, submitted (1996b).
- Quandt *et al.*, Mol Plant-Microbe Interact **6**: 699-703 (1993).
- Schröder *et al.*, submitted (1996).

Regulation of *Bradyrhizobium japonicum* *hemB*, a Heme Biosynthesis Gene.

Sarita Chauhan and Mark R. O'Brian

Department of Biochemistry, State University of New York, Buffalo, New York 14214 USA

1. INTRODUCTION

Symbiotic rhizobial bacteroids respire efficiently in the O_2-limited environment of a legume nodule to support nitrogen fixation and maintain viability, and changes in plant and bacterial heme protein synthesis are an essential feature of nodule ontogeny. Plant leghemolobin is the most abundant nodule protein, and it serves to facilitate O_2 diffusion to bacteroids (2). Accordingly, soybean heme synthesis genes are strongly induced and maintained at high levels in nodules (9, 10, 13, 18, 19) and these observations run counter to the long-held view that the host lacks the synthetic capacity for heme formation in the symbiotic organ (reviewed in ref. 15). The bacterial heme protein profile also changes during differentiation, and these changes include the expression of a terminal oxidase with a high affinity for O_2 (11, 17) and an increase in the overall quantity of cytochromes. Herein, we describe recent developments on the regulation of heme synthesis in *Bradyrhizobium japonicum* with particular attention to *hemB*, the gene encoding δ-aminolevulinic acid (ALA) dehydratase. Analysis of mutants indicates that ALA dehydratase is the first essential bacterial step for heme synthesis in nodules even though it is the second step of the heme pathway (5, 6, 8, 18). The first step, ALA synthase, is not required for heme synthesis in soybean or cowpea nodules, and evidence suggests that a *hemA* mutant of *B. japonicum* is rescued by uptake of ALA derived from the plant host (14, 18).

2. REGULATION OF *hemB* BY IRON

Heme is synthesized by the enzymatic chelation of ferrous iron into protoporphyrin. Porphyrins are toxic to cells because they catalyze the formation of reactive oxygen species, thus, a *prima facie* argument can be made for regulation of the heme pathway by iron. Iron is a limiting nutrient for rhizobia in the soil because iron is in the oxidized state and therefore nearly

NATO ASI Series, Vol. G 39
Biological Fixation of Nitrogen
for Ecology and Sustainable Agriculture
Edited by A. Legocki, H. Bothe, A. Pühler
© Springer-Verlag Berlin Heidelberg 1997

insoluble at pH 7. Within nodules, iron is unlikely to be free in solution, and it is usually complexed with other proteins in eukaryotes. In either situation, iron may be limiting to rhizobia and it is therefore a potential regulator of heme

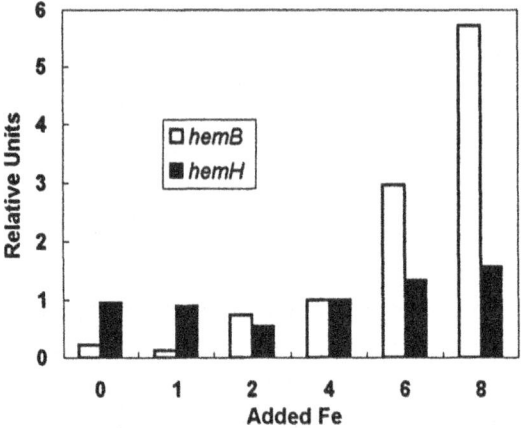

Figure 1. Expression of *hemB* mRNA in *B. japonicum* cells grown with various FeCl$_3$ concentrations. The mRNA was detected by an RNase protection assay using an antisense probe to a portion of the *hemB* or *hemH* gene. Bands from the autoradiograms were analyzed by scanning densitometry and presented as relative units. The quantity of mRNA at 4 μM Fe was arbitrarily assigned as 1 for each gene. The iron concentration in media with no exogenous iron added is 0.3 μM as determined by atomic absorption.

biosynthesis. Indeed, the *B. japonicum* strain I110 *hemB* gene is regulated by iron as seen by a 30- to 50-fold higher levels of mRNA in cells grown in 8 μM Fe compared with cells grown in media with no added iron (Fig. 1; the actual iron concentration is 0.3 μM as determined by atomic absorption). The *hemH* gene, however, was essentially unaffected by the exogenous iron concentration. The induction of *hemB* message is fast and is observed by 15 minutes after addition of 8 μM FeCl$_3$ to cells grown in iron-deprived media. Turnover of *hemB* mRNA in cells treated with rifampicin is also observed within 15 minutes in the presence or absence of exogenous iron, indicating that iron does not regulate turnover, and that the induction is likely to be due to synthesis.

3. OXYGEN CONTROL OF *hemB*

Oxygen is a key regulator of nodule development in legume nodules (20), thus we are currently studying the effects of O$_2$ on bacterial heme synthesis with an emphasis on *hemB*. Cultured cells of *B. japonicum* strain I110 subjected to anaerobiosis show a strong and rapid increase in *hemB* mRNA as observed by an RNase protection assay, whereas *hemH*, the gene encoding the heme synthesis gene ferrochelatase, shows no induction. Over a 100-fold induction of *hemB* mRNA is observed by 15 minutes after O_2 is removed and is maintained for at least two hours. Transcriptional run-off experiments show that the rate of *hemB* mRNA synthesis is increased greatly by O_2 deprivation. Furthermore, turnover of *hemB* message in cells treated with rifampicin occurs by 15 minutes after introduction of O_2 to anaerobic cultures. Thus, the induction of message in O_2-limited cells is due primarily to synthesis rather than turnover.

FixL and FixJ comprise a two-component regulatory system that allows rhizobia to sense and respond to O_2 for regulation of genes essential for development and symbiosis (reviewed in 4). FixL/FixJ is likely to be the primary system for O_2-dependent signal transduction leading to bacterial differentiation in *Rhizobium meliloti* (20), and it works in conjunction with NifA for direct O_2 sensing in *B. japonicum* (reviewed in ref. 7). We examined the induction of *hemB* in the *fixJ* strain 7360 (obtained from H. Hennecke; ref. 1) by RNase protection and western blot analyses and found that mRNA and protein are not induced in the mutant strain, implicating a role for the transcriptional activator in *hemB* expression. We do not yet know whether FixJ interacts with the *hemB* gene directly or exerts its affect through another factor such as FixK. Cultured cells of wild type strain I110 show a two-fold increase in *b*- and *c*-type cytochromes and the disappearance of cytochrome aa_3 when grown in 2% O_2. However, growth of the *fixJ* strain in 2% O_2 shows no cytochrome increase, and the aa_3 oxidase persists. We conclude that heme synthesis and heme protein expression are under developmental control in *B. japonicum*, and is important for the repression of cytochrome aa_3 as well as the induction of the symbiotic oxidase encoded by the *fixNOQP* (see ref. 7).

ALA synthase activity can be induced by restricted aeration of stationary phase *B. japonicum* cultured cells (3), but studies of its regulation using *hemA-lacZ* fusions have not yielded consistent conclusions. Reports indicating that promoter activity is not O_2 dependent in *B. japonicum* (12) is countered by data showing an increase in activity in microaerobic cells in *B. japonicum* that is fixL/fixJ-dependent (16). In our hands, transcriptional *lacZ* fusion studies using heme synthesis promoters correlate poorly with direct mRNA analysis in terms of the magnitude and time of expression. This is probably because, unlike many symbiotically relevant genes such as *nif*, *fix* or *hup*, heme synthesis genes are essential for viability *ex planta*, and expression is significant in "uninduced" cells. Thus, synthesis of a very stable protein such as β-galactosidase from a heme promoter allows substantial accumulation of the protein without induction. In the studies where an O_2 effect is observed, basal rates were minimized either by using cells in stationary phase (3), where ALA synthase activity is low (McGinnis and O'Brian, unpublished results) or by omitting iron from the growth medium (16). Therefore, ALA synthase is probably under O_2 control in *B. japonicum* under certain conditions, but indirect analysis using *lacZ* fusions should not be over interpreted.

4. FINAL COMMENTS

The conclusion that heme biosynthesis is developmentally controlled is consistent with the correlation between changes in the pattern of heme protein expression and nodule ontogeny. Heme synthesis is regulated by iron and by O_2 deprivation, which is in accordance with the physiology of the organism. The regulation of *hemB*, and probably other steps of the heme pathway, are under positive and negative control, and the level of expression in nodules is likely to be a composite of these regulatory mechanisms.

5. REFERENCES

1. Anthamatten, D. and H. Hennecke. 1991. Mol. Gen. Genet. **225**: 38-48.
2. Appleby, C.A. 1992. Sci. Progress Oxford. **76**: 365-398.
3. Avissar, Y.J. and K.D. Nadler. 1978. J. Bacteriol. **135**: 782-789.
4. Batut, J. and P. Boistard. 1994. Antonie van Leewenhoeck **66**: 129-150.
5. Chauhan, S. and M.R. O'Brian. 1993. J. Bacteriol. **175**: 7222-7227.
6. Chauhan, S. and M.R. O'Brian. 1995. J. Biol. Chem. **270**: 19823-19827.
7. Fischer, H.-M. 1994. Microbiol. Rev. **58**: 352-386.
8. Frustaci, J.M., and M.R. O'Brian. 1992. J. Bacteriol. **174**: 4223-4229.
9. Frustaci, J.M., I. Sangwan, and M.R. O'Brian. 1995. J.Biol. Chem. **270**: 7387-7393.
10. Kaczor, C, M.W. Smith, I. Sangwan and M.R. O'Brian. 1994. Plant Physiol. **104:** 1411-1417.
11. Keefe, R.G. and R.J. Maier. 1993. Biochim. Biophys. Acta **1183:** 91-104.
12. Kim, H. and R.J. Maier. 1990. J. Biol. Chem. **265**: 18729-18732.
13. Madsen, O., L. Sandal, N.N. Sandal, and K.A. Marcker. 1993. Plant Mol. Biol. **23:** 35-43.
14. McGinnis, S.D. and M.R. O'Brian. 1995. Plant Physiol. **108**: 1547-1552.
15. O'Brian, M.R. 1996. J. Bacteriol. **178**: 2471-2478.
16. Page, K.M. and M.L. Guerinot. 1995. J. Bacteriol. **177**: 3979-3984.
17. Preisig, O., D. Anthamatten and H. Hennecke. 1993. Proc. Natl. Acad. Sci. U.S.A. **90**: 3309-3313
18. Sangwan, I., and M.R. O'Brian. 1991. Science **251**: 1220-1222.
19. Sangwan, I., and M.R. O'Brian. 1993. Plant Physiol. **102**: 829-834.
20. Soupene, E., M. Foussard, P. Boistard, G. Truchet and J. Batut. 1995. Proc. Natl. Acad. Sci. U.S.A. **92**: 3759-3763.

Part 4

Bacterium-Plant Surface Interaction

Rhizobial Capsular and Lipopolysaccharides: Evidence for their Importance in *Rhizobium*-Legume Symbiosis.

Russell W. Carlson[1], L. Scott Forsberg[1], Elmar Kannenberg[1], Ben Jeyaretnam[1], and Bradley Reuhs[1].

[1] Complex Carbohydrate Research Center, University of Georgia, Athens, GA, USA

Introduction

The cell surface of rhizobia is comprised of a number of polysaccharides; extracellular (EPSs), capsular (KPSs), and lipopolysaccharides (LPSs). Each is important in forming an effective nitrogen-fixing symbiosis. Defective mutants are unable to invade the host root cortical cells in a normal manner, or the infection thread is aborted prior to cortical cell invasion (1,2).

1 Rhizobial Lipopolysaccharide Structures

The general architecture of bacterial LPSs, including those of rhizobia, is shown in Figure 1. The most characterized rhizobial LPSs are those from *R. leguminosarum* (*Rl*) strains, including that from *R. etli* (*Re*) CE3, formerly known as *R. leguminosarum* biovar phaseoli (*Rlp*) CE3 (8-10). In this short paper, it is not possible to describe the structural details reported for all rhizobial LPSs. However, structures have been reported for the LPS core regions of *B. japonicum* and *R. meliloti* (*Rm*) (3-5), the O-chain polysaccharide from *R. tropici* (6) and from *R. leguminosarum* bv. trifolii (*Rlt*) (6a), and *Rm* lipid A (7).

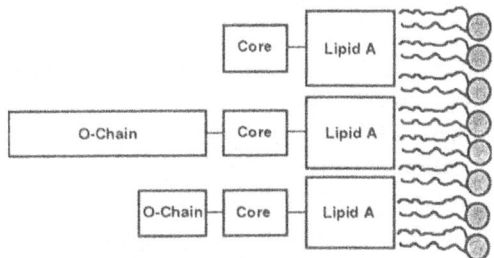

Figure 1. Schematic showing the various structural regions of LPS and that LPSs exist without O-chain polysaccharide and with truncated forms of O-chain

There appears to be a common core region among all strains of *Rl*, including that from *Re* CE3 (9). The similarity in core structures was monitored by high performance anion exchange chromatography (HPAEC) of the LPS mild acid hydrolysates (manuscript in preparation). The HPAEC profiles for the core oligosaccharides from *R. leguminosarum* biovar viciae (*Rlv*), *Rlt*, and *Re* are identical to one another (11). In addition, a panel of monoclonal antibodies to the core region of *Rl* 3841 LPS cross reacts with all other *Rl* LPSs tested, including *Re* CE3 LPS (12), supporting the conclusion that these strains share a common core structure.

The lipid A of *Rl* LPSs is very different from that of enteric LPSs. It contains (a.) a galacturonosyl residue rather than phosphate α-linked to the 4'-position of the lipid A, (b.) a 2-aminogluconate (GlcN-onate) residue at its "reducing" end rather than GlcN, (c.) a mixture of β-hydroxy fatty acyl residues instead of only β-hydroxymyristic acid, (d.) a long chain 27-hydroxyoctacosanoyl residue, and does not contain acyloxyacyl residues.

2 The biosynthesis of *R. leguminosarum* lipid A

Due to the ununusual of *Rl* lipid A structure, it was of interest to determine the unique aspects of its biosynthetic pathway compared to that of enteric lipid A. Chris Raetz and co-workers had defined the biosynthesis of *E. coli* lipid A, particularly the first six enzymatic steps leading from UDP-GlcNAc to Kdo_2-Lipid IVA (19). It was demonstrated that *Rl* contained these same enzyme activities (17). Thus, it was proposed that Kdo_2-Lipid IVA, or a very similar analog, must be a precursor to the mature *Rl* lipid A, and that there is a point of divergence between the *E. coli* and *Rl* lipid A biosynthesis which occurs after Kdo_2-Lipid IVA synthesis. This implied that unique enzymes in *Rl* process the Kdo_2-Lipid IVA precursor into the mature lipid A. One unique enzyme is a phosphatase which removes the 4'-phosphate from Kdo_2-Lipid IVA (18). This enzyme activity is dependent on the presence of the two Kdo residues and is membrane bound (18). The characterization of other unique *Rl* enzymes is in progress by Dr. Raetz's group.

These results suggest that the synthesis of Kdo_2-Lipid IVA must be a crucial step for a very wide distribution of Gram-negative bacteria. This is not unexpected since all steps leading to the synthesis of Kdo_2-Lipid IVA, as well as those involved in the synthesis of Kdo, are essential for cell viability (26). Thus, a general prediction for the synthesis of lipid A in Gram-negative bacteria is that the essential steps leading to Kdo_2-Lipid IVA synthesis are common, and that specific enzymes leading to unique lipid A structures necessarily follow after these common steps.

3 The functions of rhizobial LPSs, EPSs, and KPSs in symbiosis

As mentioned in the "Introduction", LPS mutants which lack the O-chain polysaccharide are symbiotically defective implying that an intact LPS is essential for symbiosis. In addition, it has also been shown, using monoclonal antibodies, that the LPS undergoes structural changes during symbiosis (20-23). These changes occur in the O-chain polysaccharide portion of the LPS and some can be observed when the bacteria are grown under conditions thought to mimic, in part, those found inside the nodule, *e.g.* low pH or low O_2 tension (2,22,27). It has also been reported that, in the case of *Re*, one of these changes is inducible by root or seed exudates from bean (23); the compounds from the seed coat being anthocyanins (24). In addition, phenol-water extraction of bacteroids, or of bacteria grown under low pH or low O_2, results in a different distribution of LPS in the water and phenol layers compared to that from normally cultured bacteria (12). In the case *R. fredii*, and more recently NGR234, growth of the bacteria in the presence of a *nod* gene inducer resulted in the expression of novel forms of LPS (25,26).

The functions of LPS in symbiotic infection are not known. However, they may be involved in the (a.) avoidance of the host defense response, (b.) cell invasion (endocytosis), and/or (c.) required for the exchange of metabolites between the plant host and the bacterial symbiont. Some work has been done showing that LPS defective (*i.e.* missing the O-chain) mutants result in the stimulation of certain defense responses; *e.g.* callose and lignin deposition, induction of aromatic compounds, and phytoalexin stimulation (27-32). Thus, the presence of the O-chain and the changes in its structure during infection may be important in avoiding or supressing of the plant's defenses.

Little is known concerning how the defective LPSs may stimulate the plant's defense response. However, there is considerable knowledge as to how anmimal cells respond to LPS (*i.e.* endotoxin). This process has been briefly summarized (33.34) and is depicted in Figure 2. Serum contains an LPS binding protein (LBP) which functions by delivering one molecule of LPS from an LPS aggregate to (in macrophages) a GPI-anchored membrane protein, CD14. This CD14-LBP-LPS complex delivers the LPS to a transmembrane receptor which is involved in stimulating the signal transduction cascade leading to the production of various cytokines. Whether or not plant cells act in an analogous manner to animal cells is not known. However, the animal system may serve as a model for future investigations.

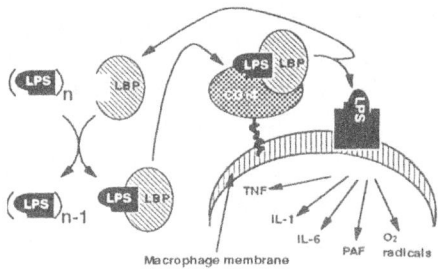

Figure 2. A schematic shown the mechanism by which a macrophage cell responds to LPS. LBP = LPS binding protein, CD14 = a GPI anchored protein, TNF = tumor necrosis factor, IL = interleukin, PAF = platlet activating factor.

As with LPS, EPS and KPS may have important functions in the avoidence or supression of the host defense response. It had been shown that relatively small amounts of EPS can complement an EPS⁻ mutant resulting in nitrogen-fixing nodules (36,37). In the case of *Rm*, the active EPS I consists of approximately a tetrameric version of the eight sugar repeat unit (36). Recently, it has been shown that the 15-20-mer version of the disaccharide repeat unit of EPS II in amounts as little as 7 picomoles also corrects the EPS⁻ *Rm* phenotype (38). Similar results have been obtained for the KPS; *i.e.* a low molecular weight version of the polymer can correct the EPS⁻ phenotype (Reuhs, personal communication). Only those *Rm* strains containing the *rpkZ* gene produce forms of KPS which can substitute for EPS (35).

The EPSs, KPSs, and intact LPSs, may act as signal molecules involved in regulating the host's defense reponse. The plant's defense system is stimulated by a variety of molecules, called elicitors; *e.g.* hepta-β-glucosides, chitin and chitosan oligomers, oligouronides, glycopeptides, etc. (for reviews see, 39,40). A single plant can respond to a variety of these structures. That any one of such a variety of structures can result in stimulation of the defense pathway suggests that the putative receptors for these different structures all have a common point of intersection with regard to activating the host defense system. This common point of intersection may include membrane proteins involved in H^+, K^+, Cl^-, and Ca^{2+} fluxes, the formation of H_2O_2, and/or the phosphorylation systems involved in signal transduction (again, see above reviews). Similarly, the EPSs, KPSs, LPSs may act as supressors (28) of the defense pathway, again with their putative receptors acting at the same common point of intersection mentioned above for elicitors. A recent presentation (41) provided data indicating that *Rm* low molecular weight EPS I or EPS II, and LPS can inhibit the alkalinization of alfafa tissue culture cells by an elicitor. Recently, it was reported that KPS can cause a transient expression of chalcone synthase mRNA in alfalfa (42). Some supressors may act simply by competitively inhibiting the binding of an elicitor to its receptor. An example of this may be the cyclic-β-3,6-linked glucan from *B. japonicum* which appears to inhibit the binding of the hepta-β-glucan elicitor to its receptor (43).

More experimentation to determine the effects of EPSs, KPSs, and LPSs on specific aspects of the defense pathway are most likely in progress. However, when such experiments are done one has to be careful to use the purified form of EPS, KPS, or LPS that is known to be important in symbiotic infection.

Acknowledgements

This work was supported, in part, by grants from the NSF (IBN-9305022), the NIH (GM89583), and the DOE (DE-FG05-93ER20097).

Citations

1. Leigh, J. A. and Walker, G. C. (1994) *Trends Genet.* **10**, 63-67
2. Noel, K. D. (1992) in *Molecular signals in plant-microbe communications* (Verma, D. P. S. ed) pp. 341-357, CRC Press, Boca Raton, Ann Arbor, London
3. Carlson, R. W. and Krishnaiah, B. S. (1992) *Carbohydr. Res.* **231**, 205-219
4. Russa, R., Bruneteau, M., Shashkov, A. S., Urbanik-Sypniewska, T., and Mayer, H. (1996) *Arch. Microbiol.* **165**, 26-33
5. Russa, R., Urbanik-Sypniewska, T., Choma, A., and Mayer, H. (1991) *FEMS Microbiol. Lett.* **84**, 337-344
6. Gil-Serrano, A. M., González-Jiménez, I., Mateo, P. T., Bernabé, M., Jiménez-Barbero, J., Megías, M., and Romero-Vázquez, M. J. (1995) *Carbohydr. Res.* **275**, 285-294
6a. Wang, Y. and Hollingsworth, R.I. (1994) *Carbohydr.Res.* **260**, 305-317.
7. Urbanik-Sypniewska, T., Seydel, U., Greck, M., Weckesser, J., and Mayer, H. (1989) *Arch. Microbiol.* **152**, 527-532
8. Bhat, U. R., Forsberg, L. S., and Carlson, R. W. (1994) *J. Biol. Chem.* **269**, 14402-14410
9. Carlson, R. W., Reuhs, B., Chen, T.-B., Bhat, U. R., and Noel, K. D. (1995) *J. Biol. Chem.* **270**, 11783-11788
10. Bhat, U. R., Bhagyalakshmi, S. K., and Carlson, R. W. (1991) *Carbohydr. Res.* **220**, 219-227
11. Kannenberg, E., Xie, S., and Carlson, R.W. (1996) *MPMI* **Knoxville, Tn**, H65 (abstract).
12. Kannenberg, E., Forsberg, L. S., and Carlson, R. W. (1996) *Plant and Soil* (in press)
13. Gil-Serrano, A. M., González-Jiménez, I., Tejero-Mateo, P., Megías, M., and Romero-Vázquez, M. J. (1994) *J. Bacteriol.* **176**, 2454-2457
14. Mayer, H., Krauss, J. H., Urbanik-Sypniewska, T., Puvanesarajah, V., Stacey, G., and Auling, G. (1989) *Arch. Microbiol.* **151**, 111-116
15. Bhat, U. R., Mayer, H., Yokota, A., Hollingsworth, R. I., and Carlson, R. W. (1991) *J. Bacteriol.* **173**, 2155-2159
16. Russa, R., Urbanik-Sypniewska, T., Lindström, K., and Mayer, H. (1995) *Arch. Microbiol.* **163**, 345-351
17. Price, N. P. J., Kelly, T. M., Raetz, C. R. H., and Carlson, R. W. (1994) *J. Bacteriol.* **176**, 4646-4655
18. Price, N. P. J., Jeyaretnam, B., Carlson, R. W., Kadrmas, J. L., Raetz, C. R. H., and Brozek, K. A. (1995) *Proc. Nat. Acad. Sci. USA* **92**, 7352-7356
19. Raetz, C. R. H. (1993) *J. Bacteriol.* **175**, 5745-5753
20. Kannenberg, E. L. and Brewin, N. J. (1989) *J. Bacteriol.* **171**, 4543-4548
21. Kannenberg, E. L. and Brewin, N. J. (1994) *Trends in Microbiol.* **2**, 277-283
22. Tao, H., Brewin, N. J., and Noel, K. D. (1992) *J. Bacteriol.* **174**, 2222-2229
23. Noel, K. D., Duelli, D. M., Tao, H., and Brewin, N. J. (1996) *MPI* **9**, 180-186
24. Duelli, D. M. and Noel, K. D. (1996) *MPMI* **Knoxville, TN**, H26(Abstract)

25. Reuhs, B. L., Kim, J. S., Badgett, A., and Carlson, R. W. (1994) *Mol. Plant Microbe Interact.* **7**, 240-247

26. Jabbouri, S., Hanin, M., Fellay, R., Quesada-Vincens, D., Reuhs, B., Carlson, R. W., Perret, X., Freiberg, C., Rosenthal, A., Leclerc, D., Broughton, W. J., and Relic, B. (1996) *MPMI* **Knoxville, TN**, (Abstract)

27. Staehelin, C., Müller, J., Mellor, R. B., Wiemken, A., and Boller, T. (1992) *Planta* **187**, 295-300

28. Niehaus, K., Kapp, D., and Pühler, A. (1993) *Planta* **190**, 415-425

29. Perotto, S., Brewin, N. J., and Kannenberg, E. L. (1994) *Mol. Plant Microbe Interact.* **7**, 99-112

30. Lawson, C. G. R., Djordjevic, M. A., Weinman, J. J., and Rolfe, B. G. (1994) *Mol. Plant Microbe Interact.* **7**, 498-507

31. Parniske, M., Schmidt, P. E., Kosch, K., and Müller, P. (1994) *Mol. Plant Microbe Interact.* **7**, 631-638

32. Van Workum, W. A. T., van Brussel, A. A. N., Tak, T., Wijffelman, C. A., and Kijne, J. W. (1995) *Mol. Plant Microbe Interact.* **8**, 278-285

33. Beutler, B. and Kruys, V. (1995) *J. Cardiovasc. Pharmacol.* **25**, S1-S8

34. Schletter, J., Holger, H., Ulmer, A. J., and Rietschel, E. T. (1995) *Arch. Microbiol.* **164**, 383-389

35. Reuhs, B. L., Williams, M. N. V., Kim, J. S., Carlson, R. W., and Côté, F. (1995) *J. Bacteriol.* **177**, 4289-4296

36. Battisti, L., Lara, J. C., and Leigh, J. A. (1992) *Proc. Natl. Acad. Sci. USA* **89**, 5625-5629

37. Djordjevic, S. P., Chen, H., Batley, M., Redmond, J. W., and Rolfe, B. G. (1987) *J. Bacteriol.* **169**, 53-60

38. Gonzales, J. E., Reuhs, B. L., and Walker, G. C. (1996) *Proc. Natl. Acad. Sci. USA* **93**, 8636-8641

39. Kombrink, E. and Somssich, I. E. (1995) *Adv. Bot. Res.* **21**, 1-34

40. Côté, F. and Hahn, M. G. (1994) *Plant Mol. Biol.* **26**, 1379-1411

41. Niehaus, K., Baier, R., and Puhler, A. (1996) *MPMI* **Knoxville, TN**, L16(Abstract)

42. Becquart-de Kozak, I., Reuhs, B. L., Buffard, D., Breda, C., Kim, J. S., Esnault, R., and Kondorosi, A. (1996) *Mol. Plant Microbe Interact.* (in press)

43. Mithöfer, A., Bhagwat, A. A., Feger, M., and Ebel, J. (1996) *Planta* **199**, 270-275

Changes in Rhizobium Lipopolysaccharide Structure Induced by Host Compounds

K. Dale Noel, Dominik M. Duelli, and Valerie J. Neumann

Department of Biology, Marquette University, Milwaukee, WI, USA

1 Host-induced Lipopolysaccharide Structural Refinements

The components of bacterial outer membranes vary according to the conditions of growth. This type of variation has been well established for the proteins and, to a lesser extent, the lipopolysaccharide (LPS) of bacteria that associate with animals. Similarly, there has long been the conviction that the conditions of root nodules induce a particular outer membrane composition in *Rhizobium* bacteroids that differs from that of the bacteria grown in laboratory media. Indeed, using monoclonal antibodies coupled with gel electrophoresis and electron microscopy, de Maagd et al (1994) and Brewin and colleagues (1991) have provided persuasive evidence that the outer membrane proteins and LPS of *Rhizobium leguminosarum* change during the course of legume infection. The importance and specific functions of such alterations in outer membrane structure remain unclear, however.

In *Rhizobium etli* CE3 the differences between bacteroid LPS and the LPS of the bacteria grown in typical laboratory media are subtle and most easily detected by monoclonal antibodies generated by Brewin (Tao et al. 1992). As with *R. leguminosarum* (Brewin 1991), certain conditions of growth ex planta also result in *R. etli* LPS alterations that can be detected with these antibodies. These conditions include low pH, oxygen and phosphate limitation, and temperature stress. In addition, exudate from host *Phaseolus vulgaris* seeds and roots induces changes in LPS structure that, at least antigenically, are similar to those found in the nodule (Noel et al 1996). Particular exudate compounds that induce this effect have been identified, and bacterial mutants unable to undergo this response have been isolated.

2 Inducing Compounds in Seed Exudate

The inducing activity released from germinating seeds was highest during the first 12 h of imbibition of water. The crude exudate collected during the first 24 h of germination was resolved by reverse-phase high performance liquid chromatography (HPLC). Several fractions that eluted consecutively, those with absorbance maxima between 530 and 560 nm, had activity. These combined fractions were hydrolyzed with acid, extracted into amyl alcohol, and resolved by HPLC conditions optimized for anthocyanidin separation. Fractions with activity were the ones having the spectral properties of anthocyanidins, with the most abundant one having the properties of delphinidin (Fig. 1). To test whether the activity might be due to a co-eluting contaminant, delphinidin was obtained also from a commercial source and purified from eggplant peels. Both sources gave specific activities that were similar to that of the material purified from bean exudate.

NATO ASI Series, Vol. G 39
Biological Fixation of Nitrogen
for Ecology and Sustainable Agriculture
Edited by A. Legocki, H. Bothe, A. Pühler
© Springer-Verlag Berlin Heidelberg 1997

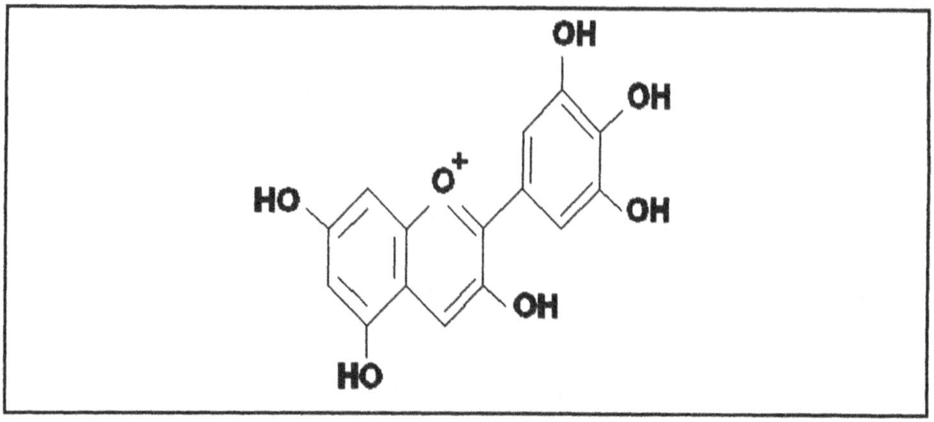

Fig. 1. Delphinidin

The anthocyanidins were found in the crude exudates as glycosylated derivatives (anthocyanins), as determined by using extraction methods and a purification scheme designed specifically for anthocyanins. The anthocyanidin in each purified anthocyanin was identified by comparison of HPLC elution with anthocyanidin standards, thin-layer chromatographic behavior, UV/visible spectra, spectral shifts in AlCl₃, and gas chromatography. The sugars were identified by derivatization and gas chromatography. The most abundant anthocyanin (57% of the total) was delphinidin-3-O-glucoside. In descending order of abundance the other major anthocyanins were petunidin-3-O-glucoside, a cyanidin diglycoside, malvidin-3-O-glucoside, and a delphinidin diglycoside. Each was active in inducing the LPS antigenic change, although the malvin had lower specific activity. The anthocyanidins exhibited half-maximal activity at 10 to 50 μM, depending on the compound.

Root exudates also triggered the LPS antigenic change (Noel et al 1996), but the major *nod*-inducers from the root, naringenin and genistein, were very poor inducers of this effect. Anthocyanins are not known to be present in roots, and HPLC analysis indicated that the active root compounds eluted from the column before seed anthocyanins would and much earlier than narigenin or genistein. A white variety of *Phaseolus vulgaris*, which does not produce anthocyanins, does not have activity in its seed exudate, but its root exudate is active.

3 Distinction between *nod*-induction and exudate-induced LPS modification

Anthocyanins from seed exudate also induce *R. etli nod* genes, but they do so at concentrations that are 1% to 10% of those needed to cause half-maximal LPS antigenic conversion. The data mentioned above indicates that in root exudate the major *nod* inducers are distinct from the compounds that induce this LPS change. Another observation that distinguishes *nod* induction from LPS modification is that mutants lacking the Sym plasmid of *R. etli* CE3 carry out the LPS antigenic conversion in a fashion that appears to be identical with with the wild type (Noel et al 1996). Therefore, *nodD* and any other genes found only on this plasmid apparently are not involved. This effect joins the few known examples of responses to legume exudates that do not involve *nodD*.

4 Mutants Defective in this Response

In order to identify the protein components required and determine what role this phenomenon plays in the physiology of *R. etli*, particularly in its symbiosis with *Phaseolus vulgaris*, modification-deficient mutants are being isolated. A screening procedure based on transposon mutagenesis and immuno-staining of colony lifts has been devised. Two types of mutants are sought. One referred to as Lpm⁻ (Lipopolysaccharide modification) still binds strongly to JIM28 antibodies after growth in the presence of exudate or purified anthocyanins, whereas the wild type no longer binds to the antibody after this treatment. The other (Lpe⁻, Lipopolysaccharide epitope) lacks the antibody epitope under all conditions (Fig. 2). In this latter type of mutant the mutations may target the modifying enzyme itself or master regulators of the decorations of the polysaccharide backbone that seem to be responsible for the epitopes recognized by the antibodies. As might be expected, since it has the bacteroid immunotype, this type of mutant is symbiotically proficient.

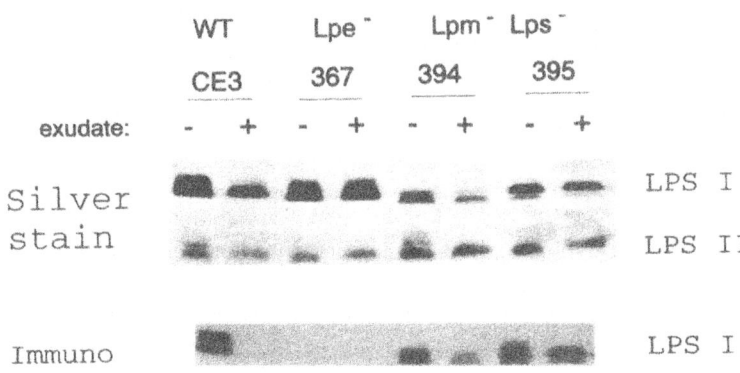

Fig. 2. Lpe⁻ mutant CE367 and Lpm⁻ Lps⁻ mutants CE394 and CE395. Each mutant and wild type CE3 were grown under standard conditions (-) or in the presence of seed exudate (+). After SDS-PAGE of SDS extracts of the bacterial cells, the nitrocellulose blot ("immuno") was reacted with JIM28 antibodies and the residual material in the gel was stained by the silver-periodate procedure.

The ideal Lpm⁻ mutant for determining whether the exudate-induced change is important in symbiosis has not been isolated to this point. Strains CE394 and CE395 exemplify the more common Lpm⁻ phenotype among the mutants thus far isolated and analyzed. This type is referred to as Lpm⁻ Lps⁻ because it exhibits obvious abnormalities in LPS banding on SDS polyacrylamide gel electrophoresis, regardless of the conditions of growth (Fig. 2). The three mutants of this type that have tested for symbiotic properties (strains CE394, CE395, and CE397) were severely defective in infection. However, the alteration in structure or the deficiency in LPS I (the O-polysaccharide-containing form of LPS) could have been responsible; i.e, it is not possible to attribute its other properties merely to the lack of response to exudate.

In strain CE396 (Fig. 2), on the other hand, the LPS I gel profile and abundance appear normal. This mutant exhibits the wild-type symbiotic proficiency, but it also is not a fair test of the importance of the LPS conversion. Although the binding to antibody JIM28 is not lost at the concentration of seed exudate used in this experiment, the binding is diminished relative to growth without inducer, and at not much higher concentrations of inducers this mutant displays the wild-type antigenic conversion. Hence, the search continues for mutants which have normal LPS I gel profiles but which do not change the LPS at all in response to exudate.

One hypothesis to explain the lack of antigenic conversion in the Lpm⁻ Lps⁻ mutants is that a putative modifying enzyme cannot recognize the aberrant basal LPS structure in these mutants and therefore the LPS remains antigenic. A corollary of this idea is that less of the complete structure is required to constitute a strong epitope for the antibody, including putative decorative groups, than is required for recognition by modifying enzymes.

Lpm⁻ mutants of both classes respond to growth at low pH in normal fashion. Therefore, the pathways by which LPS is modified in response to low pH and to exudate are at least partially independent. Moreover, it seems that less of the complete LPS structure is needed to carry out the modification at low pH than in response to exudate.

Citations

Brewin, N.J. 1991. Development of the legume root nodule. Ann. Rev. Cell Biol. 7:191-226.

de Maagd, R.A., Yang, W., Goosen-de Roo, L., Mulders, I.H.M., Roest, H.P., Spaink, H.P., Bisseling, T., and Lugtenberg, B.J.J. 1994. Down-regulation of expression of the *Rhizobium leguminosarum* outer membrane protein gene *ropA* occurs abruptly in interzone II-III of pea nodules and can be uncoupled from *nif* gene activation. Mol. Plant-Microbe Interact. 7:276-277.

Noel, K.D. 1992. Rhizobial polysaccharides required in symbiosis with legumes. Pages 341-357 in:Molecular Signals in Plant-Microbe Communications. D.P.S. Verma, ed. CRC Press, Boca Raton, FL.

Noel, K.D., Duelli, D.M., Tao, H., and Brewin, N.J. 1996. Antigenic change in the lipopolysaccharide of *Rhizobium etli* CFN42 induced by exudates of *Phaseolus vulgaris*. Mol. Plant-Microbe Interact. 4:332-340.

Tao, H., Brewin, N.J., and Noel, K.D. 1992. *Rhizobium leguminosarum* CFN42 lipopolysaccharide antigenic changes induced by environmental conditions. J. Bacteriol. 174:2222-2229.

Symbiotic suppression of the *Medicago sativa* plant defence system by *Rhizobium meliloti* oligosaccharides

Karsten Niehaus[1], Ruth Baier[1], Bodo Kohring[2], Erwin Flaschel[2] and Alfred Pühler[2]

[1]Lehrstuhl für Genetik, Fakultät für Biologie, Universität Bielefeld, P.O.Box 100131, D-33501 Bielefeld, Germany
[2]Lehrstuhl für Fermentationstechnik, Technische Fakultät, Universität Bielefeld, P.O.Box 100131, D-33501 Bielefeld, Germany

The establishment of nitrogen-fixing root nodules in the *Rhizobium*-legume symbiosis is a complex, multistep interaction between the bacterium and the specific host plant. Major early events in this plant-microbe interaction are the perception of plant borne flavonids, the synthesis of a lipooligosaccharide, the nodulation factor, and the infection of the plant via a curled root hair by the microsymbiont. The infection of a curled root hair is characterized by the formation and sustained development of the infection thread, in which the bacteria multiply and grow towards the simultaniously initiated meristem within the root cortex (Hirsch 1992). *Rhizobium meliloti* mutants that fail to synthesize the exopolysaccharide succinoglycan (EPS I) were unable to infect the plant, nevertheless, they induce non-infected pseudonodules. A detailed analysis of these pseudonodules revealed strong evidence for the induction of a plant defence response by the mutated microsymbiont (Niehaus et al. 1993). This observation gave the experimental evidence that also symbiotic bacteria have to deal with the plant defence system. In analogy to compatible and incompatible interactions between pathogens and plants, a „gene for gene" hypothesis was also formulated for symbiotic interactions (Djordjevic et al. 1987). While specific recognition is the basis for incompatibility in plant-pathogen interactions, it appears that specific recognition is necessary for compatibility in symbiosis. Failure to communicate will lead either to a complete lack of interaction or the abortion of the infection process at various stages. So it has been suggested that successful infection of the host plant may also depend on the ability of rhizobia to escape or to suppress the induction of a plant defence reaction that normally serves to prevent infection by pathogens (Vance 1983, Niehaus et al. 1993).

In this paper we describe the influence of various potential signal molecules from *R. meliloti* with respect to their ability to induce or suppress the plant defence system of the host plant alfalfa (*Medicago sativa*) or the nonhost plants tobacco (*Nicotiana tabacum*) and tomato (*Lycopersicum esculentum*). As a modell system we established plant cell cultures that reacted to the addition of elicitors with a strong transient alkalinization of their culture media, a bioassay for a plant defence related signal transduction.

NATO ASI Series, Vol. G 39
Biological Fixation of Nitrogen
for Ecology and Sustainable Agriculture
Edited by A. Legocki, H. Bothe, A. Pühler
© Springer-Verlag Berlin Heidelberg 1997

Establishment of elicitor responsive plant cell cultures as a model system to analyse the action of *Rhizobial* signal molecules

Within the complex structure of entire tissues, like a plant root, recognition of signal molecules may depend on their ability to reach the target cells. Moreover, the reaction of a few single target cells may not be detectable in the background of a majority of non-responding cells. Therefore we established elicitor responsive cell cultures of the *R. meliloti* host plant alfalfa. Selected suspension cultures of alfalfa reacted to a crude extract of bakers yeast elicitors (YE) with an accumulation of the phytoalexin medicarpin and enhanced peroxidase activity, indicating a typical plant defence reaction. In addtion, the selected cell cultures reacted to small amounts of YE with a transient alkalinization reaction of their culture medium, indicating a sensitive perception system for these compounds (Fig. 1). Pharmacological studies, using e.g. protein kinase inhibitors like K252a, showed that the elicitor induced alkalinization is the consequence of a preceding signal transduction event (Fig. 1). As a control, elicitor responsive cell cultures of tomato and tobacco, nonhosts of *R. meliloti*, were selected.

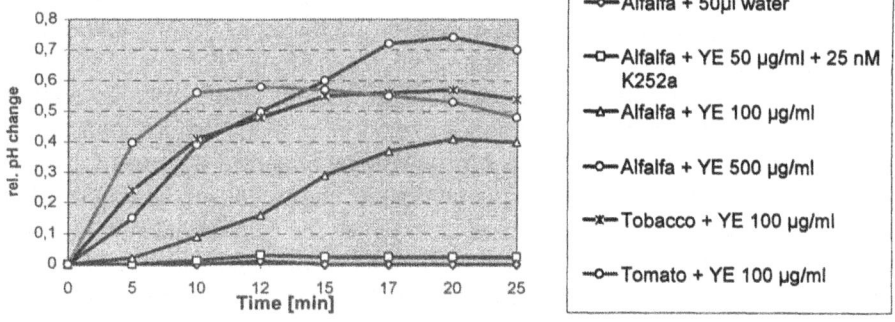

Fig. 1 Alkalinization of alfalfa, tobaco and tomato plant cell cultures in response to yeast elicitors (YE) and the protein kinase inhibitor K252a.

Rhizobium meliloti **nodulation factors induce a defence related alkalinization response in non host- but not in host- plant cell cultures**

Important early steps in the establishment of the symbiosis, like the root hair curling and induction of the root nodule meristem are dependent on the synthesis of the rhizobial nodulation factors. The rhizobial nodulation factors are structurally related to oligochitin fragments known to act as elicitors of defence related responses in different fungi-plant interactions. Staehelin et al. (1994) described that *R. meliloti* nodulation factors, as well as chitin oligomers, induce a transient alkalinization in tomato cell cultures, indicating that both substances were recognized as elicitors in this system. Therefore we compared the reaction of host- and nonhost plant cell cultures to *R. meliloti* nodulation factors. The nodulation factors were extracted from the culture by adsorption on XAD-2 resin and specific elution with ethanol. After semi-preparative reversed phase HPLC a nodulation factor extract was obtained. Analytical HPLC and mass spectroscopy showed that the extract consisted of NodRm-IV(Ac,S), -IV(S), -V(Ac,S) and -V(S). Suspension cultured tomato- and tobacco-cells, nonhosts of *R. meliloti*, reacted to the nodulation factors (10^{-6}M) and to chitin oligomers (DP 5, 10^{-5}M) with a transient alkalinization of their culture medium, indicating a sensitivity for both compounds. Alfalfa cell cultures showed in comparisiopn no alkalinization reaction when treated with nodulation factors ($10^{-4/-5/-6}$M). Treatment with chitin oligomers (DP 6, 10^{-5}M) induced the alkalinization reaction, similar to nonhost plants. These observations suggest that the host specific *R. meliloti* nodulation factors, elicit a rapid plant defence related reaction on the nonhost plants tomato and tobacco, whereas the host plant alfalfa is able to distinguish between chitin elicitors and „substituted chitin oligomers" like the nodulation factors.

The *Rhizobium meliloti* surface carbohydrates, low-molecular weight EPS I and LPS suppress the defence related alkalinization response in alfalfa but not in tobacco and tomato plant cell cultures

From *R. meliloti* mutants, defective in the synthesis of exopolysaccharides (EPS I) or lipopolysaccharides, we speculated that these compounds have a function in the suppression of plant defence reactions in the host plant alfalfa. The *R. meliloti* mutant exoP* produced only low molecular weight (LMW) EPS I. Nevertheless, this mutant is able to infect the plant (Becker et al. 1995). In accordance with this Battisti et al. (1992) reported that the defect in invasion of the *R. meliloti* EPS I non-producing mutants could be restored by the exogenious addition of purified LMW EPS I.

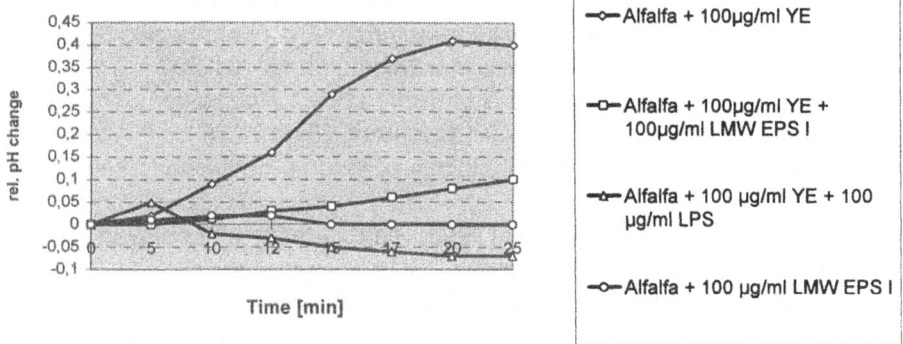

Fig. 2 Suppression of the yeast elicitor induced alkalinization reaction in alfalfa cell cultures by *R. meliloti* low molecular EPS I or LPS

We could show that in alfalfa cell cultures the yeast elicitor induced alkalinization which was suppressed by the simultanious application of LMW EPS I (Fig. 2). High molecular weight EPS I however, provoked no reduction of the elicitor response. Among other tested oligosaccharides, only the *R. meliloti* LPS, was able to suppress the elicitor action. In tobacco or tomato cell cultures no such a suppression was not observed. These data provide a strong evidence for a specific recognition of the *R. meliloti* LMW EPS I and LPS by the host plant as a suppressor of the plant defence system enabling the symbiont to infect the plant. In addition, these observations suggest that the host specific *R. meliloti* nodulation factors elicit a rapid plant defence related response on the non host plants, whereas the host plant alfalfa does not recognize them as elicitors (Tab. 1).

Tab. 1 Biological activity of rhizobial signal molecules in host- and non host-cell cultures.

Substance	Alfalfa cells	Tomato cells	Tobacco cells
yeast elicitors	E	E	E
chitinoligosaccharide	E	E	E
R. meliloti nodulation factors	N	E	E
R. meliloti LMW EPS I	S	N	N
R. meliloti LPS	S	N	N

E elicitor, S suppressor, N no influence on the   alkalinization responce

References:

Battisti L., Lara J.C. and J.A. Leigh (1992) Specific oligosaccharide form of the *Rhizobium meliloti* exopolysaccharide promotes nodule invasion in alfalfa. Proc. Natl. Acad. Sci. **89**:5625-5629

Becker A., Niehaus K. and A. Pühler (1995) Low molecular weight succinoglycan is predominantly produced by *Rhizobium meliloti* strains carrying a mutated ExoP protein characterized by a periplasmic N-terminal and a missing C-terminal domain. Mol. Microbiol. **16**:191-203

Djordjevic M.A., Gabriel D.W. and B.G. Rolfe (1987) *Rhizobium* - The refined parasite of legumes. Annu. Rev. Phytopathol. **25**:145-168

Hirsch A.M. (1992) Developmental biology of legume nodulation. New Phytol. **122**: 211-237

Niehaus K., Kapp D. and A. Pühler (1993) Plant defence and delayed infection of alfalfa pseudonodules induced by an exopolysaccharide (EPS I)-deficient *Rhizobium meliloti* mutant. Planta **190**:415-425

Staehelin C., Granado J., Müller J., Wiemken A., Mellor R.B., Felix G., Regenass M., Broughton W.J., and T. Boller (1994) Perception of *Rhizobium* nodulation factors by tomato cells and inactivation by root chitinases. Proc Natl. Acad. Sci. **91**:2196-2200

Vance C.P. (1983) *Rhizobium* infection and nodulation: a beneficial plant disease? Ann. Rev. Microbiol. **37**:399-424

Biochemical and Molecular Analyses of Rhizobial Responses to Legume Flavonoids

J. E. Cooper[1], J. R. Rao[1], L. De Cooman[2], T. M[c]Corry[1], A. J. Bjourson[1], H. L. Steele[1], W. J. Broughton[3] and D. Werner[4]

[1]Department of Applied Plant Science, The Queen's University of Belfast, Newforge Lane, Belfast BT9 5PX, Northern Ireland
[2]Laboratory of Plant Biochemistry, Rijksuniversiteit Gent, K L Ledeganckstraat 35, B-9000, Gent, Belgium
[3]LBMPS, University of Geneva, CH-1292 Chambésy, Geneva, Switzerland
[4]Department of Biology, University of Marburg, D-35032 Marburg, Germany

1. Introduction

In addition to inducing the expression of *nod* genes and the subsequent synthesis of Nod factors in rhizobia, legume flavonoids also elicit a number of other responses in the free-living form of these microsymbionts. These include degradation of the flavonoid itself (Rao and Cooper, 1994), expression of genes which share no homology with nodulation gene promoters (Sadowsky *et al.*, 1988; Perret *et al.*, 1994) and exudation of new proteins into the surrounding environment (Krishnan and Pueppke, 1993). We have used an array of biochemical, chemical analytical and molecular techniques to investigate mechanisms of flavonoid degradation exhibited in rhizobia, changes in flavonoid content of legume root exudates during incubation with rhizobia, the fate of a *nod* gene-inducing flavonoid during Nod factor synthesis and the nature of gene expression in rhizobia during exposure to flavonoids. This paper summarises the principal findings to date from our research on these four facets of the interaction between flavonoids and rhizobia.

2. Results and Discussion

2.1 Rhizobial Degradation and Modification of Flavonoids

Using direct GC-MS analyses of derivatized extracts from culture supernatants we have shown that *Rhizobium* species degrade individual flavonoids by initiating multiple fissions in the C-rings of the compounds and releasing products whose structures are based on the conserved A- and B-rings. Other open and closed C-ring modification products may also be detected among the metabolites. For example, phoroglucinol (conserved A-ring product), protocatechuic acid (conserved B-ring product) and tetrahydroxy flavanone (C-ring modification product) can all be detected in supernatants of *Rhizobium meliloti* cultures supplemented with the flavone, luteolin at a concentration of 10 μM (Rao and Cooper, 1994). Recently, it has also been

NATO ASI Series, Vol. G 39
Biological Fixation of Nitrogen
for Ecology and Sustainable Agriculture
Edited by A. Legocki, H. Bothe, A. Pühler
© Springer-Verlag Berlin Heidelberg 1997

demonstrated that *Bradyrhizobium japonicum* catabolises the isoflavonoids genistein and daidzein via similar C-ring fission mechanisms (Rao and Cooper, 1995). The products arising from the metabolism of genistein are shown in Table 1. They include several compounds with previously established functions relating to *nod* gene induction (coumestrol), *nod* gene inhibition (umbelliferone, naringenin) or chemotaxis (4 - hydroxycinnamic acid).

Table 1 Metabolites identified from supernatants of genistein-supplemented media during incubation with *B. japonicum* USDA 110 *spc*4 for 2-24 h.

Metabolites	Mode of biotransformation	Molecular ion peaks of methyl ether of the metabolite	Conc. (nM)
Naringenin chalcone	C-ring	314, 137, 105	50
Naringenin	modification	314, 180, 134	120
Coumestrol		296, 240	140
Phloroglucinol	Conservation of	168, 153, 110	280
Phloroglucinol carboxylic acid	A-ring	226, 211, 183	1,560
4-Methyl umbelliferone		174, 148, 133	190
Umbelliferone		176, 148, 133	690
4-Hydroxycinnamic acid	Conservation	192, 161, 133	1,354
Phenylacetic acid	of B-ring	150, 135, 107	550
4-Hydroxybenzoic acid		166, 135	980

A combination of separation and identification methods (polyamide-coated TLC, solid-phase extraction, GC-MS and HPLC) confirmed the presence of several mainly C-ring modification products together with unmetabolised isoflavone in *B. japonicum* cells. For example, daidzein-induced cells contained liquiritigenin (256 nM), coumestrol (390 nM), umbelliferone (279 nM) and daidzein itself (370 nM). The concurrent presence of inducers and inhibitors inside *B. japonicum* cells raises the possibility of competitive binding to NodD proteins as a mechanism for regulating *nod* gene expression.

Biotransformation of naturally occurring flavonoids in legume seed and root exudates has also been investigated. Using *Lotus pedunculatus* as the source of exudates, multiple hyphenated techniques (TLC-UV, HPLC-UV, CZE-UV, GC-MS) were employed to detect flavonoids in a manner analagous to the chemical screening of medicinal plants for bioactive compounds described by Hostettman *et al.* (1996). A photodiode array detector at the UV detection stage provides useful information on flavonoids and their substitution patterns. *L. pedunculatus* seed exudates contained catechin, naringenin, kaempferol, quercetin (aglycone) and at least four quercetin glycosides. Root exudates from sterile seedlings contained catechin, naringenin, quercetin, apigenin and kaempferol together with a number of other partially characterised flavones, flavanones and chalcones. Incubation of plant-free root exudates in the presence of *Rhizobium loti* caused a significant reduction in the concentration of quercetin and the formation of several new aromatic compounds, including 3,4 dihydroxy flavone. These results confirmed that rhizobia can effect biotransformations on mixtures of naturally occurring flavonoids as well as on individual authentic compounds.

2.2 Assimilation of ^{14}C naringenin by *Rhizobium leguminosarum* bv viciae and incorporation of labelled carbon atoms into the acyl side chain of a host-specific Nod factor

The fate of naringenin during *nod* gene induction in *R. leguminosarum* bv viciae was determined. After incubation of a wild type strain (RBL5560) or its pSym-cured derivative for 2h in a medium supplemented with A-ring labelled (4a, 6, 8) ^{14}C naringenin (sp activity 6,259 kBq/μmol) at a concentration of 2 nM (ca 12.5 kBq) and cold acetate (0.5 μM), a radio carbon inventory of cell and supernatant extracts was obtained . ^{14}C acetate (sp activity 2072 kBq/μmol) at a concentration of 0.5 μM (1036 kBq) and cold naringenin (2 nM) were also supplied to the same two strains in a separate experiment which was used to locate the positions of Nod metabolites on thin layer chromatograms of culture supernatants.

These experiments confirmed that RBL5560 cells assimilated and catabolised naringenin during the period of *nod* gene activation. The radio label from ^{14}C naringenin was located principally in cellular proteins, lipids and cell wall components, with smaller amounts in nucleic acid and low molecular weight pools. More surprisingly, radio activity analyses of Nod metabolites from RBL5560 supplied with ^{14}C naringenin detected a signal (2189 dpm) in a fraction corresponding to the Nod metabolite NodRlv IV. 95% of this activity resided in the fatty acid moiety of the molecule. These data provide the first biochemical evidence of the fate of an inducer flavonoid during *nod* gene activation and demonstrate a contribution of carbon atoms from a *nod* gene inducer to a Nod metabolite (Rao *et al.*, 1996).

2.3 Flavonoid - dependent gene expression in rhizobia

Non-*nod* gene expression during exposure to flavonoids has been investigated in *Rhizobium* sp. NGR234. The availability of an ordered cosmid library of this strain's symbiotic plasmid (Perret *et al.*, 1991) has permitted the detection of a number of flavonoid-inducible loci by means of competitive RNA hybridizations to Southern blots of cosmid digests. By combining data from competitive RNA hybridizations, subtractive DNA hybridization and shot gun sequencing it was possible to identify a small collection of cosmid restriction fragments which were induced by daidzein, and were present in NRG234 but not in the closely related strain *R. fredii* USDA257. An example was a symbiotically active gene with strong homology to the leucine responsive regulatory protein of *E. coli* (Perret *et al.*, 1994).

Flavonoid-dependent differential gene expression in NGR234 was also studied by means of the RNA arbitrarily primed polymerase chain reaction (RAP-PCR; Welsh *et al.*, 1992), a technique which is similar to differential display reverse transcription PCR (DDRT-PCR; Liang and Pardee, 1992) for eucaryotic mRNA but which also samples non-polyadenylated RNAs. Using RAP-PCR we have identified a 269 bp daidzein-induced product with 93% homology to a soybean cultivar specificity gene in *R. fredii* : *nolU* (Meinhardt *et al.*, 1993). Expression of *nolU* in *R. fredii* USDA257 is known to be flavonoid-induced and NGR234 also contains genomic DNA sequences which hybridize to a *nolU* probe from USDA257. Increased expression of a *nolU* homologue in NGR234 during exposure to daidzein might therefore be anticipated, even though this organism has very weak nodulating ability on any soybean cultivar. Another product from NGR234 with increased expression in the presence of daidzein shared 99% homology with the polyphosphate kinase gene of *E. coli (ppk)* over 302 of its 308 bp length. Expression analyses of both the *nolU* and *ppk* homologues in NGR234 are currently being undertaken and other flavonoid induced or repressed RAP-PCR products are being sequenced.

Acknowledgements

This work was supported by grants from the Leverhulme Trust, the British Council, the Deutscher Akademische Austauschdienst and the Swiss National Science Foundation.

References

Hostettman, K. *et al.* (1996). In Flavonoids and Bioflavonoids 1995, (eds. Antus, K., Gabor, M. and Vetschera, K.), 35-50, Akademiai Kiado, Budapest, Hungary.
Krishnan, H.B. and Pueppke, S.G. (1993). Mol. Plant -Microbe Interact. 6, 107-113.
Liang, P. and Pardee, A.B. (1992). Science 21, 4272-4280.
Meinhardt, L.W. *et al.* (1993). Mol. Microbiol. 9, 17-29.
Perrett, X. *et al.* (1991). Proc. Natl. Acad. Sci. USA 88, 1923-1927.
Perrett, X. *et al.* (1994). Nucleic Acids Res. 22, 1335-1341.
Rao, J.R. and Cooper, J.E. (1994). J. Bacteriol. 176, 5409-5413.
Rao, J.R and Cooper, J.E. (1995). Mol. Plant-Microbe Interact. 8, 855-862.
Rao, J.R. *et al.*, (1996). Plant Soil (in press).
Sadowsky, M.J. *et al.* (1988). J. Bacteriol. 170, 171-178.
Welsh, J. *et al.* (1992). Nucleic Acids Res. 19, 4965-4970.

Flavonoid-inducible regions in the symbiotic plasmid of *Rhizobium etli*

Lourdes Girard[1], Adriana Corvera[1], Arlette Savagnac[2], Jean-Claude Promé[2],
Esperanza Martínez-Romero[1] and David Romero[1]

[1] Departamento de Genética Molecular, Centro de Investigación sobre Fijación de Nitrógeno, UNAM Ap. Postal 565-A., 62210 Cuernavaca, Mor., México and [2] Laboratoire de Pharmacologie et Toxicologie Fondamentales, CNRS, 205 Route de Narbonne, 31077 Toulouse cedex, France.

1. Introduction

Control of gene expression by compounds present in root exudates is an important aspect for the ecology of rizospheric bacteria. Flavonoid compounds are particularly relevant for interactions between bacteria of the genus *Rhizobium* and leguminous plants. Data from different laboratories indicate that flavonoids are responsible for the induction of genes involved in the nodulation process (reviewed in ref. 1). Flavonoids also influence a variety of processes including the determination of efficiency of nodulation (2), competitivity (3), chemotaxis (4) and resistance to plant phytoalexins (5). Thus, the identification and analysis of bacterial genomic regions controlled by root flavonoids is an important step for the understanding of the molecular aspects that culminate in the establishment of a nitrogen -fixing symbiosis.

Usually, the identification of gene loci whose expression is dependent on root flavonoids has been achieved through gene fusion techniques. Although the effectivity of this strategy has been proven repeatedly, its use generally demands the generation and testing of a large number of gene insertions. Moreover, positional information on the respective loci is only achieved through the use of additional approaches.

Intensive use of molecular technologies during the past decade has lead to the establishment of complete physical maps for the symbiotic plasmids of several *Rhizobium* species (6,7). The existence of these maps has paved the way for the use of alternative approaches that allow a global evaluation of transcriptional activity in response to flavonoid induction. One of these is based on competitive hybridization, employing total RNA labeled on the 5' end as probes for hybridization over digests of an ordered cosmid collection representing a whole symbiotic plasmid. This strategy lead to the identification of several regions on the symbiotic plasmid of *Rhizobium* sp NGR 234 that are transcriptionally induced upon flavonoid stimulation (8). Powerful as this approach can be, it suffers of several limitations. The first limitation is that its sensitivity is somewhat restricted, due to the use of probes that are radioactively labeled only on the 5' end. The second limitation stems from the use of competitive hybridization. This allows the

NATO ASI Series, Vol. G 39
Biological Fixation of Nitrogen
for Ecology and Sustainable Agriculture
Edited by A. Legocki, H. Bothe, A. Pühler
© Springer-Verlag Berlin Heidelberg 1997

identification of regions whose expression is enhanced by flavonoids, but it is ineffective for the identification of loci whose expression is repressed by this stimulus. The third limitation is that this strategy does not give quantitative information on the degree of expression of a given region.

In this paper, we describe our recent results employing a new approach that circumvents these limitations. Use of this alternative on the symbiotic plasmid of *Rhizobium etli* allowed us to evaluate the extent of transcriptional activity in this plasmid under a variety of environmental conditions. Finally, we will comment some results that lead us to the identification of a new gene, called *nolL*, that is involved in modification of the nodulation factor in this species.

2. A global approach for evaluation of transcriptional activity in the symbiotic plasmid.

The strategy employed in this work is based on a technique originally employed in *Escherichia coli*, as part of the efforts for sequencing the genome of this bacterium (9). This technique uses total RNA that is isolated from bacteria exposed to the desired stimulus. This total RNA is then used as the substrate for the synthesis of complementary DNA, where synthesis is primed by a collection of random hexamers. cDNA synthesis is carried out in the presence of ^{32}P-α-dCTP, thus increasing the sensitivity of further manipulations. After cDNA synthesis, RNA is removed by alkaline hydrolisis, thus leading to the generation of ^{32}P single stranded cDNA's. These ^{32}P sscDNA's are then used as hybridization probes over digests of an ordered cosmid collection that represents the whole pSym of *R. etli* (6). Densitometric integration of the hybridization signals obtained, gives us quantitative information on transcriptional activity. Pairwise comparisons with control treatments allows then to establish induction or repression ratios for each region. Experimental details about this approach, as well as its use for the identification of genomic regions responsive to a variety of environmental stimuli will be published elsewhere (10).

3. Four sectors in the pSym of *Rhizobium etli* are strongly induced upon flavonoid stimulation.

This strategy was used to determine the extent and location of transcriptional activation in the pSym of *R. etli*, upon induction with the isoflavone genistein. Four extended sectors, named I to IV, were induced at least fourfold by genistein. These sectors colectivelly span 26 out of the 85 *Bam*HI bands that represent the whole pSym. Thus, at most one quarter of the genetic information present in the pSym is transcriptionally induced by genistein.

Most of the nodulation genes already identified in *R. etli* are located in sector I. Thus, sectors II to IV mark the location of new isoflavone-induced regions that might be involved in nodulation or other flavonoid-controlled processes. These regions are now being sequenced, to determine their role in *Rhizobium*-legume interactions.

4. One of the new regions corresponds to the *nolL* gene, which is involved in modification of nodulation factors.

It is known that after induction with flavonoids *R. etli* produces chitopentameric compounds of N-acetyl-D-glucosamine which are N-methyl-N-acylated with *cis*-vaccenic acid or stearic acid on their non-reducing end, and 4-O-acetyl-L-fucosylated at position 6 of the reducing glucosamine. A small fraction of these compounds is also carbomoylated at the C4 position on the non-reducing end (11,12). Unexpectedly, Nod factors of *R.etli* and *R. loti* are identical (13).

The production of Nod factors in a strain carrying a large deletion on the pSym of *R. etli*, that removes sectors III and IV, was studied. This strain (CFNX250) shows a different pattern of Nod factors as compared with the wild-type strain CFN42, i.e. analysis by direct-phase TLC revealed two bands corresponding to Nod factors in strain CFN42 but only one in strain CFNX250. Mass spectrometry analysis of the Nod factors produced by CFNX250 showed that they bear carbamoyl and fucosyl substituents at their non-reducing and reducing ends, respectively, but they are not acetyl-fucosylated. This phenotype has been complemented with a defined region of the pSym, located in the newly detected sector IV. This locus is located 25 kb away from sector I. Sequence analysis of the new locus shows a significant similarity with the *nolL* gene from *R. loti*, which is claimed to encode an acetyl transferase. NolL is essential for *R. loti* to nodulate *Lotus pedunculatus* and *Leucaena leucocephala* (14). Acetylation of the fucose is not essential in *R. etli* to nodulate *Phaseolus vulgaris*, since strain CFNX250 shows a normal nodulation kinetics on this plant.

Thus, these results suggest that, (i) the *nolL* gene in *R. etli* is involved in 4-O-acetylation of the fucosyl residue in the nodulation factor, (ii) that acetylation is not necessary for the addition of the fucose to the reducing glucosamine and, (iii) the acetyl fucose modification might be necessary for *R. etli* to nodulate some plants other than the common bean.

Acknowledgments.

We are indebted to Joanna Stepkowska for help in preparing the manuscript. Work in the authors laboratory was partially supported by grant IN201595 from DGAPA - UNAM (Mexico).

References

1. Carlson, R.W., N.P.J. Price and G. Stacey (1994). Mol. Plant-Microbe Interact.6:684-695.
2. Schultze, M., E. Kondorosi, P. Ratet, M. Buiré and A. Kondorosi (1994) Int. Rev. Cytol. 156:1-75.
3. Bhagwat, A.A. and D.L. Keister (1992) Appl. Environ. Microbiol. 58:1490-1495.
4. Aguilar, J.M.M., A.M. Ashby, A.J.M. Richards, G.A. Loake, M.D. Watson and C.H. Shaw (1988). J. Gen. Microbiol. 134:2741-2746.
5. Parniske, M., B. Ahlborn and D. Werner (1991) J. Bacteriol. 173:3432-3439.
6. Girard, M.L., M. Flores, S. Brom, D. Romero, R. Palacios and G. Dávila (1991) J. Bacteriol. 173:2411-2419.
7. Perret, X., W.J. Broughton and S. Brenner (1991). Proc. Natl. Acad. Sci. USA 88:1923-1927.
8. Fellay, R., X. Perret, V. Viprey, W.J. Broughton and S. Brenner (1995). Mol. Microbiol. 16:657-667.
9. Chuang, S., D.L. Daniels and F.R. Blattner (1993) J. Bacteriol. 175:2026-2036.
10. Girard, L., B. Valderrama, R. Palacios, D. Romero and G. Dávila (1996) Microbiology 142: in press.
11. Poupot, R., E. Martínez-Romero, N. Gautier and J.C. Promé (1995). J. Biol. Chem. 270:6050-6055.
12. Cárdenas, L., J. Domínguez, C. Quinto, J.M. López-Lara, B.J.J. Lugtenberg, H.P. Spaink, G.J. Rademaker, J. Haverkamp and J.E. Thomas-Oates (1995). Plant. Mol. Biol. 29:453-464.
13. López-Lara, I.M., J.D.V. van den Berg, J.E. Thomas-Oates, J. Gluschka, B.J.J. Lugtenberg and H.P. Spaink (1995). Mol. Microbiol 15:627-638.
14. Scott, D.B., C.A. Young, J.M. Collins-Emerson, E.A. Terzaghi, E.S. Rockman, P.E. Lewis and C.E. Pankhurst (1996) Mol. Plant-Microbe Interact. 9:187-197.

AZOSPIRILLUM GENES INVOLVED IN CHEMOTAXIS AND ADHESION TO PLANT ROOTS

Sara Moens[1], Els Van Bastelaere[1], Ann Vande Broek[1], Mark Lambrecht[1], Veerle Keijers[1], Luis Fernando Revers[2], Luciane M.P. Passaglia[2], Irene S. Schrank[2], Jos Vanderleyden[1]

[1]F.A. Janssens Laboratory of Genetics, K.U. Leuven, Willem de Croylaan 42, B-3001 Heverlee, Belgium,
[2]Universidade Federal Rio Grande do Sul, Departamento de Genética and Departamento de Biotecnologia, Cx. Postal 15005, Porto Alegre, RS91501-970, Brazil

Introduction

Bacteria of the genus *Azospirillum* are diazotrophs that colonize the roots of plants. Colonization patterns can be visualized by using strains equipped with a reporter gene (Vande Broek *et al.*, 1993; Arsène *et al.*, 1994). The initial steps of bacterial colonization are chemotaxis and adhesion to the root surface. In order to characterize the bacterial genes and signals that determine these processes genetic and biochemical approaches were used.

Results

Azospirillum brasilense Sp7 cells, grown in the absence and presence of wheat root exudates, revealed a 40 kDa acidic protein of which the synthesis is dependent on the presence of wheat root exudates (Van Bastelaere *et al.*, 1993). With a polyclonal antiserum, it was further demonstrated that the synthesis of this protein occurs also in other *A. brasilense* strains (Sp6, Sp245, Sp246) and in *A. lipoferum* strains tested (SpBr17, SF50, DN64). D-galactose was found to mimic the presence of plant root exudates. Using partial amino acid sequencing and DNA hybridizations with derived

NATO ASI Series, Vol. G 39
Biological Fixation of Nitrogen
for Ecology and Sustainable Agriculture
Edited by A. Legocki, H. Bothe, A. Pühler
© Springer-Verlag Berlin Heidelberg 1997

oligonucleotides, a 7 kbp *MluI* DNA restriction fragment of the *A. brasilense* Sp7 chromosome was isolated. DNA sequencing allowed us to define two ORFs. ORF1 (Mr of 37,452) shows homology (37 % of identical amino acids) with the *Agrobacterium tumefaciens* GbpR protein (galactose-binding protein regulator), the transcriptional regulator (LysR type) of *chvE* (Doty *et al.*, 1993); ORF2 (Mr of 38,247) shows homology (74 % of identical amino acids) with the *A. tumefaciens* ChvE protein (Huang *et al.*, 1990; Shimoda *et al.*, 1993). In *A. tumefaciens*, ChvE is involved in chemotaxis towards and uptake of sugars (D-galactose, D-fucose-L-arabinose) and is required for enhanced expression of *vir* genes in the presence of these monosaccharides. To study the role of these ORFs in *A. brasilense* , and the regulation of expression of the corresponding genes (*sbpA* for ORF2 and *gbpR* for ORF1), corresponding mutants and gene fusions were constructed. A translational *sbpA::gusA* fusion (pFAJ115), an *sbpA* mutant (FAJ110) and a *gbpR* mutant (FAJ111) of *A. brasilense* Sp245 were constructed. pFAJ115 was introduced in *A. brasilense* Sp245 and FAJ110. β-glucuronidase activity was measured in cells grown in the absence and presence of sugars and wheat root exudates, as shown in Table 1. Expression analysis of FAJ111 (pFAJ115) revealed that *gbpR* is not strictly required for expression of *sbpA* but enhances expression 3 to 5-fold (data not shown). These results confirm that expression of *sbpA* is dependent on the presence of sugars, some of which might be present in wheat root exudates, and is regulated by GbpR. Subsequently wild-type and *sbpA* mutant Sp245 cells were analyzed for chemotaxis towards sugars. Results are shown in Table 2.

Table 1. Induction of the *sbpA::gusA* fusion in *A. brasilense*. β-glucuronidase activity was measured after addition of the sugars or wheat root exudates to minimal (MMAB) medium. Units were calculated as defined by Miller for β-galactosidase activity (1972). The values shown are the mean of 3 replicates.

Compound added	Strains tested	
	Sp245 (pFAJ115)	FAJ110 (pFAJ115)
None	0	0
L-arabinose (10 μM)	179 (± 41)	641 (± 160)
D-fucose (10 μM)	82 (± 20)	134 (± 37)
D-galactose (10 μM)	235 (± 20)	208 (± 58)
Wheat root exudates	129 (± 27)	ND

Table 2. Chemotaxis of *A. brasilense* Sp245 and the *sbpA* mutant towards sugars, tested in semisolid H_2O-agar, containing discs impregnated with different sugars (10mM). Cells were grown either in minimal (MMAB) medium (non-induced) or in minimal medium, supplemented with 10 mM D-fucose (induced). Semisolid-agar (20 ml) was mixed with 10^9 cells *A. brasilense* cells, and chemotaxis was evaluated after 5 hours of incubation.

Strains	Sugars added					
	control	L-ara	D-fruc	D-fuc	D-gal	D-xyl
Sp245	-	-	+	-	-	-
Sp245 induced	-	+	+	+	+	-
FAJ110	-	-	+	-	-	-
FAJ110 induced	-	-	+	-	-	-

These results show that chemotaxis of *A. brasilense* towards L-arabinose, D-fucose and D-galactose is dependent on the induced expression of *sbpA*, whereas chemotaxis towards D-fructose is SbpA independent and appears to be constitutively expressed. A more detailed analysis of the *A. brasilense* SbpA protein is described by Van Bastelaere *et al.* (1996). Recently, we have identified an *A. brasilense* gene that encodes an ORF (532 AA) with significant homology to bacterial methyl-accepting chemoreceptor proteins (MCPs). This will allow us to further study the chemotaxis pathways in *A. brasilense*.

Chemotaxis not only requires chemo-attractants but is also dependent on the motility of the bacteria. *A. brasilense* shows a mixed type of flagellation: a single polar flagellum which is always present; lateral flagella, when cells are grown in semi-solid medium (Moens and Vanderleyden 1996a). Previously, we have demonstrated that the structural protein, Fla1, of the polar flagellum, is a glycoprotein that acts as a plant root surface adhesin (Moens *et al.*, 1995a). Recently, we have also isolated the structural gene, *laf1*, for the lateral flagella (Moens *et al.*, 1995b). Subsequently, the question was raised how the expression of *laf1* is regulated. Therefore a translational *laf1::gusA* fusion (pFAJ0210) was constructed. Incubating liquid grown *A. brasilense* Sp7 (pFAJ0210) cells with a monospecific polyclonal antibody raised against the polar flagellum (AS1) revealed that binding of this antibody to the polar flagellum induces expression of *laf1*.

From these results, it can be concluded that hindrance of rotation of the polar

flagellum can induce expression of the structural gene for the lateral flagella (Moens

et al., 1996b).

References

Arsène F, Katapitiya S, Kennedy IR, Elmerich C (1994) Use of *lacZ* fusions to study the expression of *nif* enes of *Azospirillum brasilense* in association with plants. Mol Plant-Microbe Interact 7: 748-757

Doty SL, Chang M, Nester W (1993) The chromosomal virulence gene, *chvE*, of *Agrobacterium tumefaciens* is regulated by a LysR family member. J Bacteriol 175: 7880-7886

Huang MW, Cangelosi GA, Halperin W, Nester E (1990) A chromosomal *Agrobacterium tumefaciens* gene required for effective plant signal transduction. J Bacteriol 172: 1814-1822

Miller JH (19972) Experiments in molecular genetics. Cold Spring Harbor Laboratory, Cold Spring Harbor, NY

Moens S, Michiels K, Vanderleyden J 1995a Glycosylation of the flagellin of the polar flagellum of *Azospirillum brasilense*, a gram-negative nitrogen-fixing bacterium. Microbiol 141, 2651-2657

Moens S, Michiels K, Keijers V, Van Leuven, F, Vanderleyden J 1995b Cloning, sequencing and phenotypic analysis of *laf1*, encoding the flagellin of the lateral flagella of *Azospirillum brasilense* Sp7. J Bacteriol 177, 5419-5426

Moens S, Vanderleyden, J 1996a Functions of bacterial flagella. Crit Rev Microbiol 22(2), 67-100.

Moens S, Schloter M, Vanderleyden J 1996b Expression of the structural gene, *laf1*, encoding the flagellin of the lateral flagella in *Azospirillum brasilense* Sp7. J Bacteriol 178, 5017-5019

Shimoda N, Toyoda-Yamamoto A, Aoki S, Machida Y (1993) Genetic evidence for an interaction between the VirA sensor protein and the ChvE sugar-binding protein of *Agrobacterium*. J Biol Chem 268: 26552-26558

Van Bastelaere E, De Mot R, Michiels K, Vanderleyden J, 1993 Differential gene expression in *Azospirillum* spp. by plant root exudates: analysis of protein profiles by two-dimensional polyacrylamide gel electrophoresis. FEMS Microbiol Lett 112, 335-342

Van Bastelaere E, Vermeiren H, Keijers, V, Proost, P, Vanderleyden, J 1995 Characterization of a sugar-binding protein from *Azospirillum brasilense* mediating chemotaxis to and uptake of sugars. Mol Microbiol, in press

Vande Broek A, Michiels J, Van Gool A, Vanderleyden J 1993 Spatial-temporal Colonization Patterns of *Azospirillum brasilense* on the wheat root surface and expression of the bacterial *nifH* gene during the association. Mol Plant-Microbe Interact 6, 592-600

Part 5

Molecular Microbial Ecology

Molecular Microbial Ecology

John E Beringer

School of Biological Sciences, University of Bristol, Woodland Road, Bristol, BS8 1UG, UK

The title of this conference is "Biological Fixation of Nitrogen for Ecology and Sustainable Agriculture" and it is appropriate, therefore, that there be a session on microbial ecology. Bacteria are the only organisms that have evolved to fix atmospheric nitrogen gas and have a very important role in the nutrition of plants. They, and their genes, are also of critical importance if we are to develop more sustainable agricultural systems in the future. By sustainable I mean systems in which inputs of fertiliser and other chemicals can be maintained almost indefinitely. For phosphate, potassium and other major nutrients sustainability will be very difficult to achieve because supplies of suitable raw materials for fertiliser manufacture are finite. On the other hand fixed nitrogen, which is often the major limiting nutrient for plant productivity, need never be a limiting factor if biological nitrogen fixation is exploited effectively.

Until we are able to construct plants carrying nitrogen fixation and assimilation genes capable of functioning efficiently, we will be dependent on fixation by bacteria. This means that for many years to come our crops will derive biologically-fixed nitrogen from bacteria already in the soil, or from bacterial inoculants, or from a combination of both. In practice it is unusual for an inoculant strain to be dominant if there are already members of the same species in the soil and thus the amount of fixed nitrogen a crop will receive is determined by the proportion of introduced to existing bacteria, which is usually highly biased towards the latter. Indeed, with *Rhizobium* and *Bradyrhizobium* it has long been recognised that inoculation is usually futile in soils with a history of inoculation or with existing populations of these bacteria. If we are to be able to exploit our existing ability to genetically modify nitrogen fixing bacteria to enhance nitrogen fixation with crop plants it will be essential to find ways of enhancing the ability of introduced strains to compete with indigenous members of the same species. The competition will need to be at many different levels, starting with competition for nutrients and the colonisation of the rhizosphere, through to interactions with the host that may result in the formation of complex structures, such as nodules.

The solution to the problem of delivering sufficient inoculant to plant roots so that it can function effectively must surely be based on a much better understanding of microbial ecology than we have at present. For example, what factors influence the ability of strains to survive? How many indigenous strains might be present? How stable are populations and can this stability be changed to facilitate the introduction of inoculants? What is needed to interact with roots of the target crop and be able to multiply sufficiently? Why do certain strains dominate, even though they may numerically be less common than others? What controls host range? Can the bacteria or plants be manipulated to provide a competitive advantage to enable the inoculant to nodulate?

Perhaps most important of all, are there methods now available to begin to answer these questions and to design inoculants that are very competitive for the host in question, but will not themselves dominate and impede the introduction of other strains in the future, as has happened when inoculants have been used in the USA and Australia? Of course the answer is yes in terms of strain construction because molecular genetic techniques allow us to make strains carrying new genes, or with existing genes removed or modified, almost at will. The major limiting factor being insufficient understanding of ecology and interactions with hosts to enable much progress to be made in designing better inoculants. Molecular genetics also offers the opportunity to study microbial ecology in ways that were not previously possible by allowing us to look at the genomes of strains in the soil and thus to obtain very precise information about strain diversity.

The papers in this session range from studies of populations of *Rhizobium* in the soil that indicate that bacteriocin production and resistance may lead to dramatic changes in the diversity of strains, through work identifying metabolic pathways in *Rhizobium* that are stimulated by host plants and as a result enhance the ability of the bacteria to colonise roots, to the identification of a gene that restricts the range of hosts nodulated by a strain of *Bradyrhizobium*.

PLANT REGULATION OF ROOT COLONIZATION BY *Rhizobium meliloti*

Donald A. Phillips, Wolfgang R. Streit, Hanne Volpin, Cecillia M. Joseph

Dept. of Agronomy & Range Science, Univ. of California, Davis, CA 95616 USA
DAPhillips@ucdavis.edu

Keywords. biotin, carbon dioxide, respiration

1 Introduction

Rhizobial colonization of roots is affected by diverse processes, including survival in soil, chemotaxis and attachment (2, 9, 13), antibiotic production (12), and growth (10). Our group is studying the possibility that bacterial processes which are regulated by the host plant may be especially important for root colonization. That concept was supported by data showing that mutations in a stachydrine-inducible gene involved in stachydrine uptake or in the flavonoid/betaine-inducible *nodC* gene in *Rhizobium meliloti* 1021 (Rm1021) impaired competitive colonization by 65% and 94%, respectively (6). Those contributions to competitiveness are substantial, but we now believe that stimulatory effects on growth produced by plant-derived biotin may contribute even more significantly to root colonization. This report explores how rhizobial colonists may use biotin and CO_2 released from roots.

2 The Role of Biotin in Root Colonization

Biotin is a component of alfalfa root exudate (7), and its stimulatory effect on rhizobial growth in laboratory media has long been known (15). We were surprised to find, however, that adding trace amounts of biotin to the alfalfa rhizosphere promoted root colonization by Rm1021 (10) (Fig. 1A). Saturation of the response near 50 nmol biotin/plant indicates the primary role of biotin in the rhizosphere is as a catalyst rather than as an energy substrate. Tests with biotin auxotroph Rm1021-B3 established that *R. meliloti* grows significantly on plant-derived biotin in the rhizosphere, and the importance of exogenous biotin for colonization of an alfalfa root also was supported by a significant decrease in growth measured for Rm1021 in the presence of the biotin-binding protein avidin (10). On the basis of those results we initiated work designed to enhance biotin synthesis in *R. meliloti*, to increase uptake of external biotin, and to identify biotin-regulated processes that contribute to root colonization.

To increase biotin availability in Rm1021, we introduced biotin synthesis genes from *Escherichia coli* and observed that the transconjugants overproduced biotin and grew faster than Rm1021 in lab medium (11). The recombinant cells showed a slightly extended lag phase in their growth and a significant decrease in viability, which was partially alleviated by an unknown element present near the *E. coli* biotin synthesis genes. Genetic solutions to these impairments are being sought.

NATO ASI Series, Vol. G 39
Biological Fixation of Nitrogen
for Ecology and Sustainable Agriculture
Edited by A. Legocki, H. Bothe, A. Pühler
© Springer-Verlag Berlin Heidelberg 1997

Our plans for increasing biotin uptake in Rm1021 and thereby helping the bacteria benefit more from plant-derived biotin have produced unexpected and interesting results. Biotin uptake has not been defined at the molecular level in any bacteria including *E. coli*, so facts from Rm1021 will contribute basic knowledge on this process. Our data comparing ^{14}C-biotin transport in biotin auxotrophs of *E. coli* and Rm1021 show that Rm1021 takes up biotin at a 50-fold lower concentration than *E. coli*. This sensitivity undoubtedly helps *R. meliloti* respond to biotin in root exudate, and it has complicated our efforts to define the nature of a biotin-inducible mutation isolated as Rm1021-B5 after introducing Tn5-B30(*nptII*), a transposable promoter probe (8). Assays show that the promoter on the mutated gene in Rm1021-B5 responds to 1 nM biotin and produces a six-fold increase in NPTII protein at 40 nM biotin. Rm1021-B5 competes equally with Rm1021 in defined medium without biotin, but if the two strains are coinoculated with low biotin, Rm1021 grows better than the mutant. That phenotype is consistent with impaired biotin transport, but when avidin-treated cells in stationary phase are assayed with ^{14}C-biotin, the mutant takes up 50% more biotin than Rm1021. Sequence analyses indicate the mutated gene is quite small and that no homologous gene has been reported in any bacteria. Adjacent open-reading frames, however, show high homologies to several previously reported genes associated with stationary phase processes. Based on these data, we speculate that Rm1021-B5 is mutated in a gene involved in sensing internal biotin.

Rm1021 responds to exogenous biotin with a remarkable increase in growth rate (Fig. 1B). Tests with Rm1021-B5 under the same conditions show the rapid growth is delayed but not prevented. This response is consistent with our interpretation that the gene mutated in Rm1021-B5 is involved in sensing biotin, and it reinforces the concept that plant-derived biotin may trigger *R. meliloti* growth in soil. How the biotin stimulation of growth in Rm1021 (Fig. 1B) relates to similar effects reported previously for flavonoids (4) remains to be determined.

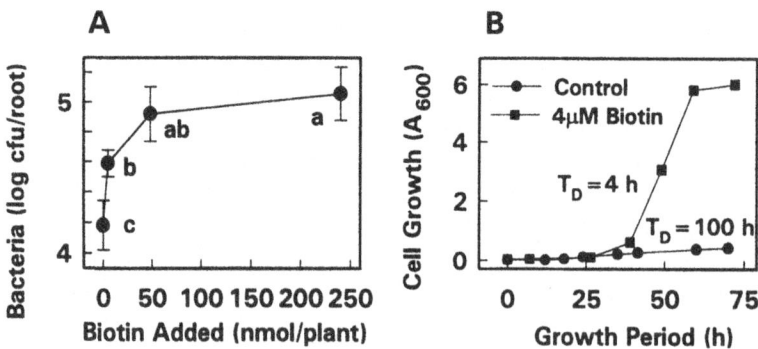

Figure 1. Effect of biotin on Rm1021 colonizing alfalfa roots (A) or growing in defined liquid media (B). Root colonization tests (A) were conducted with bacteria that had grown with optimum exogenous biotin. Biotin was added on day 0, 2 and 4 to supply the indicated total amounts. Bacteria were inoculated on each seed at day 0 (<100 cfu/seed) and recovered on day 6 from roots. The liquid growth test (B) was done in defined GTS medium after three subcultures in the same biotin-free medium.

3 Carbon Dioxide and Competitive Root Colonization

The strongly promotive effects of biotin on rhizobial root colonization must also reflect the catalytic role of this vitamin as a cofactor in several carboxylase reactions that use CO_2 as a substrate, including acetyl-CoA carboxylase, pyruvate carboxylase and propionyl-CoA carboxylase. Rhizobia require CO_2 for growth (5), and acetyl-CoA carboxylase undoubtedly contributes to rapid growth through its role in formation of membranes which help bacteria compete effectively for diffusible substrates.

Because CO_2 is required for rhizobial growth, one can ask if rhizosphere CO_2 levels are optimum for these bacteria. It is known that rhizosphere CO_2 concentrations are quite high (3) and that roots evolve more CO_2 in the presence of microorganisms (1). Some workers have interpreted this response as an indication that soil microbes stimulate root exudation and then respire carbon compounds in the exudate (3). Our results, however, show that both living and dead Rm1021 cells increase CO_2 evolution from roots (Table 1). This finding suggests that bacterial products affect root respiration, and it is consistent with the fact that fungal elicitors increase CO_2 evolution from cultured plant cells (14). Rm1021 promotes root respiration by interacting with the root-hair containing segment of roots (Table 1), but Nod factors do not completely explain the results. For example, one compound released by Rm1021 has no root-hair curling activity, but it is very effective at eliciting increases in root CO_2 evolution. The structure of the compound is being determined and mutants incapable of producing the compound are being sought. Whether such mutants will show altered root colonization is not known. Knowing the structure of the molecule may permit external regulation of plant respiration and possibly have affects on plant productivity and/or survival.

Table 1. *R. meliloti* effects on CO_2 evolution by alfalfa roots. Intact roots were exposed to bacteria for 24 h and then cut into a 1-cm tip and a 3-cm subtending region to measure CO_2. Values in a column followed by different letters showed significant ($P \leq 0.01$) treatment effects. Roots exposed to dead Rm1021 remained sterile throughout the 24-h incubation period.

			Entire root	
Treatment	Segment with root hairs	Root tip	Uncut control	Sum of segments
		(mmol CO_2/g/h)		(calculated)
Sterile control	0.66 a	3.2 a	1.10 a	1.17 a
Living Rm1021	1.26 b	3.0 a	1.54 b	1.60 b
UV-killed Rm1021	1.35 b	3.3 a	1.65 b	1.74 b

Acknowledgments: Work described here was funded by US National Science Foundation grant IBN-92-18567. W.R. Streit is supported in part by a Feodor von Lynen stipend from the Alexander von Humboldt Foundation. H. Volpin has a postdoctoral award from BARD, the US-Israel Binational Agricultural Research and Development Fund.

4 References

1. Barber, D. A., and J. K. Martin. 1976. The release of organic substances by cereal roots into soil. New Phytol. 76:69-80.
2. Bauer, W. D., and G. Caetano-Anollés. 1990. Chemotaxis, induced gene expression and competitiveness in the rhizosphere. Plant and Soil. 129:45-52.
3. Cheng, W. X., D. C. Coleman, C. R. Carroll, and C. A. Hoffman. 1993. In situ measurement of root respiration and soluble C-concentrations in the rhizosphere. Soil Biol. Biochem. 25:1189-1196.
4. Hartwig, U. A., C. M. Joseph, and D. A. Phillips. 1991. Flavonoids released naturally from alfalfa seeds enhance growth rate of *Rhizobium meliloti*. Plant Physiol. 95:797-803.
5. Lowe, R. H., and H. J. Evans. 1962. Carbon dioxide requirement for growth of legume nodule bacteria. Soil Sci. 94:351-356.
6. Phillips, D. A., W. R. Streit, H. Volpin, J. D. Palumbo, C. M. Joseph, E. Sande, F. J. de Bruijn, and C. I. Kado. 1996. Plant regulation of bacterial root colonization. In press. *In* G. Stacey, B. Mullin, P. Gresshoff (eds.), 1996 IS-MPMI Symposium Proceedings. IS-MPMI, St. Paul, MN. In press.
7. Rovira, A. D., and J. R. Harris. 1961. Plant root excretions in relation to the rhizosphere effect V. The exudation of B-group vitamins. Plant and Soil. 14:199-214.
8. Simon, R., J. Quandt, and W. Klipp. 1989. New derivatives of transposon Tn*5* suitable for mobilization of replicons, generation of operon fusions and induction of genes in gram-negative bacteria. Gene. 80:161-169.
9. Smit, G., S. Swart, B. J. J. Lugtenberg, and J. W. Kijne. 1992. Molecular mechanisms of attachment of *Rhizobium* bacteria to plant roots. Molec. Microbiol. 6:2897-2903.
10. Streit, W. R., C. M. Joseph, and D. A. Phillips. 1996. Biotin and other water-soluble vitamins are key growth factors for alfalfa rhizosphere colonization by *Rhizobium meliloti* 1021. Molec. Plant-Microbe Interact. 5:330-338.
11. Streit, W. R., and D. A. Phillips. 1996. Recombinant *Rhizobium meliloti* strains with extra biotin synthesis capability. Appl. Environ. Microbiol. In press.
12. Triplett, E. W. 1988. Isolation of genes involved in nodulation competitiveness from *Rhizobium leguminosarum* bv. *trifolii* T24. Proc. Natl. Acad. Sci. USA. 85:3810-3814.
13. Vande Broek, A., and J. Vanderleyden. 1995. The role of bacterial motility, chemotaxis, and attachment in bacteria-plant interactions. Molec. Plant-Microbe Interact. 8:800-810.
14. Vera-Estrella, R., E. Blumwald, and V. J. Higgins. 1992. Effect of specific elicitors of *Cladosporium fulvum* on tomato suspension cells - evidence for the involvement of active oxygen species. Plant Physiol. 99:1208-1215.
15. West, P. M., and P. W. Wilson. 1939. Growth factor requirements of the root nodule bacteria. J. Bacteriol. 37:161-185.

HOST-CONTROLLED RESTRICTION OF NODULATION BY BRADYRHIZOBIUM JAPONICUM STRAIN USDA 110 AND CHARACTERIZATION OF A GENE REGULATING NODULATION

Michael J. Sadowsky[1], Scott M. Lohrke[1], Bradly Day[2], V. Kumar Kolli[3], Rob Hancock[3], Russell Carlson[3], Gary Stacey[2], James H. Orf[4], and Zhaokun Tong[1]

[1]Department of Microbiology and Department of Soil, Water, and Climate, University of Minnesota, St. Paul, MN 55108, [2]Department of Microbiology, University of Tennessee, Knoxville, TN 37996, [3]Complex Carbohydrate Research Center, University of Georgia, Athens, GA 30602, and [4]Department of Agronomy and Plant Genetics, University of Minnesota, St. Paul, MN 55108.

Keywords: Bradyrhizobium japonicum, soybean, genotype-specific nodulation gene, nodulation factors

1 Introduction

A major emphasis of recent dinitrogen fixation research has been the development of genetic engineering techniques for the construction of rhizobial strains with enhanced nitrogen fixation. However, faced with competition from indigenous rhizobial and bradyrhizobial populations, even highly efficient N_2 fixing strains are destined to form few nodules and have a minimal effect upon N_2 fixation. While most studies of competition among *B. japonicum* strains have focused on soil - rhizosphere ecology, the selection of more competitive strains, or improved inoculation techniques, comparatively little research has considered utilizing the genetic components of both *Bradyrhizobium* and the host plant to alter nodule occupancy.

2 Nodulation Restricting Soybean Genotypes

Several studies have demonstrated that soybean genotypes are differentially nodulated by strains of *B. japonicum* (2,12,16). Soybean genotypes restricting nodulation by specific strains or serogroups of *Bradyrhizobium* are well documented (16). In several instances, the plant alleles controlling nodulation restriction have been identified. The single dominant gene Rj_2 conditions restricted nodulation with all available strains of the 122 and c1 serogroups. The single dominant gene Rj_3 conditions restricted nodulation with USDA 33. A third single dominant gene Rj_4 conditions restricted nodulation with strain USDA 61. Two new single recessive soybean genes, rj_5 and rj_6, which restrict nodulation by *B. japonicum* have also been described (9). These alleles appear to function like rj_1 to restrict nodulation by all bradyrhizobia. More recently, we determined that a single recessive host gene (RJ 110) in PI 417566 conditions restriction of nodulation by USDA 110 (7).

In 1989, we reported the identification of a soybean genotype, PI 417566, which restricts nodulation and reduces the competitiveness of strain strains USDA 430, USDA 129, and USDA 110 (3). We subsequently showed that PI 417566 restricted nodulation by some, but not all, *B. japonicum* serogroup 110 strains (8) and that the nodulation process is not aborted until sometime after the formation of nodule primordia (11). Through reciprocal grafting, we showed that restricted nodulation conditioned by PI 417566 is due to soybean root factors and plant growth temperature (11).

3 Genetics of Nodulation Restriction

Common nodulation, host-specific nodulation, and genotype-specific nodulation genes all play a major role in host-controlled nodulation restriction (15). While many *hsn* genes have been localized in *Rhizobium* and *Bradyrhizobium*, there has been only one report of the identification of genotype-specific nodulation (GSN) genes in the bradyrhizobia (10). Specifically, GSN genes refer to those bacterial sequences which allow nodulation of specific plant genotypes within a given legume species. Genes involved in the genotype-specific nodulation genes of legumes can act in either a positive or negative manner to control nodulation specificity (15). An insertion in a negatively-acting nodulation gene extends host-range, while disruption of a positively-acting GSN gene limits host range. The *B. japonicum nol*A gene, identified in strain USDA 110, is a positively-acting gene that allows serogroup 123 strains to nodulate PI 377578 (10). An example of another positively-acting GSN gene is *nod*X found in *Rhizobium leguminosarum* bv. *viceae* strains (4). This gene mediates the O-acetylation of nod factors. Taken together, results from all of these studies indicate that both positively acting and negatively acting GSN determinants can control specificity in legume-*Rhizobium* (*Bradyrhizobium*) symbioses (15).

3.1 Extracellular nodulation factors in bradyrhizobia.

Nodulation factors from several strains of *B. japonicum* have been characterized (1,13). The basic structure of the *B. japonicum* nodulation factor is five, β-1,4-linked N-acetylglucosamine residues with a methyl fucose moiety and a fatty acid side chain. The ability of *B. japonicum* to nodulate soybean, but not siratro, depends on the presence of methylfucose at the terminal reducing residue, and requires the activity of the nodulation gene *nodZ* (14). The fatty acid chains present on nodulation factors from *B. japonicum* or *B. elkanii* are either C16:0, C16:1, or C18:1. Acetyl groups are present at the reducing terminal end of some *B. japonicum* and *B. elkanii* nodulation factors, and some *B. elkanii* strains have a carbamyl group at the terminal reducing residue (1). None of these modifications of nodulation factors have been shown to be involved in host specificity, with the exception of the methylfucose group. *B. japonicum* strain USDA 110, synthesizes only one major nodulation factor, which does not contain an acetyl group, and contains a C18:1 fatty acid.

3.2 A neagtively-acting GSN gene from Bradyrhizobium japonicum

In 1995, we reported the isolation of a *Tn5*-induced mutant of *B. japonicum* strain USDA 110, D4-2.5 that had the ability of overcome nodulation restriction conditioned by soybean PI 417566 (8). This mutant was unchanged in its nodulation host range, relative to the wild-type strain, with the exception of the ability to nodulate genotype PI 417566 (8). Moreover, wild-type USDA 110 could competitively block nodulation by the mutant. To our knowledge, this is the first report of a negatively-acting, *avr*-like genotype-specific nodulation gene in *Bradyrhizobium*. It should be noted, however, that while we previously isolated a *Tn5*-induced serogroup 123 mutant (671-21) which could nodulate PI 377578, subsequent studies indicated that the site of the *Tn5* mutation was not responsible for the host range extension phenotype (5).

More recently, we isolated, cloned, and characterized the wild-type DNA region corresponding to the *Tn5* insertion site in mutant D4-2.5 (6). This DNA region was subsequently cloned in plasmid pSL5.2. To verify that the host-range extension phenotype of mutant D4.2-5 was due to the *Tn5* insertion, plasmid pSL5.2 was subjected to random *Tn5* mutagenesis using λ::*Tn5*. Independent *Tn5* insertions in pSL5.2, that were located in and around the original *Tn5* insertion site in mutant D4.2-5, were homologously recombined into wild-type USDA 110 and transconjugants were tested for nodulation ability on soybean PI 417566. One of the new *Tn5* insertion mutants, 110-36, had gained the ability to nodulate the PI genotype, indicating that the nodulation phenotype of mutant D4-2.5 was due to *Tn5*. The 110-36 insertion was located in the same 0.9 kb *PstI-HindIII* fragment of pSL5.2 as was the original insertion in D4-2.5.

3.3 Characterization of noeD

We have sequenced the defined gene region and found that it contains a single open reading frame (ORF) of 474 nucleotides. We have named this ORF, *noeD*. The *noeD* encodes a potential protein of 158 amino acids. No similarity was detected between the *noeD* DNA sequence or the predicted amino acid sequence and any known genes in the nucleotide and protein data bases. No *nod* box consensus sequence was located upstream of noeD, however, a region upstream of *noeD* had high similarity (about 76% similarity and 62% identity) to the N-terminal region of the *R. meliloti* and *R. leguminosarum* bv. viceae *nodM* genes, which have been postulated to be D-glucosamine synthase. This is the first report of the presence of the *nod*M gene in *B. japonicum*. Southern hybridization analysis of the *noe*D gene region indicated that it is not closely linked to the main or auxiliary nodulation gene clusters in *B. japonicum*. Transcriptional analyses done using *nodC::lacZ* gene fusions and isoflavonoid inducers indicated that *noeD* positively regulates the expression of the common nodulation genes, a mutation in *noeD* reduced *nod*C-lacZ activity to about 50% of wild-type level.

To determine if *noeD* influences expression of nodulation factors, nod factors were isolated and purified from USDA 110 and mutant D4-2.5 and separated by HPLC. The HPLC elution profiles from strain USDA 110 and the *noeD* mutant differed in respect to the retention times and quantity of eluting material. While the elution profile of partially purified nodulation factors from strain USDA 110 had 5 peaks, with peak 2 in the largest amount, the *noeD* mutant, strain D4.2-5, had an elution profile with 5 major peaks, with peaks 2, 3, and 5 in the largest amounts. While there was more of the peak 2 than the peak 3 in the USDA 110 elution profile, that from the *noeD* mutant contained more of the peak 3 than peak 2 material. In addition, the *noeD* mutant had a greater amount of peak 5 than did USDA 110. The structures of Nod factors from USDA 110 and the *noeD* mutant were determined by chemical composition analysis, NMR spectroscopy, FAB, GC-MS and MS-MS analyses. The major nod factor produced by USDA 110 is NodBj-V(C18:1,MeFuc) (1,13), which is reflected in the HPLC elution profile Peak 2. Results indicated that in addition to NodBj-V(C18:1,MeFuc) (Peak 2), USDA 110 also produces lesser amounts of NodBj-V(C16:0,MeFuc) (Peak 1), and the acetylated nod factor, NodBj-V (Ac,C18:1,MeFuc) (Peak 3). The mutant, however produced a much greater amount of NodBj-V(Ac,C18:1,MeFuc) than did the wild-type strain and the relative ratios of acetylated:non-acetylated nod factors in the mutant is reversed from that seen with the wild-type strain. Moreover, the *noeD* mutant produced more of peak 5 material than did USDA 110. Preliminary results suggest that the peak 5 material is NodBj-V(C20:1,NMe,MeFuc).

In summary, in this study we provide evidence that a single, negatively-acting gene, noeD, allows B. japonicum strain USDA 110 to overcome nodulation restriction conditioned by the BJ 110 gene in PI 417566. Since mutations in noeD only affect nodulation of PI 417566, we conclude that this locus constitutes a bonafide GSN gene. In addition, results of this study strongly suggest that the ability of the noeD- mutant to nodulate PI 417566 is dependent on an elevated level of the acetylated nod factor, NodBj-V(Ac,C18:1, MeFuc) or on the presence of NodBj-V(C20:1,NMe,MeFuc). Taken together, our results also suggest that host and bacterial genes are interacting in a gene-for-gene manner to control nodulation specificity in this symbiotic system.

4 References Cited

1. Carlson, R.W., Sanjuan, J., Bhat U.R., Glushka, J., Spaink, H.P., Wijfjes, A.H.M., van Brussel, A.A.N., Stokkermans, T.J.W., Peters, N.K., and Stacey, G., 1993. *J. Biol. Chem.* **268**:18372-18381.
2. Cregan, P.B., and Keyser, H.H., 1986. *Crop Sci.* **26**:911-916.
3. Cregan, P.B., Keyser, H.H,, and Sadowsky, M.J., 1989. *Crop Sci.* **29**:307-312.
4. Firmin, J.L., Wilson, K.E., Carlson, R.W., Davies, A.E., and Downie, J.A., 1993. *Mol. Microbiol.* **10**:351-360.
5. Judd, A.K., Sadowsky, M.J., Bhagwat, A., Cregan, P.B., and Liu, R.-L., 1993. *FEMS Microbiol. Lett.* **106**:205-210.
6. Lohrke, S.M., Day, B., Kolli, V.K., Hancock, R., Carlson, R.W., Stacey, G., Orf, J.H., Tong, Z., and Sadowsky, M.J., 1996. *J. Bacteriol.* (Submitted).
7. Lohrke, S.M., Orf, J.H., and Sadowsky, M.J., 1996. *Crop Sci.* (In Press).
8. Lohrke, S.M., Orf, J.H., Martínez-Romero, E., and Sadowsky, M.J., 1995. *Appl. Environ. Microbiol.* **61**:2378-2383.
9. Pracht, J.E., Nickell, C.D., and Harper, J.E., 1993. *Crop Sci.* **33**:711-713.
10. Sadowsky, M.J., Cregan, P.B., Gottfert, M., Sharma, A., Gerhold, D., Rodriguez-Quinones, F., Keyser, H.H., Henneke, H., and Stacey, G., 1991. *Proc. Natl. Acad. Sci.* (USA). **88**:637-641.
11. Sadowsky, M.J., Kosslak, R.M., Madrzak, C.J., Golinska, B., and Cregan, P.B., 1995. *Appl. Environ. Microbiol.* **61**:832-836.
12. Sadowsky, M.J., Tully, R.E., Cregan, P.B., and Keyser, H.H., 1987. *Appl. Environ. Microbiol.* **53**:2624-2630.
13. Sanjuan, J., Carlson, R.W., Spaink, H.P., Bhat, U.P., Barbour, W.M., Glushka, J., and Stacey, G., 1992. *Proc. Natl. Acad. Sci.* (USA) **89**:8789-8793.
14. Stacey, G., Luka, S., Sanjuan, J., Banfalvi, Z., Nieukoop, A.J., Chun, J.Y., Forsberg, L., and Carslon, R., 1994. *J. Bacteriol.* **176**:620-633.
15. Triplett, E.W., and Sadowksy, M.J., 1992. *Ann. Rev. Microbiol.* **46**:399-428.
16. Vest G., Weber, D.F., and Sloger, C., 1973. *In* B.E. Caldwell (ed.), Soybeans: Improvement, Production, and Uses. *Agronomy* **16**:353-390. Am. Soc. of Agron., Madison, Wisc.

Enhancing the potential of microbial inoculants through molecular microbial ecology.

[1]O'Gara, F., [2]Peruch, U., [3]Barea, J.M., [4]Nuti, M.P., [5]Bonfante, P., [1]Moënne-Loccoz, Y., [1]McCarthy, J., [1]Lohrke, S., [6]Powell, J.

[1]Microbiology Department, University College, Cork, Ireland, [2]Agronomica Srl, Ravenna, Italy, [3]Department of Microbiology, CSIC, Grenada, Spain, [4]Dipartimento di Biotecnologie Agrarie, Universita di Padova, Padova, Italy, [5]Dipartmento di Biologia Vegetale, Universita degli Studi di Torino, Torino, Italy, [6]Irish Sugar plc, Carlow, Ireland.

Introduction

The current interest associated with the use of microbial inoculants is in part a response to concerns associated with the use of synthetic chemical pesticides and fertilisers in current conventional agricultural practices. Due to increasing concerns with regard to the safety of synthetic chemical pesticides, restrictions have been placed on the use of a number of fungicides used to control fungal root diseases with more strict regulations including total bans expected in the future (e.g. methyl bromide). Therefore, viable alternatives must be made available to replace current chemical control methods. Strains of *Pseudomonas fluorescens* which exhibit antifungal activities against a number of soil-borne fungal pathogens under laboratory conditions have been isolated from soils which are naturally suppressive to fungal disease. These strains have been the focus of investigations into protection of plants from soil-borne fungal pathogens (3,4,8,10,17). Other microbial species have recieved increased attention as inoculants as a result of their ability to stimulate plant performance. These include the use of biofertiliser strains such as rhizobia, which establish N_2-fixing symbioses with leguminous plants, thereby alleviating the need for exogenous applications of nitrogen-based fertilisers. Other important biofertiliser strains include the arbuscular mycorrhizal (AM) fungi. AM fungi establish symbiotic associations with many crop species and benefit the host plant mainly by increasing access to soil nutrients such as phosphorous (12). In order for biocontrol strains to realise their full potential in sustainable environmentally friendly agriculture, they must be compatible with other microbial inoculants (i.e. rhizobia and AM fungi) designed to combine biocontrol and biofertiliser functions in a single inoculum. This need for compatiblity is demonstrated by observations that even chemical control agents can have negative effects on microbial inoculants. For example, Thiram, which is used as a seed coat to control fungal disease, has been demonstrated to negatively affect survival of *Bradyrhizobium japonicum* inoculants and subsequent nodulation of soybean (13).

Examination of the mechanisms whereby biocontrol strains exert their antifungal properties has revealed a close association between biocontrol and the production of antifungal secondary metabolites including phenazines and phloroglucinols (3,7,9,11,14,16). A number of additional metabolites have also been implicated in biocontrol including iron-chelating siderophores, HCN, and salicylic acid (7,9). The use of microbial inoculants in effective biocontrol and biofertiliser applications involve the introduction of large numbers of bacteria into the environment. The ecological impact associated with the deliberate release of large numbers of microbial inoculants, particularly biocontrol strains genetically modified to overproduce antifungal metabolites such as Phl needs to be evaluated. The soil microbial community is complex and consists of a number of groups which are important from an agronomic perspective (i.e. rhizobia and AM fungi). Within the EU-funded IMPACT project, WT and GM biocontrol strains were evaluated for effects on compatibility with recognised biofertiliser strains. Another aspect of this project was to evaluate biocontrol inoculants for effects on selected resident soil microbial communities. Ecological effects were determined at a number of levels, including the soil microbial community as a whole, pathogenic microorganisms (i.e. *P.ultimum*), and selected beneficial microbial strains (*Rhizobium, P. fluorescens* and arbuscular mycorrhizal (AM) fungi). In addition, biocontrol strains were evaluated for compatibility with the biofertiliser inoculant *Bradyrhizobium japonicum*.

P. fluorescens strain F113 was isolated from the rhizosphere of sugarbeets and which exerts biological control activity against *Pythium ultimum*, the causative agent of "damping off" of sugarbeet (8). The biocontrol capability of F113 has been linked to the production of the secondary metabolite 2,4-diacetylphloroglucinol (Phl) (2,7,8). A mutant derivative, F113G22, is deficient in Phl production and has lost the ability to protect sugarbeets from *P. ultimum* induced damping-off in soil microcosms. Introduction of a plasmid containing the

NATO ASI Series, Vol. G 39
Biological Fixation of Nitrogen
for Ecology and Sustainable Agriculture
Edited by A. Legocki, H. Bothe, A. Pühler
© Springer-Verlag Berlin Heidelberg 1997

Phl biosynthetic genes into F113G22 restores Phl production and biocontrol ability (8). A major focus of research with regard to biological control is the generation of genetically modified inoculant strains with improved biocontrol abilities. The antifungal nature of secondary metabolites such as phenazines and phloroglucinols have made them prime targets for genetic manipulation. Inoculation of cucumber with a genetically modified *P. fluorescens* strain CHAO resulted in a significant increase in protection against *P. ultimum* in soil microcosms (15).

Figure 1: Integration of biocontrol and biofertiliser inoculants in compatible corsortia

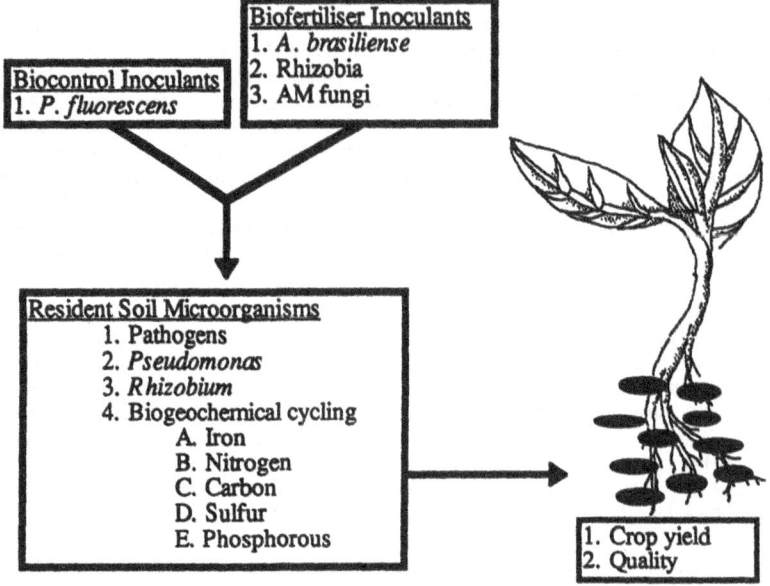

Effect of biocontrol strains in inoculant consortia containing biofertiliser (*Bradyrhizobium japonicum*) strains.

Within the IMPACT project, the use of mixtures of microbial inoculants with distinct functions (i.e. biocontrol and biofertiliser) in agronomic conditions were evaluated. Previous work by de Freitas, et.al. (6) demonstrated that coinoculation of bean with *R. phaseoli* and the wheat-growth promoting strain *P. syringae* R25, resulted in reductions in nodulation and plant biomass in non-sterile soil microcosms. Against this background, in 1994 and 1995 the effect of inoculating soybean with an inoculant mixture of *B. japonicum* and *P. fluorescens* F113 was evaluated in field trials in Italy (5). There were no significant differences with regard to crop yield or quality between coinoculated plants and plants which were singly inoculated with either *B. japonicum* or F113. A significant improvement in nodulation was observed in 1994 with coinoculated plants compared to the singly inoculated plants. No negative effects were associated with any of the inoculants used with regard to crop yield or quality in 1994 or 1995. These results indicate that the biocontrol strain F113 does not adversely affect *B. japonicum* inoculants and under certain circumstances may even stimulate biofertiliser functions. The requirement for multiple evaluations of microbial inoculants is emphasized by results obtained in 1995 in which there was no observed stimulation of nodulation by coinoculation with F113 and *B. japonicum*. These results demonstrate the requirement for multiple field trials to take into account variations in environmental and field conditions from year to year which may affect inoculant performance.

Impact of inoculants on indigenous populations of *Rhizobium leguminosarum* bv *trifolii*.

To evaluate biocontrol strains for possible negative effects on the indigenous microbial community of soil, rhizobia were chosen as potentially useful indicator organisms. These strains are of particular importance in agriculture as a consequence of their N_2-fixing symbiotic associations with legume crops. Negative effects on the indigenous population of rhizobia should be detected through plant parameters including nodulation, plant yield, and nitrogen content. In 1995 uninoculated red clover was utilised as a "biosensor" crop to detect perturbations in the indigenous community of *R. leguminosarum* bv. *trifolii* at a field site in which sugarbeet had been inoculated the previous year with *P. fluorescens* F113. No significant differences were detected with regard to crop yield or quality between those plots inoculated with F113 and those which had received the uninoculated controls. Additionally, no significant difference in the degree of nodulation of the clover root system was detected between the three treatments. Taken together, these results indicate that the use of F113 as an inoculant does not appear to adversely affect the resident soil *Rhizobium* community as measured by plant yield, health and nodulation. The use of crop rotation involving leguminous crop species is an integral aspect of low-impact, sustainable agricultural practices and the apparent lack of negative effects of biocontrol inoculants on rhizobial species is a highly significant finding.

Effect of biocontrol strains on beneficial arbuscular mycorrhizal fungi

Arbuscular mycorrhizal (AM) fungi are obligate biotrophic fungi which establish symbiotic associations with a majority (80%) of land plants. They are important for overall plant health, nutrient cycling, and conservation of overall soil structure (12). The selection of biocontrol strains such as F113 have been made as a consequence of their antifungal activities toward fungal pathogens. Due to the importance of AM symbiosis in agroecosystems, evaluations of biocontrol inoculants (e.g. F113) for effects on AM fungi was a major aspect of research within the IMPACT project. Evaluations of WT and GM biocontrol inoculants, including strains overproducing Phl, on the representative AM fungus *Glomus mosseae* were carried out *in vitro* and in soil microcosms. Strains evaluated included WT F113 and GM derivatives F113G22 (Phl-) and F113 pCU203 (Phl overproducer). While there was no effect on spore germination associated with any of the strains, F113 and F113G22 resulted in a significant stimulation of mycelial development, whereas F113 pCU203 did not, compared to the uninoculated control (1). In microcosm evaluations, none of the biocontrol strains exhibited any negative effects on plant biomass compared to the uninoculated control. In fact, a significant increase in mycorrhizal formation was observed with each strain. In studies designed to evaluate attachment of biocontrol strains to AM fungi, differences in binding were detected between the biocontrol strains, but that this was not related to the ability to produce Phl (1).

Conclusions

Within the IMPACT project it was the stated goal to develop and evaluate WT and GM microbial inoculants, particularly biocontrol strains for use within agricultural systems. A particularly important aspect of this project was to evaluate the compatibility of microbial inoculants within sustainable, environmentally friendly agricultural practices. The overall results from this project has demonstrated the lack of any significant detrimental effects with regard to the indigenous microbial community of soil associated with the use of biocontrol strains. This includes microorganisms which are of particular importance in maintaining the natural fertility and productivity of soil (e.g. rhizobia and AM fungi). The lack of negative effects on these microbial groups indicates that application of inoculants containing corsortia with biocontrol and biofertiliser functions may be a viable option to promote environmentally friendly agriculture. In this respect it is significant that some of these results suggest that under certain circumstances, biocontrol strains may actually improve the function and performance of biofertiliser strains. Additional studies which are designed to identify environmental factors which play a role in inoculant performance will further increase the reliability and function of microbial inoculants.

Table 1: Bacteria currently utilised as agricultural inoculants

Microbial inoculant	Crop inoculated	Function
Anabaena-Azolla	Rice	Biofertiliser
Azospirillum brasilense and *A. lipoferum*	Cereals	Biofertiliser
Bradyrhizobium japonicum	Forage and grain legumes	Biofertiliser
Rhizobium spp.	Forage and grain legumes	Biofertiliser
Frankia spp.	Non-leguminous trees (*Alnus*)	Biofertiliser
Bacillus subtilis and *B. thuringensis*	Different Crops	Biopesticide
Pseudomonas fluorescens	Sugarbeet	Biopesticide
Pseudomonas spp.	Various crops	Biopesticide

References

1. **Barea, J.M., Andrade, G., Bianciotto, V., Dowling, D., Lohrke, S., Bonfante, P., and F. O'Gara.** 1996. Impact on arbuscular mycorrhiza formation of *Pseudomonas* strains used as inoculants for the biocontrol of soil borne plant fungal pathogens. Appl. Environ. Microbiol. (Submitted).

2. **Carrol, H. Moënne-Loccoz, Y., Dowling, D.N. and F. O'Gara.** 1995. Mutational Disruption of the biosynthesis genes coding for the antifungal metabolite 2,4-diacetylphloroglucinol does not influence the ecological fitness of *Pseudomonas fluorescens* F113 in the rhizosphere of sugarbeets. **61:** 3002-3007.

3. **Cook, R.J.** 1993. Making greater use of introduced microorganisms for biological control of plant pathogens. Annu. Rev. Phytopathol. **31:** 53-80.

4. **Cook, R.J., Thomashow, L.S., Weller, D.M., Fujimoto, D., Mazzola, M., Bangera, G. and D.-S. Kim.** Molecular mechanisms of defense by rhizobacteria against root disease. 1995. Proc. Natl. Acad. Sci.USA. **92:** 4197-4201.

5. **Corich, V., Giacomini, A., Concherti, G., Ritzerfeld, B., Vendramin, E., Struffi, P., Basaglia, M., Squartini, A., Casella, S., Nuti, M.P., Peruch, U., Poggiolini, S., DeTroch, P., Vanderleyden, J., Fedi, S., Fenton, A., Moënne-Loccoz, Y., Dowling, D. and F. O'Gara.** 1995. Environmental impact of genetically modified *Azospirillum brasiliense*, *Pseudomonas fluorescens*, and *Rhizobium leguminosarum* released as soil/seed inoculants. In: 3rd Symposium on biosafety results of field tests of genetically modified plants and microorganisms. Monterey, Ca. USA, November 1994. pp.371-388.

6. **de Freitas, J.R., Gupta, V.V.S.R. and J.J. Germida.** 1993. Influence of *Pseudomonas syringae* R25 and *P. putida* R105 on the growth and N_2 fixation (acetylene reduction activity) of pea (*Pisum sativum* L.) and field bean (*Phaseolus vulgaris* L.). Biol. Fertil. Soils 16: 215-220.

7. **Dowling, D.N. and F. O'Gara.** 1994. Metabolites of *Pseudomonas* involved in the biocontrol of plant disease. Trends in Biotechnol. 12: 133-141.

8. **Fenton, A.M., P.M. Stephens, J. Crowley, M. O'Callaghan, and F. O'Gara.** 1992. Exploitation of gene(s) involved in 2,4- diacetylphloroglucinol biosynthesis to confer a new biocontrol capability to a *Pseudomonas* strain. Appl. Environ. Microbiol. **58:** 3873-3878.

9. **Keel, C., Schnider, U., Maurhofer, M., Voisard, C., Laville, J., Burger, U., Wirthner, P., Haas D. and G. Defago.** 1992. Suppression of root disease by *Pseudomonas fluorescens* CHA0: Importance of the bacterial secondary metabolite 2,4-diacetylphloroglucinol. Mol. Plant-Microbe Interact. **5:** 4-13.

10.   Maurhofer, M., Keel, C., Haas, D. and G. Defago. 1994. Pyoluteorin production by
       Pseudomonas fluorescens strain CHA0 is involved in the suppression of of *Pythium* damping-off of
       cress but not of cucumber. Eur. J. Plant Pathol. **100**: 221-232.

11.   O'Sullivan, D.J. and F. O'Gara. 1992. Traits of fluorescent *Pseudomonas* spp. involved in
       suppression of plant root pathogens. Microbial Rev. **56**: 662-676.

12.   Read, D.J., Lewis, D.H., Fitter, A.H. and I.J. Alexander. 1992. Mycorrhizas in
       ecosystems. CAB International, Oxford.

13.   Rennie, R.J. and S. Dubetz. 1984. Effects of fungicides and herbicides on nodulation and N_2-
       fixation in soybean fields lacking indigenous *Rhizobium japonicum*. Agron. J. **76**: 451-454.

14.   Russo, A. Moënne-Loccoz, Y., Fedi, S., Higgins, P., Fenton, A., Dowling, D.N.,
       O'Regan, M., and F. O'Gara. 1996. Improved delivery of biocontrol *Pseudomonas* and their
       antifungal metabolites using alginate polymers. Appl. Microbiol. Biotechnol. **44**: 740-745.

15.   Schnider U., Keel. C., Blumer, C., Troxler,J., Defago, G., and D. Haas. 1995.
       Amplification of the housekeeping sigma factor in *Pseudomonas fluorescens* CHA0 enhances antibiotic
       production and improves biocontrol abilities. J. Bacteriol. **177**: 5387-5392.

16.   Shanahan, P., O'Sullivan, D.J., Simpson, P., Glennon, J.D. and F. O'Gara. 1992
       Isolation of 2,4-Diacetylphloroglucinol from a fluorescent pseudomonad and investigation of
       physiological parameters influencing its production. Appl. Environ. Microbiol. **58**: 353-358.

17.   Thomashow, L.S. and D. Weller. 1988. Role of phenazine antibiotic from *Pseudomonas
       fluorescens* in biological control of *Gaeumannomyces graminis* var. *tritici*. J. Bacteriol. **170**: 3499-
       3508.

Part 6

Nitrogen Fixing Systems

Nitrogen Fixation in *Rhodospirillum rubrum*: Regulation of Activity and Generation of Reductant

Stefan Nordlund, Magnus Johansson, Anders Lindblad, and Agneta Norén

Department of Biochemistry, Arrhenius Laboratories for Natural Sciences, Stockholm University, S-106 91 Stockholm, Sweden.

1. INTRODUCTION

Nitrogen fixation in phototrophic bacteria has been studied in greatest detail at the molecular level in *Rhodospirillum rubrum* and *Rhodobacter capsulatus*. The general characteristics of nitrogenase are very similar to those of the enzymes from e.g. *Klebsiella pneumoniae* and *Azotobacter vinelandii* (reviewed in 1). There is however one very interesting difference: nitrogen fixation in *R. rubrum*, *Rb. capsulatus* and a number of other phototrophic bacteria, is regulated at the metabolic level in addition to the transcriptional control demonstrated in all diazotrophs studied. This metabolic regulation is manifested as an inhibition of nitrogenase activity when cells are subjected to "switch-off" effectors, such as ammonium ions, glutamine or darkness. At the molecular level this inhibition is due to the conversion of dinitrogenase reductase (the Fe-protein) from an active to an inactive form by ADP-ribosylation of an arginine residue in one of the subunits. The modified form can not interact with dinitrogenase (the MoFe-protein). ADP-ribosylation is catalyzed by ADP-ribosyl transferase (DRAT) with NAD^+ as ADP-ribose donor and dinitrogenase reductase activating glycohydrolase (DRAG) catalyzes the reverse reaction.

In all diazotrophs studied nitrogenase expression is under transcriptional control in response to the availability of fixed nitrogen (reviewed in 2). In *K. pneumoniae* and a number of other organisms the P_{II} protein has a central role in this control system. P_{II} can exist in two forms, unmodified and uridylylated. The unmodified form interacts with NTRB, enhancing the phosphatase activity of this bifunctional enzyme, which then catalyzes the hydrolysis of phosphate from NTRC-P. NTRC can not as the phosphorylated form activate transcription from σ^{54} dependent promoters, e.g. the one of the *nifLA* operon. On the other hand, unmodified P_{II} can not interact with NTRB, which then acts as a kinase, phosphorylating NTRC, which leads to transcription of *nifLA* and thus nitrogenase expression.

The two forms of P_{II} are furthermore involved in the regulation of glutamine synthetase activity. The modified form activating the deadenylylating activity of the *glnE* product, whereas the unmodified stimulates the adenylylating activity of this enzyme, producing the less active form of glutamine synthetase.

The sensor in the P_{II} regulatory cascade is the *glnD* product, a bifunctional enzyme which catalyses the addition of UMP to P_{II} at high α-ketoglutarate/glutamine ratios and the removal of UMP at low ratios. The transcriptional control of *nif* genes in

NATO ASI Series, Vol. G 39
Biological Fixation of Nitrogen
for Ecology and Sustainable Agriculture
Edited by A. Legocki, H. Bothe, A. Pühler
© Springer-Verlag Berlin Heidelberg 1997

photosynthetic bacteria has been studied in greatest detail in *Rb. capsulatus*. Although there are general similarities with the *K. pneumoniae* system there are also significant differences, e.g. NTRC-P activated promoters studied are not σ^{54} dependent (reviewed in 3).

The electron transport system to nitrogenase has only been elucidated in great detail in *K. pneumoniae*. In this organism pyruvate is oxidized by an pyruvate: flavodoxin oxidoreductase, the product of *nifJ*, and the electrons transferred to dinitrogenase reductase by a flavodoxin, encoded by *nifF* (4). Although a flavodoxin and its gene, *nifF*, have been characterized in *Rb. capsulatus* (5), there are convincing evidence showing that the major electron donor to nitrogenase in this organism is a ferredoxin, encoded by *fdxN* (6).

2. NITROGEN FIXATION IN *R. rubrum*

2.1. Metabolic regulation

Although DRAG and DRAT, and the corresponding genes have been characterized in great detail, the signal(s) turning DRAT on and DRAG off when a "switch-off" effector is added, is not yet identified. We have previously demonstrated that addition of NAD^+ to nitrogen fixing *R. rubrum* will lead to inhibition of nitrogenase activity (7). Furthermore, we have reported that addition of "switch-off" effectors will cause an increase in $[NAD^+]$, as measured by fluorescence (8).

We now report that addition of NAD^+ indeed does lead to a modification of dinitrogenase reductase which does not take place in a DRAT⁻ mutant (9) and we believe that "switch-off" by NAD^+ is due to the same events as ammonium "switch-off". We suggest that the signal to DRAT involves the increase in $[NAD^+]$ and that DRAG is regulated through its binding to the chromatophore membrane, which we believe is dependent on the energization of the membrane. Such a model would explain why DRAG is found associated to the membranes when *R. rubrum* cells are broken.

This model implies that there is a specific binding site for DRAG on the membranes and to study this we have investigated the effect of different nucleotides on the association of DRAG to chromatophores. Addition of MgGDP specifically dissociated DRAG from the membrane, but there was no effect when GTP, ATP or ADP was added. It is tempting to speculate that guanin nucleotides could have a regulatory role, as they have in eukaryotes. It still remains to be demonstrated if there are variations in the concentration of GDP during "switch-off" and if there is a specific GTPase responding to "switch-off" conditions.

2.2. The role of P_{II}

We have previously shown that *glnB*, the gene encoding P_{II}, is located just upstream of *glnA* in *R. rubrum* and that there is both a *glnBA* and a *glnA* mRNA (10). We have identified sequences similar to σ^{54} and σ^{70} consensus promoters just upstream of *glnB* and a probable NTRC binding sequence overlapping the latter. We have not been able to identify a promoter directly upstream of *glnA* and believe that the *glnA* mRNA is produced by specific processing of the *glnBA* transcript. There is in fact a sequence similar to a RNase E cleavage site in between *glnB* and *glnA*.

In this communication we report on further studies on the effect of "switch-off" conditions on the P_{II} protein, as assayed by SDS-PAGE and Western Blot. Addition of ammonium ions, glutamine or NAD^+ leads to the conversion of the slower migrating form present under growth on N_2, to a faster migrating species. After addition of glutamate, both forms are present. There was no change in migration when cells were subjected to darkness. We have not yet been able to demonstrate that the change in migration is due to modification by UMP. Addition of methionine sulfoximine (MSX) prevented the conversion of P_{II} by ammonium ions and also nitrogenase "switch-off" as has been shown before. On the other hand MSX has no effect on either when glutamine was added as effector. Interestingly, the was no change in migration of P_{II} when MSX was added before NAD^+.

From these studies we conclude that the molecular events leading to modification of dinitrogenase reductase and those involved in the change in P_{II} migration (modification) are not regulated by the same mechanism. This would also indicate that P_{II} probably does not have a role as a signal to DRAG/DRAT. The molecular basis for the two forms of P_{II} still remains to be identified.

2.3. Electron transport to nitrogenase

We have previously isolated and purified an enzyme with properties very similar to those of NIFJ from *K. pneumoniae*, the major difference being that it is expressed not only under diazotrophic conditions but also under repressing conditions. Purified enzyme together with a ferredoxin fraction from *R. rubrum* and pyruvate could support nitrogenase activity *in vitro* (11). Subsequent studies led to the characterization of the gene encoding this enzyme and the generation of a mutant by kanamycin cassette insertion. The mutant, lacking the enzyme, grew as well as the wild-type under all conditions studied, including diazotrophy. From these studies we concluded that generation of reductant for nitrogenase in *R. rubrum* does not involve NIFJ as in *K. pneumoniae* and in agreement with this we could not demonstrate the presence of a *nifF* gene in *R. rubrum* (12).

We suggest a model for electron transport to nitrogenase in *R.rubrum* that involves a membrane bound enzyme (complex) which catalyzes the reduction of ferredoxin in a reaction requiring a membrane potential. The electron donor to this enzyme is suggested to be the reduced Q-pool. There have been previous reports on the involvement of a membrane potential in photosynthetic bacteria (13) and our model is supported by the identification of the *rnf* genes in *Rb. capsulatus*. These genes encode proteins shown to be involved in electron transport to nitrogenase and from sequence predictions some of them are membrane proteins (14).

In studies using fluoroacetate we have shown that nitrogenase activity in *R. rubrum* requires an operating TCA-cycle and we believe that its role is to generate NADH which would then reduce the Q-pool. In order to demonstrate the dependence of the membrane potential we have studied the effect of varying light intensities and of

154

uncouplers, on nitrogenase activity. In addition changes in the ATP/ADP ratio were monitored. These studies were done both in the wild-type and a mutant devoid of DRAT (9), to eliminate the influence of the metabolic regulation. Our results clearly show that variations in the ATP/ADP ratio have very little effect on nitrogenase activity. When light intensity is decreased there is very little change in the ATP/ADP ratio and nitrogenase activity decreases only at very low intensities. Addition of uncouplers led to a decrease in the ATP/ADP ratio at low uncoupler concentrations at which there was very little effect on nitrogenase activity. At higher concentrations however, activity dropped drastically. The results are clearly in agreement with our model and we are presently trying to identify membrane proteins expressed specifically under nitrogen fixing conditions.

3. Acknowledgement

This work was supported by grants from the Swedish Natural Science Research Council to S.N.

4. References

1.   Ludden, P. W. and Roberts, G. P. (1995) in "Anoxygenic Photosynthetic Bacteria", (Blankenship, R. E., Madigan, M. T. and Bauer, C. E., eds.) pp. 929-947, Kluwer Academic Publishers, Dordrecht.
2.   Merrick, M. J. and Edwards, R. A. (1995) Microbiol. Rev., 59, 604-622
3.   Kranz, R. G. and Cullen, P. J. (1995) in "Anoxygenic Photosynthetic Bacteria", (Blankenship, R. E., Madigan, M. T. and Bauer, C. E., eds.) pp. 1191-1208, Kluwer Academic Publishers, Dordrecht.
4.   Ludden, P. W. (1991) Curr. Top. Bioenerg., 16, 369-390.
5.   Gennaro, G., Hübner, P., Sandmeier, U., Yakunin, A. F. and Hallenbeck, P. C. (1996) J. Bacteriol., 178, 3949-3952.
6.   Jouanneau, Y., Meyer, C., Naud, I. and Klipp, W. (1995) Biochim. Biophys. Acta, 1232, 33-42.
7.   Soliman, A. and Nordlund, S. (1992) Arch. Microbiol., 117, 53-60.
8.   Norén, A. and Nordlund, S. (1994) FEBS Lett., 356, 43-45.
9.   Liang, J., Nielsen, G. M., Lies, D. P., Burris, R. H., Roberts, G. P. and Ludden, P. W. (1991) J. Bacteriol., 173, 6903-6909.
10.   Johansson, M. and Nordlund, S. (1996) Microbiol., 142, 1265-1272.
11.   Brostedt, E. and Nordlund, S. (1991) Biochem. J., 279, 155-158
12.   Lindblad, A., Jansson, J., Brostedt, E., Johansson, M., Hellman, U. and Nordlund, S. (1996) Mol. Microbiol., 20, 559-568.
13.   Haaker, H., Laane, C., Hellingwerf, K., Houwer, B., Konings, W. N. and Veeger, C. (1982) Eir. J. Biochem., 127, 639-645.
14.   Schmehl, M., Jahn, A., Meyer zu Visendorf, A., Hennecke, S., Masepohl, B., Schuppler, M., Marxer, M., Oelze, J. and Klipp, W. (1993) Mol. Gen. Genet., 241, 602-615.

MOLECULAR STUDIES OF THE ELECTRON TRANSPORT PATHWAY TO NITROGENASE IN *RHODOBACTER CAPSULATUS*

JOUANNEAU, Y., JEONG, H.-S., HUGO, N., MEYER, C., and WILLISON, J. C.

CEA/Grenoble, Laboratoire de Biochimie Microbienne, CNRS URA 1130, Département de Biologie Moléculaire et Structurale, 38054 Grenoble cedex 9, France

1 Introduction

Nitrogenase reduces N_2 to ammonia in a multiple-electron reaction that requires a source of low-potential reductant. In nitrogen-fixing bacteria, a ferredoxin or a flavodoxin, operating at a redox potential of - 400 mV or below, is thought to donate electrons to the nitrogenase complex. However, the electron carrier that specifically reduces nitrogenase has, in very few species, been identified and little is known of the metabolic pathways generating reductants for nitrogen fixation.

Our studies are focused on the identification of the protein components involved in electron transport to nitrogenase in the photosynthetic bacterium *Rhodobacter capsulatus*. Six ferredoxins have been isolated and characterized from this bacterium. Three of these, named FdI, FdII and FdIII, were found to be [4Fe-4S]-containing dicluster ferredoxins (1-5), while the other three ferredoxins contain a single [2Fe-2S] cluster (6-8). Based on *in vitro* experiments, only FdI and FdII could mediate electron transfer to nitrogenase and either one appeared as a potential electron donor (1,3). Genetic studies indicated that FdII has some essential metabolic function, because its structural gene could not be disrupted, but its function appears to be unrelated to nitrogen fixation (9,10). On the other hand, the *fdxN* gene encoding FdI is localized within a *nif* operon (2), and the deletion of *fdxN* resulted in a drastic reduction of N_2 fixation (11,12). These findings, as well as additional evidence given below, indicate that FdI serves as the physiological electron donor to nitrogenase in *R.capsulatus*. The molecular interaction between FdI and dinitrogenase reductase has been studied by using site-directed mutants of the ferredoxin, as well as by cros-linking experiments. Our work has also been directed towards the biochemical characterization of the products of the *rnf* genes, which may code for a specific electron transport system (12). The properties of some components of this system are presented. It is proposed that the *rnf* gene products form a membrane complex which might function as a ubiquinol-ferredoxin oxidoreductase.

2 Role of FdI and studies on its interaction with nitrogenase

In order to assess the physiological role of FdI, we have studied the phenotype of a *fdxN*-deleted mutant. Growth on N_2 was markedly affected and nitrogenase activity was reduced 20-fold. The lack of FdI affected not only electron transport to nitrogenase, but also the stability of the nitrogenase enzyme itself (11). Introduction of a plasmid carrying a copy of *fdxN* restored a wild-type phenotype. None of the other ferredoxins could functionally replace FdI. However, the residual nitrogenase activity observed in

NATO ASI Series, Vol. G 39
Biological Fixation of Nitrogen
for Ecology and Sustainable Agriculture
Edited by A. Legocki, H. Bothe, A. Pühler
© Springer-Verlag Berlin Heidelberg 1997

the *fdxN* mutant suggested that an alternate electron carrier could partially subtitute for FdI. A flavodoxin recently isolated from *R.capsulatus* might be responsible for this activity (13). As the synthesis of the flavodoxin was stimulated under iron deprivation, it might replace FdI when iron becomes limiting.

Cross-linking experiments were performed to investigate the interaction between FdI and dinitrogenase reductase (Rc2). A covalent complex between FdI and one Rc2 subunit was obtained upon incubation of the two purified proteins with ethyl-(dimethylaminopropyl)-carbodiimide(EDC). Cross-linking was prevented by raising the ionic strength, indicating that electrostatic bonds are involved in the interaction. The molecular interaction between the two redox partners was also studied by site-directed mutagenesis (14). To help choose target residues, a three-dimensional model of FdI was constructed, based on the structure of a homologous ferredoxin. Three mutant proteins were purified and found to have spectroscopic and redox properties similar to wild-type FdI. They also proved to be competent electron donors to nitrogenase as they could restore the wild-type phenotype when expressed in a *fdxN*-deleted strain. However, two variants bearing substitutions of charged residues predicted to be at the surface of the molecule exhibited altered kinetic parameters. A D36H variant showed a marked decrease of its affinity towards Rc2, compared to natural FdI. Interestingly, this mutant also failed to form a cross-linked complex with dinitrogenase reductase when incubated with EDC. It is concluded that Asp36 in FdI plays an important role in the Rc2-FdI molecular interaction.

3 Properties of a *nif*-specific electron transport system in *R.capsulatus*

A set of genes, showing no homology to known *nif* genes, has been identified within a major *nif* region of the *R.capsulatus* chromosome (12). These genes, called *rnf*, are organized in two divergently transcribed operons, controlled by typical *nif* promoters (Fig.1). One of the operons also includes two ferredoxin-encoding genes, *fdxC* and *fdxN*, coding for FdIV, a [2Fe-2S] ferredoxin (6), and FdI, respectively. The *rnf* gene products are predicted to be either membrane-bound polypeptides, or proteins containing Fe-S centers (Table1). Based on these predictions and on analyses of the phenotypes of *rnf*-deleted mutants, it was proposed that the *rnf* gene products assembled to form a membrane complex involved in electron transfer to nitrogenase(12). Recently, we have found errors in the initially released sequences. Corrections resulted in the identification of a new gene (*rnfG*) between *rnfD* and *RnfE*, that may code for a periplasmic protein (Fig.1 and Table1).

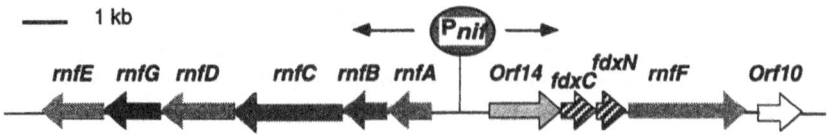

Fig.1 : Organization of the genes encoding a *nif*-specific electron transport system in *R.capsulatus*

Table 1. Predicted properties of the *rnf* gene products

Gene	Phenotype of mutants	Polypeptide size (Da)	Expected properties
rnfA	Nif⁻	20,424	Membrane protein
rnfB	Nif⁻	19,090	Fe-S protein
rnfC	Nif⁻	55,587	Fe-S protein
rnfD	Nif⁻	37,086	Membrane protein
rnfE	Nif⁻	25,826	Membrane protein
rnfF	Nif⁻	54,142	Membrane protein
rnfG	?	22,959	Periplasmic protein
fdxC	Nif⁺	10,156	Ferredoxin
fdxN	Nif⁺/⁻	6,733	Ferredoxin
Orf14	Nif⁺	48,050	Flavoprotein

A database search has revealed that the *rnfA*, *rnfD* and *rnfE* gene products share sequence similarities with three subunits of a membrane-bound enzyme that functions as a NADH-ubiquinone oxidoreductase (B. Berks, pers. commun.). This enzyme also has Na^+-translocating activity and generates a sodium gradient across the membrane in a reaction coupled to NADH oxidation (15). By analogy, it is proposed that the *rnf*-encoded complex of *R.capsilatus* might function as a Na^+-translocating ubiquinol-ferredoxin oxidoreductase.

3.1 Expression of the *rnf* genes in *E.coli* and purification of the *rnfB* gene product

As the *rnf* gene products are synthesized at low level in *R.capsulatus* , we have attempted to overexpress the *rnf* genes them in *E.coli.*. The RnfB product was overproduced in *E.coli* as a 27 kDa His-tagged polypeptide, that formed insoluble inclusion bodies. After solubilisation in urea, the recombinant RnfB was purified as a brown protein, showing an absorbance spectrum typical for a Fe-S protein. However, the instability of the protein precluded further characterisation of the cluster(s). As we suspected that the instability may have arisen from the His-tag extension fused at the N-terminus, a protein carrying a His-tag at the C-terminus was designed. The latter fusion protein was produced in soluble form in *E.coli*, but it contained no detectable Fe-S cluster. Antibodies against recombinant RnfB were used to localize the protein in *R.capsulatus*. It was found that RnfB was mainly associated with the membrane fraction, consistent with the proposed model for the Rnf complex. The RnfC protein was also produced in *E.coli* but in relatively minor amounts. On the other hand, the *rnfA*, *rnfD*, *rnfE*, and *rnfG* genes could not be expressed to significant levels in *E.coli*. As heterologous expression of the *rnf* genes had some unexplained limitations, we are now attempting to overproduce the Rnf products in the natural host.

3.2 Overexpression and purification of a 48 kDa flavoprotein

An open reading frame, called Orf14, precedes *fdxC* and *fdxN*, in an operon that also includes *rnfF* (Fig. 1). Although Orf14 is apparently not essential for nitrogen fixation (12), it potentially encodes a flavoprotein with unusual sequence features. For example,the C-terminal domain of the deduced polypeptide shows striking sequence similarities with flavodoxins. The protein was overproduced in *E.coli* as a His-tagged fusion and could be purified in good yield. The protein was yellow in colour and exhibited a spectrum indicating the presence of a flavin cofactor. A FMN group was identified at the active site. The flavoprotein did not react with pyridine nucleotides, but was readily reduced by methyl viologen or dithionite. Its mid-point redox potential was about - 200 mV. Interestingly, FdIV could serve as electron donor to the flavoprotein *in vitro*. Since the two proteins are coded by contiguous genes, they may belong to the same electron transfer pathway. Recently, a flavoprotein associated with a hydrogenase in a *Methanobacterium* species, was found to share striking sequence similarity with the *R.capsulatus* flavoprotein (16). The possible function of the latter protein is under investigation.

References

1 Hallenbeck, P.C., Jouanneau, Y. and Vignais, P.M. (1982) Biochim. Biophys. Acta 681, 168-176
2 Schatt, E., Jouanneau, Y. and Vignais, P.M. (1989) J. Bacteriol. 171, 6218-6226
3 Jouanneau, Y., Meyer, C., Gaillard, J. and Vignais, P.M. (1990) Biochem. Biophys. Res. Commun. 171, 273-279
4 Jouanneau, Y., Meyer, C., Gaillard, J., Forest, E. and Gagnon, J. (1993) J. Biol. Chem. 268, 10636-10644
5 Armengaud, J., Gaillard, J., Forest, E. and Jouanneau, Y. (1995) Eur. J. Biochem. 231, 396-404
6 Grabau, C., Schatt, E., Jouanneau, Y. and Vignais, P.M. (1991) J. Biol. Chem. 266, 324-329
7 Armengaud, J., Meyer, C. and Jouanneau, Y. (1994) Biochem. J. 300, 413-418
8 Naud, I., Vinçon, M., Garin, J., Gaillard, J., Forest, E. and Jouanneau, Y. (1994) Eur. J. Biochem. 222, 933-939
9 Duport, C., Jouanneau, Y. and Vignais, P.M. (1992) Mol. Gen. Genet. 231, 323-328
10 Saeki, K., Suetsugu, Y., Tokuda, K.I., Miyatake, Y., Young, D.A., Marrs, B.L. and Matsubara, H. (1991) J. Biol. Chem. 266, 12889-12895
11 Jouanneau, Y., Meyer, C., Naud, I. and Klipp, W. (1995) Biochim. Biophys. Acta 1232, 33-42
12 Schmehl, M., Jahn, A., Meyer zu Vilsenforf, A., Hennecke, S., Masepohl, B., Schuppler, M., Marxer, M., Oezle, J. and Klipp, W. (1993) Mol. Gen. Genet. 241, 602-615
13 Yakunin, A.F., Gennaro, G. and Hallenbeck, P.C. (1993) J. Bacteriol. 175, 6775-6780
14 Naud, I., Meyer, C., David, L., Breton, J., Gaillard, J. and Jouanneau, Y. (1996) Eur. J. Biochem. 237, 399-405
15 Beattie, P., Tan, K., Bourne, R. M., Leach, D., Rich, P. R. and Ward, F. B. (1994) FEBS Lett. 356, 333-338
16 Wasserfallen, A., Huber, K. and Leisinger, T. (1995) J. Bacteriol. 177, 2436-2441

Regulation of Alternative Nitrogenase Systems by Environmental Factors in the Cyanobacterium *Anabaena variabilis*

Teresa Thiel, Eilene M. Lyons, and Marta Zahalak

Department of Biology, University of Missouri-St. Louis,
St. Louis, MO 63130, USA

Keywords. nitrogenase, alternative nitrogenase, environmental regulation, *nif*, *vnf*, cyanobacteria, heterocyst, *Anabaena*

1 Introduction

Anabaena variabilis ATCC 29413 is a filamentous cyanobacterium that fixes dinitrogen under a variety of environmental conditions. Under aerobic conditions, nitrogen fixation occurs exclusively in specialized cells called heterocysts. Heterocysts are differentiated cells that develop in a pattern from vegetative cells in the filament in response to limitation in fixed nitrogen (reviewed, 11). The primary Mo-dependent nitrogenase in this organism is encoded by a contiguous cluster of *nif1* genes that are transcribed only in heterocysts (1, 9). These genes are very similar to the *nif* cluster of *Anabaena* sp. strain PCC 7120 (reviewed, 3), which is also transcribed exclusively in heterocysts (2). We have recently characterized alternative nitrogenase gene clusters in this strain that are absent in *Anabaena* sp. strain PCC 7120. One system, *vnf*, encodes a vanadium-dependent nitrogenase that functions in heterocysts (7); whereas the other system, *nif2*, encodes a Mo-nitrogenase that functions only under strictly anaerobic conditions in vegetative cells and in heterocysts (9). We describe here the regulation of these alternative nitrogenase systems, which respond to changes in environmental factors.

2 Characterization of the *vnf* genes

The *vnfDGK* genes of *A. variabilis* (Fig. 1) encode a V-dependent nitrogenase that is transcribed in the absence of Mo, whether or not V is present (7). These genes are similar to the genes that encode the V-nitrogenase in *Azotobacter vinelandii* (1). In *A. variabilis*, the *vnfG* gene, which encodes the δ subunit, is fused with *vnfD* into a single ORF. The promoter region contains a putative -10 region, but no identifiable -35 region and has at least one NtcA binding site (7).

NATO ASI Series, Vol. G 39
Biological Fixation of Nitrogen
for Ecology and Sustainable Agriculture
Edited by A. Legocki, H. Bothe, A. Pühler
© Springer-Verlag Berlin Heidelberg 1997

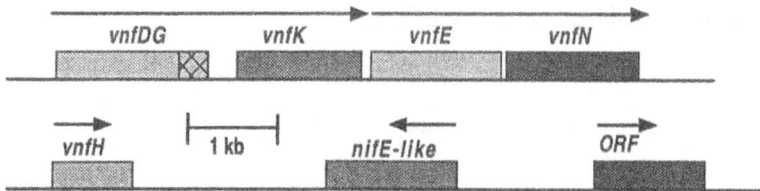

Fig 1. Genes associated with the V-nitrogenase system of *A. variabilis*. Arrows indicate direction of transcription.

The *vnfEN* genes (Fig. 1) show very weak similarity to other *nifEN/vnfEN* genes and show greater similarity to the *vnfDK* genes just upstream than to the *A. vine-landii vnfEN* genes (8). A mutant with an insertion in the *vnfN* gene lacks V-nitrogenase activity; thus, the *vnfEN* genes are essential for the V-nitrogenase in *A. variabilis* (8). In contrast, in *Azotobacter vinelandii* NifEN can substitute for VnfEN (10).

The *vnfH* gene, which is not linked to the other *vnf* genes (Fig. 1) is transcribed in the absence of Mo, with or without V (as are the other *vnf* genes). In addition, two other ORFs have been found near *vnfH*. One has similarity to *nifE* genes: a mutation in this ORF has no phenotype. The third ORF has about 100 nucleotides that are very similar to the 5' end of *vnfD*; however, the rest of this ORF has no similarity to any sequences in the GenBank database. The function of these ORFs is unknown.

3 Characterization of the *nif1* genes

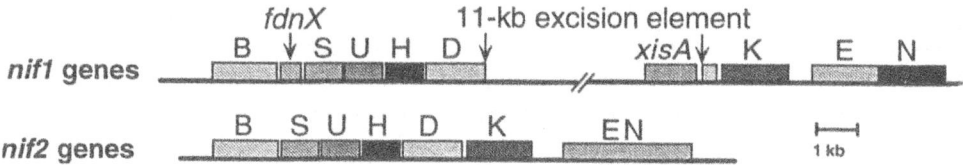

Fig. 2. Genes encoding the two Mo-nitrogenases of *A. variabilis*

The *nif1* genes of *A. variabilis* are the homologs of the *nif* genes of *Anabaena* PCC 7120 (reviewed, 3). They are about 95% identical to the *Anabaena* PCC 7120 genes at the nucleotide level and differ primarily in the absence of a 55-kb excision element in the *fdxN* gene. We subcloned, partially sequenced, and mutagenized the *nifB1*, *nifS1*, and *nifU1* genes of *A. variabilis* (Fig. 2). These genes are part of the system that encodes the heterocyst-specific Mo-nitrogenase that functions under aerobic conditions. The *nifB1* gene is essential for both the *nif1* nitrogenase and for the *vnf* nitrogenase, but neither *nifS1* nor *nifU1* is required for nitrogen fixation, with or without Mo or V (5).

4 Regulation of the *vnf* and *nif1* genes

Northern analysis and assays of β-galactosidase in a *nif1::lacZ* fusion strain have demonstrated that both the *vnf* and *nif1* genes are transcribed in the absence of Mo, with or without V (7, 9). Addition of either Mo or V to metal-starved cells resulted in a rapid increase in acetylene reduction, within 1-2 hours after metal addition, which suggests that aponitrogenases are made even in the absence of Mo and V. Acetylene reduction and slow growth of Mo- and V-starved cells further indicate that one or more nitrogenases are made and probably scavenge trace amounts of metal to provide some low level of functional nitrogenase. The relatively high rates of acetylene reduction in these poorly growing cells indicate that a Mo- and V-deficient nitrogenase may reduce acetylene much better than it reduces N_2 (8).

While both nitrogenases appear to be transcribed in Mo- and V- starved cells, we have evidence that it is the Mo-nitrogenase that is preferentially used. A mutation in the *nifDK1* genes (5) completely abolishes growth and acetylene reduction in cells starved for Mo and V; whereas, a *vnf* mutant grows and reduces acetylene in medium lacking Mo and V as well as the wild-type strain (7, 8). It appears that in the absence of Mo and V, both sets of genes (*nif1* and *vnf*) are transcribed and probably translated. However, Mo alone exerts transcriptional control by repressing the *vnf* genes; V does not have a similar effect on the *nif1* genes.

5 Characterization of the *nif2* gene cluster

The *nif2* genes of *A. variabilis* encode a Mo-dependent nitrogenase that functions only under strictly anaerobic conditions in vegetative cells and in heterocysts (9). These genes are not present in *Anabaena* PCC 7120. This cluster comprises about 13 kb, including *nifB2*, *nifS2*, *nifU2*, *nifH2*, *nifD2*, *nifK2*, *nifEN2*, *nifX2*, and *nifW2* (Fig. 2). In addition to this major cluster, there are ferredoxin genes downstream of *nifW2* that are part of the cluster (6). Unlike the *nif1* genes, this cluster lacks both the *fdxN* gene and excision elements, and the *nifE* and *nifN* genes are fused into a single ORF. A *lacZ* transcriptional reporter, fused to *nif1* and to *nif2* genes in the chromosome, was used to demonstrate, at the single cell level, that the *nif1* genes are expressed only in heterocysts under aerobic or anaerobic conditions; whereas, the *nif2* genes are expressed only under anaerobic conditions in vegetative cells and in heterocysts (9). Further characterization of mutants in each system has confirmed that the *nif2* genes are essential for anaerobic nitrogen fixation in the absence of heterocysts. Sequence comparisons indicate that the *nif2* cluster is more closely related to the *nif* genes of non-heterocyst forming cyanobacteria, which fix only under anaerobic conditions, than to the *nif1* genes.

Table 1. Summary of expression of the nitrogenase systems of *Anabaena*

	Anaerobic		Aerobic (Heterocysts)		
System	Het.	Veg. cells	-Mo -V	+Mo	+V
nif1	+	-	+	+	+
nif2	+	+	-	-	-
vnf	N.D.*	-	+	-	+

* Not determined

6 Summary of factors that regulate expression of the nitrogenase systems

I. Developmental Factor - Heterocyst differentiation
   A. controlled by availability of fixed nitrogen
   B. controls expression, probably activation, of *nif1* and *vnf*
   C. primary level of control: "default"

II. Environmental Factors
   A. Mo
      1. *nif1* - expression is independent of Mo
      2. *vnf* - repressed by Mo, possibly by a Mo-induced repressor
   B. Fixed nitrogen (NH_4Cl)
      1. represses heterocyst differentiation and thus, indirectly, represses activation of *nif1* and *vnf*
      2. represses *nif1*, *vnf*, and *nif2* transcription
      3. induces posttranslational modification of *nifH1* and *vnfH*
   C. Oxygen
      1. heterocysts are anaerobic, thus *nif1* and *vnf* systems are generally shielded from oxygen
      2. high O_2 induces posttranslational modification of NifH1
      3. *nif2* nitrogenase is very oxygen sensitive, but nothing is known about regulation by oxygen

Citations
1. Bishop, P.E., and R. Premakumar. 1992. p. 736-762. *In* G. Stacey, R.H. Burris, and H.J. Evans, ed. Biological nitrogen fixation. Chapman and Hall. New York.
2. Elhai, J., and C.P. Wolk. 1990. EMBO J. 9:3379-3388.
3. Haselkorn, R. 1992. p. 166-190. *In* G. Stacey, R.H. Burris, and H.J. Evans, ed. Biological nitrogen fixation. Chapman and Hall, Inc. New York. .
4. Herrero, A., and C.P. Wolk. 1986. J. Biol.Chem. 261:7748-7754.
5. Lyons, E.M. and T. Thiel. 1995. J. Bacteriol. 177:1570-1575.
6. Schrautemeier, B., Neveling, U., and S. Schmitz. 1995. Mol.. Microbiol. 18:357-369.
7. Thiel, T. 1993. J. Bacteriol. 175:6276-6286.
8. Thiel, T. 1996. J. Bacteriol. 178: 4493-4499.
9. Thiel, T., Lyons, E.M., Erker, J.C., and A. Ernst. 1995. Proc. Natl. Acad. Sci.USA 92:9358-9362.
10. Wolfinger, E.D. and P.E. Bishop. 1991. J. .Bacteriol. 173:7565-7572.
11. Wolk, C.P., Ernst, A., and J. Elhai. 1994. p. 769-823. *In*: D.A Bryant, ed. The Molecular Biology of the Cyanobacteria. Kluwer Academic Press. The Netherlands.

Heterocyst Differentiation and Nitrogen Fixation in Cyanobacteria

Robert Haselkorn and William J. Buikema

Department of Molecular Genetics & Cell Biology, University of Chicago,
920 East 58 Street, Chicago IL 60637 USA

Keywords. *Anabaena*, cyanobacteria, gene expression, green fluorescent protein,
heterocyst differentiation, *hetR* gene, spacing pattern

Introduction

Cyanobacteria face the difficult problem of fixing nitrogen while generating oxygen.
Some solve this problem by storing carbohydrate during the day and using the stored
material to fuel nitrogen fixation during the night. Others, notably *Trichodesmium*, a
filamentous organism that is one of the major nitrogen fixers in the ocean, fixes only
during the day, precisely at the same time that it is fixing CO_2 and generating
oxygen. How this is accomplished is still mysterious. There is no obvious
morphological differentiation in *Trichodesmium* filaments. But there is now an
acknowledged role for respiration in cyanobacterial photosynthesis, so possibly
nitrogenase is protected by respiration in *Trichodesmium*. The heterocystous
cyanobacteria, such as *Anabaena*, solve the problem of oxygen protection, at least in
part, by differentiating specialized cells for nitrogen fixation at regular intervals along
the filaments. We have been interested in the regulatory interactions among genes
and their products, starting with the environmental cue of nitrogen depletion and
concluding with a fully functional anaerobic factory for nitrogen fixation.
    The conversion of a carbon-fixing, oxygen-evolving vegetative cell into a
heterocyst requires the orderly expression of many hundreds of genes in a cell-
specific manner. In the strain *Anabaena* PCC 7120, heterocysts generally
differentiate at the ends of filaments and at regular intervals, approximately every
tenth cell, along each filament. These cells are readily distinguished by their double-
layered envelope, consisting of novel polysaccharides and crystalline glycolipids, and
prominent polar granules comprised of cyanophycin. Within the heterocyst, synthesis
of proteins for carbon fixation stops while the enzymes for nitrogen fixation and
generation of ATP and reductant for nitrogen fixation and assimilation increase.
Several DNA elements interrupting *nif* and *hup* operons in vegetative cell DNA are
excised. Systems for the orderly transport of fixed nitrogen out of the heterocyst and
carbohydrate into it are established (1).
    To learn how these conversions are accomplished, we have isolated mutants
blocked in the development of functional heterocysts and the genes complementing
the mutations. We have concentrated on regulatory mutations, but some results
obtained for other types of mutations shed light on heterocyst physiology and oxygen
protection. Two new experimental tools have been important in elucidating the
relationships among regulatory genes and their products. These are two kinds of gene

NATO ASI Series, Vol. G 39
Biological Fixation of Nitrogen
for Ecology and Sustainable Agriculture
Edited by A. Legocki, H. Bothe, A. Pühler
© Springer-Verlag Berlin Heidelberg 1997

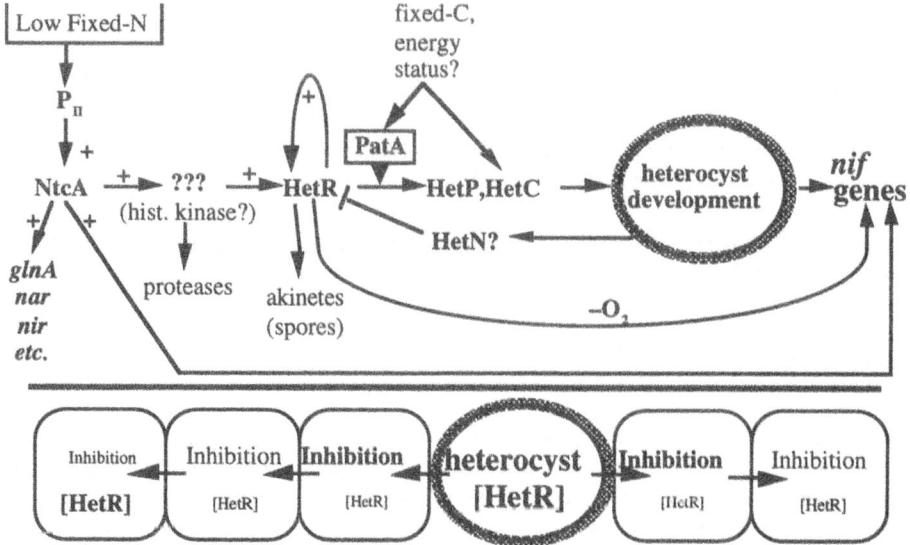

Figure 1. Model of the molecular interactions involved in heterocyst differentiation and pattern formation. The upper part of the figure attempts to show the temporal relationships among the gene products, while the lower part shows the spatial relationships. The environmental clue that starts the process is nitrogen step-down. This clue is perceived by an unknown sensor, resulting in the modification of NtcA, which (probably indirectly) activates transcription of *hetR*. Subsequent steps can be accomplished simply by expression of *hetR*, for example, from the Cu-regulated *petE* promoter. Nitrogen fixation cannot be accomplished by this expedient, however, because the rearrangement catalyzed by XisA requires NtcA for the transcription of *xisA*. The lower part of the figure indicates that HetR is responsible for the pattern in part, by providing an inhibitor of its own expression, of which there is a gradient. Experiments with extra or controlled copies of the *hetR* gene show that HetR is responsible for interpreting the gradient as well.

fusions: one fuses the Cu-regulated *petE* promoter to a regulatory gene, allowing the expression of the regulatory gene to be controlled by the experimenter. The second fuses the promoter of a regulatory gene to the structural gene for the green fluorescent protein (GFP) of the jellyfish, allowing the detection of regulatory gene expression with respect to timing and cellular pattern.

Results and Discussion

Results obtained in several labs allow the construction of a preliminary scheme for heterocyst differentiation, shown in Figure 1. The drawing attempts to show the sequence of gene actions that lead to heterocyst differentiation, above, and one of the spatial relationships below. We believe that there are many steps missing from the sequence shown and, possibly, that there are parallel, redundant paths to the same end.

The first gene product known to respond to the external cue of nitrogen starvation is NtcA. This protein was discovered independently by Jim Golden, who was studying proteins that bind to the promoter region of the *xisA* gene, and Enrique Flores, who was studying control of transcription of genes for nitrogen assimilation (2, 3). A knockout mutation of the *ntcA* gene of *Anabaena* PCC 7120 results in a

A   **B**

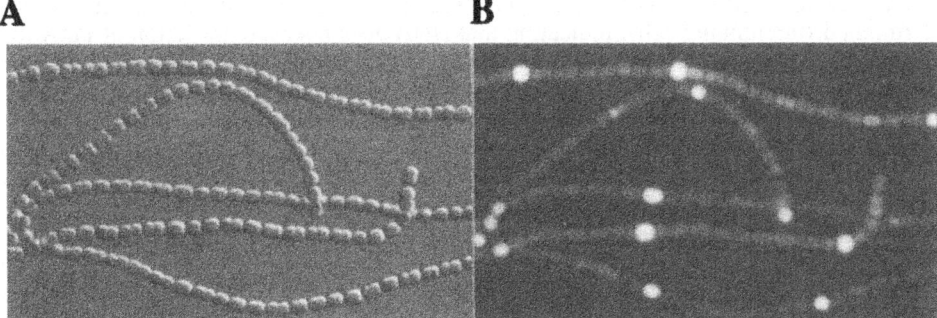

Figure 2. Expression of GFP from the *hetR* promoter in wild-type *Anabaena* 7120. The *gfp* gene was fused to the *hetR* promoter and transferred by conjugation into *Anabaena* 7120. Cells were induced by nitrogen step-down for 25 hrs and visualized by DIC (A) or fluorescence using a modified FITC filter set (B).

strain that cannot activate transcription of the *nar, nir, glnA, xisA, nifH* or *hetR* genes. Failure to activate *hetR* results in the classical HetR phenotype, which is failure to initiate heterocyst differentiation (except for the induction of proteases, which occurs independently of HetR action) (4).

NtcA is present constitutively, but its activity is controlled by nitrogen status. In ammonia-containing medium, it does not activate transcription of any of the genes listed above. The NtcA amino acid sequence is related to that of the CRP protein of *E. coli*. Originally named Bif by Golden (for Binding Factor), NtcA binds to the specific site $TGT-N_8-ACA$ found in the 5'-flanking region of most of the genes it regulates. The one exception is *hetR*, all of whose promoters defined by primer extension lack the consensus NtcA binding site. Thus it is likely that NtcA activation of *hetR* expression is indirect. We show histidine kinases between NtcA and HetR, but there is no evidence that the two-component signalling systems function there. M. Fath and E. Koshy have isolated seven histidine kinase genes by PCR but none of them has been shown to have a differentiation role yet. C. C. Zhang has also found a family of serine/threonine kinases in *Anabaena* PCC 7120 with possible functions at this level of the cascade (5).

The pattern of expression of the *hetR* gene in wild-type *Anabaena* is revealed by the fluorescence of the GFP reporter, shown in Figure 2. This pattern can be seen before morphological differentiation is observed. It is possible to activate *hetR* expression without NtcA. A fusion of the *petE* promoter to *hetR* allows *hetR* transcription to be regulated by addition of Cu^{++} to the medium. In this way, we expressed HetR in media containing ammonia, with the result that morphologically normal heterocysts differentiated. Moreover, the heterocyst frequency depended on the amount of Cu^{++} added, i.e. on the level of expression of HetR.

This experiment shows that HetR is a key protein in the process of commitment to morphological differentiation. NtcA can be bypassed for that process, but NtcA is still required for functional nitrogen fixation in this system, because both *nif* and *xisA* gene transcription require it. Nor is HetR production, alone, sufficient for heterocyst differentiation. Cu^{++}-driven expression of HetR in a *patA* mutant does not result in the differentiation of internal heterocysts (W.J. Buikema, unpublished data). Thus, at least in those cells not at the ends of filaments, PatA is required for the positive actions of HetR. PatA could be a modifier, direct or indirect, of HetR. Alternatively, PatA might be responsible for destruction of a diffusible inhibitor that acts on HetR. When a *hetR::gfp* fusion is put into a *hetR* mutant strain, fluorescence increases in all the cells, indicating that HetR participates in its own repression in vegetative cells.

When the same fusion is put into the *patA* mutant, fluorescence increases only in the terminal differentiating cells, indicating that HetR does not need the action of PatA to repress its own transcription. These results are consistent with two possibilities: either PatA modifies HetR to convert it from a repressor to an activator of *hetR* transcription, or PatA destroys an inhibitor that prevents activation of transcription of *hetR* and other downstream genes.

HetP and HetC are gene products discovered in the laboratory of Peter Wolk. They have been shown by him to work downstream of HetR. Extra copies of the *hetP* gene produce the same phenotype as extra copies of *hetR*, namely too many heterocysts (1).

HetN is a great mystery, requiring further analysis. The *hetN* gene was found first by Todd Black in the Wolk lab (6) and later by Chris Bauer in Chicago (7). Bauer isolated a cosmid with a wild-type *Anabaena* DNA insert that inhibited heterocyst differentiation. By looking at revertants and Tn*5* inserts in the cosmid, Bauer identified a gene that, alone, was responsible for preventing heterocyst differentiation. When he sequenced it, he found it was identical to the *hetN* gene that Black had discovered by way of Tn*5* mutagenesis of the chromosome. In many ways, the *hetN* gene is the opposite of the *hetR* gene. Extra copies of *hetN* prevent, rather than provoke, heterocyst differentiation. So we conclude that HetN is responsible for synthesis of the inhibitor that prevents HetR from initiating differentiation. The mystery comes from the structure of HetN: the protein seems to be an enzyme of fatty acid elongation, possibly involved in heterocyst glycolipid synthesis. Transcripts are not found until 12 hours after nitrogen step-down (7), although precocious synthesis of HetN when extra copies of the gene are present is possible. Upstream of the *hetN* gene are four ORFs of which three are clearly enzymes of glycolipid synthesis, expression of which occurs late in heterocyst differentiation and mutation of which leads to glycolipid-free and oxygen-sensitive heterocysts (7).

Our current work exploits the two fusions described, in order to understand better the elements of the regulatory network that governs heterocyst differentiation and nitrogen fixation in *Anabaena*. The critical immediate questions are (1) how NtcA responds to nitrogen step-down, (2) which players link NtcA to *hetR* expression, (3) how HetR activates genes required for heterocyst differentiation, and (4) how HetN is connected to the diffusible inhibitor of differentiation.

References

1. Wolk, C. P., Ernst, A. & Elhai, J. (1995) in *The Molecular Biology of Cyanobacteria*, eds. Bryant, D. A. (Kluwer Academic Publishers, Dordrecht, The Netherlands), pp. 769-823.
2. Wei, T.-F., Ramasubramanian, T. S., Pu, F. & Golden, J. W. (1993) J. Bacteriol. **175,** 4025-4035.
3. Frias, J. E., Flores, E. & Herrero, A. (1994) Mol. Microbiol. **14,** 823-832.
4. Buikema, W. J. & Haselkorn, R. (1991) Genes & Development **5,** 321-330.
5. Zhang, C. (1993) Proc. Natl. Acad. Sci. USA **90,** 11840-11844.
6. Black, T., A. & Wolk, C. P. (1994) J. Bacteriol. **176,** 2282-2292.
7. Bauer, C. C. (1994) PhD dissertation (University of Chicago).

TRICHODESMIUM HAS CELLS SPECIALIZED FOR NITROGEN FIXATION BUT LACKS HETEROCYSTS

Bergman, B.[1], Fredriksson C.[1], Janson S.[1], Carpenter E.J.[2], Paerl H.[3] & Lugomela C.[4]

1) Department of Botany, Stockholm Univ., S-106 91 Stockholm, Sweden; 2) Marine Science Research Center, State Univ. of New York, Strony Brook, NY11794, USA; 3) Institute of Marine Sciences, Univ. of North Carolina, Morehead City, NC28557, USA; 4) Department of Botany, Univ. of Dar es Salaam, Tanzania, Africa

Keywords: cyanobacteria, nitrogen-fixation, nitrogenase, *nif* genes, cell differentiation, CO_2 fixation

1. Introduction

Two features make research on the marine cyanobacterium *Trichodesmium* of particular relevance. The first, relates to its common occurrence and important role in a global perspective. Although *Trichodesmium* is restricted to coastal habitats and open oceans with temperatures above about $18-20^{\circ}C$, it is probably quantitatively one of the most common cyanobacterium in nature. Recent estimates also indicate that it supports the pelagic zone of the oligotrophic oceans with considerable amounts of fixed nitrogen (Carpenter & Romans 1991). Furthermore, *Trichodesmium* is unique in that it fixes nitrogen aerobically in light, a feature only known from cyanobacteria differentiating heterocysts, a specific cell type for the oxygen sensitive nitrogen-fixing enzyme nitrogenase.

The second feature is the topic of the present report. A short summary of background and consequences are summarized with emphasis on our own recent data. The reader is referred to Carpenter & Capone (1992), Bergman et al. (1994), Gallon et al. (1996), Bergman et al. (1996), Capone et al. (1996) for more comprehensive reviews on the ecology and physiology of *Trichodesmium*.

2. Results and Discussion

2.1 *Trichodesmium* contains nitrogenase

Nitrogen fixation by *Trichodesmium* colonies was demonstrated by Dugdale et al. already in 1961: however, it took another thirty years before this finding was substantiated by molecular and immunological data. These analyses not only demonstrated the occurrence of one of the cyanobacterial structural *nif* genes in natural *Trichodesmium* colonies (Zehr et al. 1989) but also that high levels of the nitrogenase protein is synthesized in cells of *Trichodesmium* (Paerl et al. 1989;

NATO ASI Series, Vol. G 39
Biological Fixation of Nitrogen
for Ecology and Sustainable Agriculture
Edited by A. Legocki, H. Bothe, A. Pühler
© Springer-Verlag Berlin Heidelberg 1997

Bergman & Carpenter 1991). The major reason for the reluctance of the biological community to accept that *Trichodesmium* possessed nitrogenase were the unusual patterns and conditions required for nitrogen fixation. In spite of being non-heterocystous, fixation of nitrogen had repeatedly been shown to take place in colonies inhabiting well aerated ocean waters, to be light dependent and maximal at mid day, while neglible at night or in the dark (Saino & Hattori 1978; Carpenter et al. 1993; Zehr et al. 1993; Paerl 1994). This behaviour is considered as atypical of non-heterocystous, but typical for heterocystous cyanobacteria. In the latter category, differentiation of heterocysts is a prerequisite for fixation of nitrogen as these cells protect the nitrogenase from oxygen inactivation. Nitrogenase is therefore exclusively found in this cell type while in the non-heterocystous cyanobacteria *Oscillatoria limosa*, *Plectonema boryanum* and *Lyngbya aestuarii* nitrogenase is synthesized in all cells, and nitrogen fixation occurs only in darkness or under anaerobiosis (Villbrandt et al. 1992; Rai et al. 1992; Paerl et al. 1991).

Our continued immunocytochemical investigations on nitrogenase in *Trichodesmium* have clearly demonstrated that colonies collected in day time from natural *Trichodesmium* populations growing in the Caribbean and Sargasso Seas, including several species collected from various locations, different years and seasons, all synthesize nitrogenase. Also, that the nitrogenase Fe-protein is almost exclusively found in *Trichodesmium* and that little (if any) is being synthesized in associated bacteria or cyanobacteria (Siddiqui et al. 1992).

2.2 *Trichodesmium* synthesizes nitrogenase in a small proportion of the cells

There are two peculiarities in the pattern of the nitrogenase synthesis in *Trichodesmium* that are noteworthy. The first is that nitrogenase is not only detected within the central areas of the *Trichodesmium* colonies as hypothesized previously (Carpenter 1983; Carpenter et al. 1990) but that it may occur in cells located in any position of the colony (Bergman & Carpenter 1991; Fredriksson & Bergman 1995). It has been assumed that cells or filaments in central areas of the colonies, i.e. aggregations of up to hundred or more of parallel filaments, were shaded enough by surrounding filaments to prevent photosynthetic oxygen evolution. The second peculiarity is that nitrogenase is not present in all cells like in the other non-heterocystous genera but restricted to comparatively few cells within each colony (Bergman & Carpenter 1991; Bergman et al. 1993; Fredriksson & Bergman 1995). On no occasion has nitrogenase been detected in all cells, and a detailed screening of a number of natural colonies (three species) collected at day time, suggested that less than about 24% of the cells contained nitrogenase (Fredriksson & Bergman 1995). However, the frequency may be lower, e.g. in colonies collected at or just before dawn which may lack nitrogenase, as the protein is synthesized *de novo* each morning (Capone et al. 1990; Fredriksson & Bergman 1995). Collectively, our data suggest that cells with nitrogenase at mid day amounts to on average 13-14% of the total cell population. This resembles the situation in heterocystous cyanobacteria in which a small fraction (5-10%) of the photosynthetic vegetative cells are differentiated into nitrogenase harbouring non-photosynthetic heterocysts.

2.3 Nitrogenase containing cells are contiguously arranged into short "zones"

What has recently become more apparent is that *Trichodesmium* may have an unusual arrangement of their nitrogenase containing cells. These cells are not spread individually along the filaments, as is the case in most heterocystous cyanobacteria. In fact, cells with nitrogenase are clustered into short stretches or "zones" within the filaments. This was first discovered in longitudinally sectioned filaments of *T. contortum* collected from Indian Ocean (Janson et al. 1994) and later shown for the cultured isolate *Trichodesmium* IMS 101 (Fredriksson, Paerl and Bergman, unpublished data), isolated by Prufert et al. (1993). Even though these two species have experienced extremely different external growth conditions there is a high congruence in the frequency of cells containing nitrogenase, and both show an identical clustering of the nitrogenase containing cells. Our current hypothesis is therefore that this is a widespread feature of the filaments of the genus *Trichodesmium*, whether growing under natural conditions or in the laboratory. Such an arrangement would make individual *Trichodesmium* filaments self supportive in terms of nitrogen and possibly more fit and competetive in waters disturbed by pronounced wave actions, which may disrupt the integrity of the *Trichodesmium* colonies.

Janson et al. (1994) were also the first to correlate the centrally located "lighter" cells of the filaments sometimes seen using light microscopy (LM), with the nitrogenase containing cells. However, the occurrence of regions of cells with a different "texture" had been commented upon already in some earlier LM analyses of natural colonies (Carpenter & Price 1976; Li & Lee 1990). It is therefore possible that these cells were also sites for nitrogenase. A preliminary, more detailed, LM investigation of *T. erythareum* from the Indian Ocean suggests that "lighter" zones are common in the filaments, that there may be more than one in each filament and that they are often centrally located (Lugomela, Fredriksson, Semesi & Bergman, unpublished data). It is now of interest to examine whether cells in these zones are the nitrogenase containing cells.

2.4 Sequestering of nitrogenase and photosynthesis into different subsets of cells ("spatial separation")

Previous data suggested that cells in *Trichodesmium* with nitrogenase may not fix carbon dioxide. This was concluded from the fact that stretches of consecutively arranged cells, often located centrally in natural and cultured *Trichodesmium* filaments, were devoid of $^{14}CO_2$-incorporation in micro-autoradiography analyses (Carpenter & Price 1976; Paerl 1994). Experiments have now for the first time verified this hypothesis. Analyses of the patterns on the isolate IMS 101 demonstrated that not only the location but also the proportion of cells with ^{14}C label, constituting approximately 85% of the total cell population, matched closely the population of cells lacking nitrogenase, about 87% (Fredriksson, Paerl & Bergman, unpublished data).

It therefore appears that *Trichodesmium* devote certain cell regions for nitrogen fixation while cells in the remaining, flanking regions continue to be photosynthetically competent. Although such data make it tempting to speculate that the same may be true for the photosynthetic light reaction, a by-product of which is oxygen, this now needs to be proven experimentally. In any case, the situation in *Trichodesmium,* a distinctly non-heterocystous cyanobacterium, in certain functional respects starts to resemble heterocystous cyanobacteria, well known to practice a spatial separation to protect nitrogenase from oxygen, but also from competition for e.g. ATP (Fay 1992). Add to this that cytochrome oxidase and glutamine synthetase are over-expressed in cells with nitrogenase in *Trichodesmium* spp. (Bergman et al. 1994; Carpenter et al. 1992) and that these cells show structural modifications, although no additional cell wall layers (Fredriksson & Bergman, unpublished data), the similarities are further strengthened.

2.5 Cell differentiation in a non-heterocystous cyanobacterium

Cyanobacteria classified as non-heterocystous are by definition not capable of cell differention and hence *Trichodesmium* may not belong to this category, i.e. section III cyanobacteria (Rippka et al. 1979). On the other hand, the cells with nitrogenase are not true heterocysts and *Trichodesmium* can therefore neither be classified as belonging to section IV cyanobacteria.

Trichodesmium cells with nitrogenase also show a gene expression which in many respects differs from the rest of the cells and may therefore be regarded as a second cell type in *Trichodesmium,* and a new cell type in cyanobacteria. Whether the cells containing nitrogenase are predestined genetically to express *nif* genes or whether all cells of the filaments have the capacity to synthesize nitrogenase is an open question. The *nifKDH* genes encoding nitrogenase, are contiguously arranged as in non-heterocystous cyanobacteria (Zehr et al. 1991; Sroga et al. 1996) which suggest that all cells may express *nif* genes. As opposed to heterocysts, however, the cells with nitrogenase in *Trichodesmium* spp. divide and retain nitrogenase in both daughter cells while doing so (Fredriksson & Bergman, unpublished data). If not permanently differentiated, there is the possibility that a new subset of cells express nitrogenase in the filaments, e.g. each morning when the synthesis of nitrogenase re-commences (Capone et al. 1990; Fredriksson & Bergman 1995). These aspects are now under consideration and we have also initiated a study on the expression of *hetR* in some non-heterocystous cyanobacteria. This gene encodes a protein required for heterocyst differentiation (Buikema & Haselkorn 1991).

3. Conclusions

Taken together we therefore suggest that, although the genus *Trichodesmium* has hitherto being classified as a non-heterocystous (non-differentiating) cyanobacterium, it should perhaps be reclassified as it differentiates certain photosynthetic cells into a novel non-^{14}C-fixing cell type in which nitrogen fixation takes place, i.e. *Trichodesmium* seems to practices division in labour through cell differentiation although it lacks heterocysts.

4. Acknowledgements

The financial support from the Swedish Natural Science Research Council, the C. Trygger´s and Bergwall´s Foundations is acknowledged.

5. References

Bergman, B. and Carpenter, E.J. (1991). Nitrogenase confined to randomly distributed trichomes in the marine cyanobacterium *Trichodesmium. J. Phycol.* **27**: 158-165.

Bergman, B., Siddiqui, P.J.A., Carpenter, E.J. & Peschek, G.A. (1993). Cytochrome oxidase: subcellular distribution and relationship to nitrogenase expression in the nonheterocystous marine cyanobacterium *Trichodesmium thiebautii. Appl. Environ. Microbiol.* **59**: 3239-3244.

Bergman, B., Carpenter, E.J., Janson, S., Sroga, G. and Fredriksson, C. (1994). Nitrogenase in the marine non-heterocystous cyanobacterium *Trichodesmium.* In *Nitrogen Fixation with Non-Legumes* (Hegazi, N.A., Fayes, H. & Monib, M., editors), 85-92. The American University Press, Cairo.

Bergman, B., Gallon, J.R, Rai, A.N. & Stal, L.J. (1996). N$_2$ fixation by non-heterocystous cyanobacteria. FEMS Microbiol. Rev. (submitted).

Buikema , W.J. & Haselkorn, R. (1991). Characterization of a gene controlling heterocyst differentiation in the cyanobacterium *Anabaena* 7120. *Genes Dev.* **5**: 321-330.

Capone, D.G., O´Neil, J.M., Zehr, J. & Carpenter, E.J. (1990). Basis for diel variation in nitrogenase activity in the marine planktonic cyanobacterium *Trichodesmium thiebautii. Appl.Environm. Microbiol.* **56**: 3532-3536.

Capone, D.G., Zehr, J.P., Paerl, H.W., Bergman, B. & Carpenter, E.J. (1996). *Trichodesmium*: A biochemical and ecological enigma. *Science* (submitted).

Carpenter, E.J. & Price, C. (1976). Marine *Oscillatoria* (*Trichodesmium*): explanation for aerobic nitrogen fixation without heterocysts. *Science* **191**: 1278-1280.

Carpenter, E.J. (1983). Physiology and ecology of marine planktonic *Oscillatoria* (*Trichodesmium*). *Mar. Biol. Lett.* **4**: 69-85.

Carpenter, E.J., Chang, J., Cottrell, M., Schubauer, J., Paerl, H.W., Bebout, B.M. & Capone, D.G. (1990). Re-evaluation of nitrogenase oxygen-protective mechanisms in the planktonic marine cyanobacterium *Trichodesmium. Mar. Ecol. Prog. Ser.* **65**: 151-158.

Carpenter, E.J. and Romans, K. (1991). Major role of the cyanobacterium *Trichodesmium* in nutrient cycling in the North Atlantic Ocean. *Science* **254**: 1356-1358.

Carpenter, E.J., Bergman, B., Dawson, R., Siddiqui, P.J.A., Söderbäck, E. & Capone, D.G. (1992). Glutamine synthetase and nitrogen cycling in colonies of the diazotrophic cyanobacteria *Trichodesmium* spp. *Appl. Environ. Microbiol.* **58**: 3122-3129.

Carpenter, E.J., O'Neil, J.M., Dawson, R., Capone, D.G., Siddiqui, P.J.A., Roenneberg, T. & Bergman, B. (1993). The tropical diazotrophic phytoplankter *Trichodesmium*: biological characteristics of two common species. *Marine Ecol. Prog. Rep.* **95**: 295-304.

Dugdale, R. C., Menzel, D.W. & Ryther, J.A. (1961). Nitrogen fixation in the Sargasso Sea. *Deep-Sea Res.* **7**: 297-300.

Fay, P. (1992). Oxygen relations of nitrogen fixation in cyanobacteria. *Microbiol. Rev.* **56**: 340-373.

Fredriksson, C. & Bergman, B. (1995). Nitrogenase quantity varies in a subset of cells within colonies of the non-heterocystous cyanobacteria *Trichodesmium* spp. *Microbiol.* **141**: 2471-2478.

Gallon, J.R., Jones, D.A. and Page, T.S. (1996). *Trichodesmium*, the paradoxical diazotroph. *Arch. Hydrobiol.* (in press).

Janson, S., Carpenter, E.J. & Bergman, B. (1994). Compartmentalisation of nitrogenase in a non-heterocystous cyanobacterium: *Trichodesmium contortum*. *FEMS Microbiol. Lett.* **118**: 9-14.

Li, C.W. & Lee, M. (1990). Cellular differentiation in the trichome *Trichodesmium thiebautii* (Cyanophyta). *Bot. Mar.* **33**: 347-353.

Paerl, H.W. (1994). Spatial segregation of CO_2 fixation in *Trichodesmium* spp.: linkage to N_2 fixation potential. *J. Phycol.* **30**: 790-799.

Paerl, H.W., Priscu, J.C. & Brawner, D.L. 1989. Immunocytochemical localization of nitrogenase in marine *Trichodesmium* aggregates: relationship to N_2 fixation potentials. *Appl. Environm. Microbiol.* **55**: 2965-2975.

Paerl, H.W., Prufert, L.E. & Ambrose, W.W. 1991. Contemporaneous N_2 fixation and oxygenic photosynthesis in the nonheterocystous, mat-forming cyanobacterium *Lyngbya aestuarii*. *Appl. Environm. Microbiol.* **57**: 3086-3092.

Prufert-Bebout, L., Paerl, H.W. and Lassen, C. (1993). Growth, nitrogen fixation and spectral attenuation in cultivated *Trichodesmium* species. *Appl. Environ. Microbiol.* **59**: 1367-1375.

Rai, A.N., Borthakur, M. & Bergman, B. (1992). Nitrogenase derepression, its regulation and metabolic changes associated with diazotrophy in the non-heterocystous cyanobacterium *Plectonema boryanum* PCC 73110. *J. Gen. Microbiol.* **138**: 481-491.

Rippka , R., Deruelles, J., Waterbury, J.B., Herdman, M. & Stanier, R.Y. (1979). Generic assignments, strain histories and properties of pure cultures of cyanobacteria. *J. Gen. Microbiol.* **111**: 1-61.

Saino, T. & Hattori, A. (1978). Diel variation in nitrogen fixation by a marine blue-green alga, *Trichodesmium thiebautii*. *Deep-Sea Res.* **25**: 1259-1263.

Siddiqui, P.J.A., Bergman, B. & Carpenter, E.J. (1992). Filamentous cyanobacterial associates of the marine planktonic cyanobacterium *Trichodesmium*. *Phycologia.*.**31**: 326-337.

Sroga, G.E., Landegren, U., Bergman, B. & Lagerström-Fermér, M. (1996). Isolation of nifH and part of nifD by modified capture polymerase chain reaction from a natural population of the marine cyanobacterium *Trichodesmium* sp. *FEMS Microbiol. Lett.* **136**: 137-145.

Villbrandt, M., Stal, L.J. ,Bergman, B. & Krumbein, W.E. (1992). Immunolocalization and western blot analyses of nitrogenase in *Oscillatoria limosa* during a light-dark cycle. *Bot. Acta* **105**: 90-96.

Zehr, J.P. & McReynolds, L.A. (1989). Use of degenerate oligonucleotides for amplification of the *nifH* gene from the marine cyanobacterium *Trichodesmium thiebautii*. *Appl. Environm. Microbiol.* **55**: 2522-2526.

Zehr, J.P., Ohki, K. & Fujita, Y. (1991). Arrangements of nitrogenase structural genes in an aerobic filamentous nonheterocystous cyanobacterium. *J. Bact.* **173**: 7055-7058.

Zehr, J.P., Wyman, M, Miller, V., Duguay, L & Capone, D.G. (1993). Modification of the iron protein of nitrogenase in natural populations of *Trichodesmium thiebautii*. *Appl. Environ. Microbiol.* **59**: 669-676.

Part 7

Nitrogen Fixation
in Sustainable Agriculture

Biological Nitrogen Fixation for Sustainable Agriculture

A. van Kammen

Department of Molecular Biology, Wageningen Agricultural University, The Netherlands

The availability of a useful nitrogen source for the plant is, except for water, the major limiting factor in agricultural productivity. That has commonly resulted in the heavy use of chemical nitrogen fertilizer to replenish the soil nitrogen, an approach that suffers from high costs and severe, environmental effects. Sustainable agriculture involves the successful management of agricultural resources to satisfy the changing human needs, while maintaining or enhancing the environmental quality and conserving natural resources. Consequently sustainability considerations demand that alternatives to nitrogen fertilizer are sought. Biological nitrogen fixation (BNF) can offer this alternative as it uses the capacity of certain, nitrogen fixing, bacteria to convert atmospheric nitrogen into, for the plant usable, ammonia. In that way BNF may have an important role in sustaining productivity of soils.

The potential of BNF to be applied in sustainable farming systems has recently been thoroughly discussed as is witnessed by the frequent meetings, which have specifically addressed this question (1, 2, 3). The reports give evidence that, indeed BNF has a large potential for sustainable agriculture. Particularly symbiotic systems such as that of legumes and Rhizobium can be a major source of nitrogen in most cropping systems, while that of Azolla and Anabeana can be of special value in the cultivation of flooded rice.

Besides, there is a strong awareness that in the next 35 years the world population will grow with 3 billion people and that will require an increase of agricultural production of 75-100% to provide for food for the world population. If such increase in production should be realized with the current agricultural management, that would similarly require a doubling of the use of fossil fuel energy for fertilizer production and cause a surpassing damage to the environment.

The topic of food production and food supply in a world with a rapidly growing population will be the main theme at the FAO World Food Conference to be held in November of this year in Rome, together with the feasibility of sustainable agricultural production.

If sustainable farming using BNF as an alternative for the application of chemical fertilizers, will be able to meet the challenges of these comprehensive and serious problems of agricultural production in the near future, a huge effort is required both for changing the farming systems and for research to optimize the amount and efficiency of BNF. Roughly 80% of BNF is the result of symbiotic nitrogen fixation by rhizobia which have the ability to induce the formation of nodules on the roots of legumes. Therefore, a major objective will be optimizing the nitrogen fixing potential of these systems.

In addition, the possibility should be investigated to extend symbiotic nitrogen fixation to crop plants, especially cereals, which are, at present, not able to fix

NATO ASI Series, Vol. G 39
Biological Fixation of Nitrogen
for Ecology and Sustainable Agriculture
Edited by A. Legocki, H. Bothe, A. Pühler
© Springer-Verlag Berlin Heidelberg 1997

nitrogen in symbiotic association with Rhizobium. This possibility is strongly advocated by the International Rice Research Institute (IRRI) at the Philippines and IRRI has established an international BNF working group of experts for assessing the opportunities for nitrogen fixation in rice and, at the same time, launched a project proposal for donor support to explore the potential opportunities (4). It is argued that the research advances in understanding Rhizobium-legume interaction at the molecular level and, furthermore, the ability to introduce new genes into rice by transformation, have created an excellent opportunity to investigate the possibility for nitrogen fixation in rice. I think this proposal deserves full support as the right time has come to explore the possibilities seriously. The elucidation of the molecular basis of the interaction of rhizobia and their legume hosts has given some understanding of the specificity of these interactions in the different steps involved in the process of nodulation in legumes. These studies have indicated that the developmental program controlling root nodule formation uses elements derived from processes common to all plants (5). Besides, some genes homologous to the early nodulin genes in legumes, have been detected in the rice genome. In the next decade great progress will be made in the genetic dissection of the formation and functioning of root nodules, for recently two model legumes *Medicago truncatula* and *Lotus japonicus*, by which transgenic can routinely be obtained, have become available, and now, provide the tools for the identification and characterization of the plant genes involved in nodule formation. That will increase our understanding of the mechanisms controlling root nodule development and the extent in which these are modifications of common developmental programs. Subsequently it might result in the design of strategies by which non-legumes, like rice, can be given the ability to enter symbiosis with nitrogen fixing bacteria.

Even if it involves processes of exceptional complexity we should not keep ourselves from assessing the opportunities for nitrogen fixation in rice. Rice is an excellent model plant for this research. As an important food crop rice is intensively studied; it has moreover a high transformation ability and transgenic rice plants can easily be obtained. At least it can be determined by carefully selected experiments, if at all it will be possible, or impossible, to establish symbiotic nitrogen fixation in rice. Similarly it can be determined whether it will perhaps be necessary to transfer some specific genes to make rice compatible for symbiotic nitrogen fixation.

It is a major challenge how to materialize by sustainable farming the increase in agricultural production for the growing world population. In that BNF can have an important role. I should want that the policy for sustainable agriculture which will be set out at the FAO World Food Conference in November, will include additional support for BNF research.

References

1. Plant and Soil, Vol. 174, 1995.
2. Soil Biology and Biochemistry 27, 4/5, 1995.
3. Proc. of the Conference on Legumes in Sustainable Farming Systems, Sept. 1996, Aberdeen, Scotland.
4. Assessing Opportunities for Nitrogen Fixation in Rice, IRRI project proposal, October 1995.
5. Mylona, P., Pawlowski, K., and Bisseling, T. (1995), Symbiotic Nitrogen Fixation, The Plant Cell 7, 869-885.

ENHANCED AGRICULTURAL SUSTAINABILITY THROUGH BIOLOGICAL NITROGEN FIXATION

Carroll P. Vance[1]

[1] USDA/ARS, Plant Science Research Unit, Department of Agronomy and Plant Genetics, 411 Borlaug Hall, University of Minnesota, St. Paul, MN 55108, USA

NEED

The Earth's population, increasing exponentially, is expected to reach 10 billion, nearly double its present status, by 2035 (Bockman et al., 1990). Of this projected population, 90% is expected to reside in tropical and subtropical regions of the developing countries in Asia, Africa, and Latin America (Waggoner, 1994). Plant sources currently provide 80% of the caloric and dietary protein needs for tropical countries, and this is not expected to change in the near future. In 1910 human beings used about 10% of the total carbon fixed through photosynthesis (Golley et al., 1992). Currently humans use 40% of that carbon, and it is estimated that by 2030 humans will require 80%. Individual protein and caloric consumption of the Earth's current 5.7 million people averages 70 g·protein·d^{-1} and 2,400 calories·d^{-1}, respectively (Waggoner, 1994). The range of protein consumption varies from 38 to 125 g protein·d^{-1}, while the range of caloric intake varies from 1,800 to 3,500·d^{-1}, with the low range values associated with developing countries (Bongaarts, 1994). The anticipated doubling of Earth's population will exacerbate the current inequalities in nutritional intake. Clearly, to maintain the current level of protein and caloric intake over the next 40 years will necessitate unprecedented increases in crop production. This enhanced production will need to be achieved despite a significant deterioration of much prime agricultural land and will require the utilization of large areas now considered marginal.

Nitrogen is the major limiting nutrient for most crop species. Acquisition and assimilation of N is second in importance only to photosynthesis for plant growth and development (Newbould, 1989). Production of high-quality, protein-rich food is extremely dependent upon the availability of necessary N. The striking rise in cereal grain yields in developed countries between 1950 and 1990 is directly attributable to a 10-fold increase in N fertilizer use. The "Green Revolution" has been spurred by the development of cereal crops that respond favorably to high N fertilization rates. A typical cereal yield of 7T·ha^{-1} requires the uptake of 200 to 300 kg N·ha^{-1} (Bockman et al., 1990; Waggoner, 1994). Concomitant with high application of N fertilizer in developed countries are volatilization of N oxides (greenhouse gases) into the atmosphere, depletion of nonrenewable resources, an imbalance in the global N cycle, and leaching of NO_3^- into groundwater (Kinzig and Socolow, 1994). By contrast, in developing countries the high cost of N fertilizer, the energy requirements for

NATO ASI Series, Vol. G 39
Biological Fixation of Nitrogen
for Ecology and Sustainable Agriculture
Edited by A. Legocki, H. Bothe, A. Pühler
© Springer-Verlag Berlin Heidelberg 1997

production, and the suboptimal transportation capabilities limit its use, especially for small farms.

Sustainable agriculture is broadly defined as agriculture that is managed toward greater resource efficiency and conservation while maintaining an environment favorable for evolution of all species (Bohlool et al., 1992; Golley et al., 1992). More simply, it is meeting the needs of the present without compromising the needs of the future. One of the driving forces behind agricultural sustainability is effective management of N in the environment. Moreover, judicious management of N inputs into cropping systems is a prerequisite for land stewardship. Successful manipulation of N inputs through the use of biologically fixed N results in farming practices that are economically viable and environmentally prudent (Bohlool et al., 1992; Vance and Graham, 1995). For example, use of N-fixing species in cropping systems reduces the need for N fertilizers and increases soil tilth. Additionally, biologically fixed N is bound in soil organic matter and thus is much less susceptible to soil chemical transformations and physical factors that lead to volatilization and leaching. Although many diverse associations contribute to symbiotic N_2 fixation (Sprent, 1984), in most agricultural settings the primary source (80%) of biologically fixed N is through the soil bacteria *Rhizobium, Bradyrhizobium, Sinorhizobium,* and *Azorhizobium*-Legume symbiosis (Vance, 1996). Legumes provide 25-35% of the worldwide protein intake. Approximately 250 million hectares of legumes are grown world wide and they fix about 90 Tg N·yr^{-1} (Kinzig and Socolow, 1994). The amount of N fixed by legumes is quite amazing since the total amount of nitrogenase in the world amounts to only a few kilograms (Delwiche, 1970). To replace the N fixed by legumes with anhydrous ammonia produced by the Haber-Bosch process would require 288 Tg of fuel and cost approximately $30 billion annually. Obviously, important goals for agriculture are enhancing the use of and improving the management of legume biologically fixed N for both humanitarian and economic reasons.

BIOLOGICALLY FIXED N: SUSTAINS CROPPING SYSTEMS

Nitrogen fixing species have played an integral role in cropping systems since the domestication of plants and have been prominently featured in rotations and intercropping, as alley crops, in pasture systems, as green manures, in agroforestry, and cover crops (Heichel, 1987; Peoples and Craswell, 1992; West and Mallarino, 1996). Mixed cropping with N_2 fixing species as a major component was the accepted norm prior to the development of the Haber Bosch process (Fujiata et al., 1992). In the U.S. and Europe the use of legumes to provide N for subsequent crops reached a maximum in the 1940s (Heichel, 1987). Since then implementation of legumes into cropping systems has declined due to the low cost of N fertilizer, high yielding cereal crops, and other economic and political forces (Heichel, 1987; Bohlool et al., 1992). A return to the widespread use of N_2 fixing plants in cropping systems of developed countries will require convincing evidence of the economic and environmental advantages accrued due to their use. Alternatively, social and/or political pressures may force changes favorable for the enhanced use of N_2 fixing species in management schemes. By contrast, more than 50% of the crops grown in Africa, India, and Latin America are either intercropped or rotated with N_2 fixing species (Fujiata et al., 1992). However, new and improved strategies need to be developed and transmitted to growers to more effectively use biological N_2 fixation in developing countries.

Symbiotically fixed N may become available to an intercrop or subsequent crop through several avenues including: (i) release from root exudates; (ii) vesicular arbuscule mycorrhizal (VAM) mediated transfer between species; (iii) N leaching from leaves and leaf litter decomposition; (iv) plow down of green manures; (v) decomposition of roots and nodules, and (vi) release of animal waste (Peoples et al., 1995a). The quantity and availability of symbiotically fixed N available to a companion or subsequent crop are governed by genotypic, environmental, and management factors. Accurate assessment of the legume contribution almost invariably requires the ^{15}N methodologies (Heichel, 1987). However, with the use of new non-N_2 fixing genotypes representative estimates of symbiotically fixed N available for other species can be obtained.

The long-held assumption is that planting legumes with companion crops, either interspersed or between rows, provides additional N to and improves the performance of the non-N_2 fixing species. However, most evidence supporting N transfer to a companion crop is indirect (Heichel, 1987; Russelle, 1996; West and Mallarino, 1996). In more exact studies where ^{15}N has been used, the amount of N in the companion crop that is derived from the N_2-fixing legumes varies from 0 to 70% (Heichel, 1987; Ledgard and Steele, 1992; West and Mallarino, 1996). Although there is a wide range in the amount of N transferred and the % of companion crop N that is derived from the legume, these figures are generally less than 40 kg N·ha^{-1} and 30%, respectively (Ledgard and Steele, 1992; Peoples et al., 1995b; West and Mallarino, 1996). Because N transfer is a function of plant growth and development, soil N content, and plant proximity, the most reliable data and maximum amounts of N transfer are obtained when the growth habit of the legume and companion species is similar, the experiment is conducted in low soil N, and plant spacing is optimal (Brophy et al., 1987; Ledgard and Steele, 1992; Russelle, 1996; Thomas, 1995). Studies conducted in the presence of high soil N frequently fail to show transfer of fixed N due to the inhibitory effect of soil N on N_2 fixation and the greater competitiveness of the non-legume species (Heichel, 1987; Fujiata et al., 1992; West and Mallarino, 1996). Brophy et al. (1987) elegant study with alfalfa and trefoil clearly demonstrated that at suboptimal densities and populations N transfer was hard to detect, while at optimal populations and spacing striking amounts of N transfer could occur.

Although most studies of N transfer do not differentiate what proportion of above and below ground components contribute to transfer, experiments with alfalfa and trefoil provide insight into this question (Ta and Faris, 1987; Lory et al., 1992; Dubach and Russelle, 1994). The quantity of symbiotically fixed N deposited in the rhizosphere of alfalfa was less than 3 kg·ha^{-1}. Additionally, the amount of N potentially available for a companion crop from the turnover of roots and nodules was approximately 13 kg N·ha^{-1} and 2 kg N·ha^{-1}, respectively. Heichel (1987) estimated that nodule turnover would contribute less than 6 kg N·ha^{-1}. These data indicate that above ground N contributions are of similar if not greater importance than below ground parts toward N transfer because total N transfer is usually around 30 to 40 kg N·ha.

Legume N as a replacement for fertilizer N derives from the practice of comparing the yield of a non-legume grown after a legume to that of a non-legume grown with fertilizer. This assessment provides an estimate of the amount of N in the non-legume that is derived from the legume. Replacement N is usually thought of in terms of rotational systems in which either the legume leaf debris and roots or the entire plant is returned to the soil for the subsequent crop. In most instances the replacement N value represents the response of the non-legume to the total available soil N and does not distinguish between symbiotically fixed N and N from other sources (Heichel, 1987; Peoples et al., 1995a; Wani et al., 1995). Because only the amount the N derived from fixation and acquired by the subsequent crop is gain of N to be credited against fertilizer use, the fertilizer N replacement value frequently is an overestimate of that contributed by fixation.

The fertilizer replacement value for legumes when grown preceding a grain crop varies from 0 to 110 kg N·ha^{-1}. This range represents composite data taken from several sources. As expected, there is substantial variation between species for N replacement values. Such variation reflects the impact of: (i) environment; (ii) moisture; (iii) plant composition, particularly C:N and phenolic content; (iv) soil texture and organic content; and (v) amount of legume incorporated into soil. Heichel (1987) and Hesterman et al. (1987) have suggested that about 40 to 70% of the fertilizer N replacement value of legumes is due to legume N. The remainder of the beneficial effect of legumes is due to increased soil health, improved disease and pest control, reduced allelopathic compounds, and enhanced water retention (Peoples and Craswell, 1992).

With the advent of increased population pressures, farming of marginal land, and degradation of the environment, research on N_2 fixation must include renewed efforts to deliver biologically fixed N to cropping systems. Research must identify the best species to incorporate into cropping systems, know the amount of N fixed by these species, and accurately predict the availability of fixed N as replacement for fertilizer N within intercrops and in rotations. Extension must be able to show on-farm success with enhanced use of biologically fixed N in practical management schemes. Lastly, we must develop new and/or value-added uses for N_2 fixing crops that make them attractive for production agriculture in developed countries.

INCREASED SUSTAINABILITY THROUGH NEW USES

Earlier sections of this article have documented the current agronomic significance of N_2 fixing species, but what about the future? Undoubtedly the use of legumes for food and fiber sources will continue at current to increased levels in the foreseeable future. New uses, however, may alter how we perceive N_2 fixing plants in agriculture and their importance to sustainable systems. Visions for the use of biological N_2 fixation in the future are numerous. Adoption of new uses could result in substantial increases in land sown to N_2 fixing species and increased diversity for cropping systems. Three projected new uses exemplify realistic possibilities: (i) develop N_2 fixing plants to decontaminate soil and water (phytoremediation); (ii) grow legumes for the generation of electrical energy; and (iii) produce industrial, pharmaceutical, and natural products in N_2 fixing plants. Nitrogen-fixing crops are ideally suited for these purposes because they use the sun's energy to achieve the desired outcome, they fix N_2 and require less fertilizer than other crops, and lastly many can be transformed with and express foreign genes. A description of the use of plants in phytoremediation will provide insight into projected new roles for legumes in sustainable agriculture.

While the use of biological agents to restore contaminated soil and water has been recognized for years (Salt et al., 1995) and plants have been used to reclaim mine spills and purify water moving into wetlands, the concept of phytoremediation of specific toxic compounds is a fairly recent development (Anderson et al., 1993; Schnoor et al., 1995). Plants may facilitate detoxification of compounds indirectly by stimulating the metabolism of rhizosphere microorganisms and through the release of enzymes into the rhizosphere. Alternatively, direct uptake of contaminants and subsequent sequestration and/or degradation can contribute to remediation (Shimp et al., 1993; Stomp et al. 1994).

The rhizosphere is characterized as the zone of soil under the immediate influence of plant roots (Anderson et al. 1993). In the rhizosphere, plant root exudates can affect soil pH, O_2 concentration, redox potential, C:N ratio, and microbial growth (Marschner and Romheld, 1996). Release of root exudates into the rhizosphere has been shown to increase microbial biomass and metabolic activity. Plant enzymes released from roots are also important components of rhizosphere biochemistry (Schnoor et al., 1995; Anderson et al., 1993). Thus, the rhizosphere has striking metabolic diversity, and this diversity is mediated through plant genotype x microbe interaction.

Legumes have been shown to facilitate the degradation (mineralization) of pesticides and other soil contaminants. Walton and Anderson (1990) have shown that rhizosphere soil of bush clover (*Lespedeza cuneata* L.) and soybean have greater rate of trichloroethylene degration than does bulk soil. Mineralization of diazinon and parathion was 18% in the rhizosphere of edible bean as compared to about 7% in root-free soil (Hsu and Bartha, 1979). Rhizosphere soil of pea containing diazinon supports substantially higher microbial populations than soil lacking vegetation

(Anderson et al., 1993). Legume exudates stimulate growth and expression of novel genes in *Rhizobium* and *Bradyrhizobium* (Denarie and Cullimore, 1993). Moreover, these bacteria degrade numerous phenolics (Parke and Ornston, 1984; Tepfer et al., 1988) and are amenable to genetic modification (Fischer and Long, 1992). Thus, rhizobia might be engineered to degrade any number of toxic compounds in soils planted with legumes. An alternative strategy might be to screen for and select legume genotypes that release enzymes from roots that degrade contaminants. Plant roots are known to exude a wide range of enzymes including esterases, peroxidases, hydroxylases, laccases, nitrilases, and dehalogenases (Schnoor et al., 1995; Waisel et al., 1996).

Not only can plants affect decontamination through the rhizosphere but also they can either sequester or degrade soil and water pollutants through direct uptake in the xylem stream. Salt et al. (1995) have demonstrated that *Brassica juncea* and *Thalspi caerulescens* can remove substantial quantities of heavy metals from soils. They also showed that sunflower (*Helianthus annulus* L.) roots rapidly absorbed and precipitated heavy metals from solution. Stomp et al. (1994) reported that several laboratories have transformed plants with the genes encoding metallothioneins, metal binding proteins, and the resultant transgenic plants grew normally in the presence of heavy metals. They suggest that the development of plants for bioremediation of heavy metals through biotechnology will occur in the near future.

Although it is well known that plants can take up and degrade herbicides and insecticides, they can also accumulate other organic aromatic compounds that pose as hazardous waste such as trinitrotoluene, dioxin, nitrobenzene, and pentachlorophenol (Paterson et al., 1990; Schnoor, 1995). Studies of herbicide resistance show that legumes and other species can develop the capacity to degrade or inactivate organic toxins (Harrison, 1992). This may occur through either natural selection, mutagenesis, or genetic engineering. We have recently initiated a project to bioremediate atrazine and enhance the use of alfalfa in sustainable agriculture (Sadowsky and Smith, 1996). Alfalfa will be transformed with the bacterial *atzA* gene, atrazine halidohydrolase, under the control of the cauliflower mosaic 35S (CaMV 35S) promoter to obtain high levels of *atzA* gene expression throughout the plant. The transgenic alfalfa should be able to detoxify atrazine. Ground water contaminated with atrazine, which is not uncommon in the U.S., will be irrigated onto the transgenic alfalfa and atrazine degradation monitored.

In the U.S. more than 32,000 toxic waste sites have been targeted for remediation, and this does not include rural areas showing contaminants such as pesticides or herbicides in ground water. Cleanup of these sites by conventional strategies could cost more than $200 billion (Salt et al., 1995). On a world-wide basis these figures are substantially higher. Legumes in phytoremediation offer a sustainable and less costly approach for cleanup of these sites.

CONCLUSIONS

Although we have made striking advances in understanding the molecular and biochemical components regulating symbiotic N_2 fixation, this has yet to be translated into applied improvements. In fact, over the last 40 years the importance of legume symbiotic N_2 fixation to agriculture has been overlooked, if not forgotten. The doubling of Earth's population expected early in the new millennium necessitates that we reaffirm the significance of legume N_2 fixation in cropping systems. Moreover, we must redefine the economic and social benefits of N_2 fixing crops in sustainable agriculture. This will require accurate documentation of reduced fertilizer N expenses, improved water quality and soil health, new uses and value-added components, and reduced pesticide inputs accruing due to the use of N_2 fixing species. Nontraditional and novel roles for legumes in cropping systems will contribute to their enhanced use in sustainable agriculture. Both traditional and biotechnological approaches will be required to increase yield and maintain high rates of N_2 fixation in crops targeted for use in sustainable agriculture.

REFERENCES

Anderson, T.A., Guthrie, E.A., and Walton, B.T., 1993, *Environ. Sci. Technol.* 27:2629-2636.

Bockman, O.-C., Kaarstad, O., Lie, O.H., and Richards, I., 1990, *Agriculture and Fertilizers: Fertilizers in Perspective*, Norsk Hydro, Drammen, Norway, pp. 245.

Bohlool, B.B., Ladha, J.K., Garrity, D.P., and George, T., 1992 , *Plant Soil* 141:1-11.

Bongaarts, J., 1994, *Sci. Amer.* 237:36-42.

Brophy, L.S., Heichel, G.H., and Russelle, M.P., 1987, *Crop Sci.* 27:753-758.

Delwiche, C.C., 1970, *Sci. Amer.* 223:136-146.

Denarie, J., and Cullimore, J., 1993, *Cell* 74:951-954.

Dubach, M., and Russelle, M.P., 1994, *Agron. J.* 86:259-266.

Fischer, R.F., and Long, S.R., 1992, *Nature* 357:655-660.

Fujiata, K., Ofosu-Budu, K.G., and Ogata, S., 1992, *Plant Soil* 141:155-175.

Golley, F. Baudry, J., Berry, R., Bornkamm, R., Dahlberg, K., Jansson, M., King, V., Lee, J., Lenz, R., Sharitz, R., and Svedin, U., 1992, *INTECOL Bulletin* 20:15-20.

Harrison, H.F., 1992, *Weed Technol* 6:613-614.

Harwood

Heichel, G.H., 1987, *In Energy in Plant Nutrition and Pest Control,* pp. 63-80, (Helsel, Z.R., ed.), Elsevier Science, Amsterdam.

Hesterman, O.B., Russelle, M.P., Sheaffer, C.C., and Heichel, G.H., 1987, *Agron J.* 79:726-731.

Hsu, T.S., and Bartha, R., 1979, *Appl. Environ. Microbio.* 37:36-41.

Kinzig, A.P., and Socolow, R.H., 1994, *Physics Today* 47:24-35.

Ledgard, S.F., and Steele, K.W., 1992, *Plant Soil* 141:137-152.

Lory, J.A., Russelle, M.P., and Heichel, G.H., 1992, *Agron. J.* 84:1023-1040.

Marschner, H., and Romheld, V., 1996, *In Plant Roots: The Hidden Half*, pp. 581-606, (Waisel, Y., et al., eds.), Marcel Dekker, Inc., New York.

Newbould, P., 1989, *Plant Soil* **115**:297-311.

Nisbet, G. S., and Webb, K. J., 1990, *In Biotechnology in Agriculture and Forestry*, pp. 38-48, Springer-Verlag, Berlin.

Parke, D., and Ornston, L. N., 1984, *J. Gen. Microbiol.* **130**:1743-1747.

Paterson, S., Mackay, D., Tam, D., and Shiu, W.Y., 1990, *Chemosphere* **21**:297-331.

Peoples, M. B., and Craswell, E. T., 1992, *Plant Soil* **14**:13-39.

Peoples, M.B., Herridge, D.F., and Ladha, J.K., 1995a, *Plant Soil* **174**:3-28.

Peoples, M.B., Ladha, J.K., and Herridge, D.F., 1995b, *Plant Soil* **174**:83-102.

Russelle, M.P., 1996, *In Nutrient Cycling in Forage Systems*, pp. 125-166, (Joost, R.E., and Roberts, C.A., eds.), Potash and Phosphate Institute and Foundation for Agronomic Research, Manhattan, Kansas.

Sadowsky, M. J., and Smith, D. R., 1996, *In Proceedings of 8th International Congresss on Molecular Plant-Microbe Interactions*, (Stacy, G., et al., eds.), Kluwer Academic Publishers, Dordrecht, (in press).

Salt, D.E., Blaylock, M., Kumar, N.P.B.A., Dushenkov, V., Ensley, B.D., Chet, I., and Raskin, I., 1995, *Biotechnology* **13**:468-474.

Schnoor, J.L., Licht, L.A., McCutcheon, S.C., Urolfe, N.L., and Carreira, L.H., 1995, *Environ. Sci. Technol.* **7**:318-323A.

Shimp, J.F., Tracy, J.E., Davis, L.C., Lee, E., Huang, W., Erickson, L.E., and Schnoor, J.L., 1995, *Crit. Rev. Envir. Sci. Technol.* **23**:41-57.

Sprent, J. I., 1984, *In Advances in Plant Physiology*, pp. 249-276, (Wilkins, M.B., ed.), Pitman, London.

Stomp, A.M., Han, K.H., Wilbert, S., Gordon, M.P., and Cunningham, S.D., 1994, *Ann. N.Y. Acad. Sci.* **721**:481-491.

Ta, T.C., and Faris, M.A., 1987, *Agron J.* **79**:820-824.

Tepfer, D., Goldmann, A., Pamboukdijian, N., Maille, M., Lepingle, A., Chevalier, D., Denarie, J., and Rosenberg, C., 1988, *J. Bacteriol.* **170**:1153-1157.

Thomas, R.J., 1995, *Plant Soil* **174**:103-118.

Vance, C.P., 1996, *In Plant Roots: The Hidden Half*, pp. 723-756, (Waisel, Y., et al., eds.), Marcel Dekker, Inc., New York.

Vance, C.P., and Graham, P.H., 1995, *In Nitrogen Fixation: Fundamentals and Applications*, pp. 77-86, (Tikhonovich, I.A., et al., eds.), Kluwer Academic Publishers, Dordrecht.

Waggoner, P.E., 1994, *Council for Agricultural Science and Technology Task Force Report 121*, Ames, Iowa, pp. 64.

Waisel, Y., Eschel, A., and Kafkafi, U., 1996, *Plant Roots: The Hidden Half*, Marcel Dekker, Inc., New York, pp. 1002.

Walton, B.T., and Anderson, T.A., 1990, *Appl. Environ. Microbiol.* **56**:1012-1016.

Wani, S.P., Rupela, O.P., and Lee, K.K., 1995, *Plant Soil* **174**:29-50.

West, C.P., and Mallarino, A.P., 1996, *In Nutrient Cycling in Forage Systems*, pp. 167-175, (Joost, R.E., and Roberts, C.A., eds.), Potash and Phosphate Institute and Foundation for Agronomic Research, Manhattan, Kansas.

TOWARDS THE APPLICATION OF NITROGEN FIXATION RESEARCH TO FORESTRY AND AGRICULTURE.

E. Martínez-Romero[1], U. Oswald-Spring[2], M. Miranda[2], J. L. García[2], L. E. Fuentes-Ramírez[1], L. López-Reyes[3], P. Estrada[1], and J. Caballero-Mellado[1].
[1]Centro de Investigación sobre Fijación de Nitrógeno, UNAM; [2]Secretaría de Desarrollo Ambiental; [3]Universidad Autónoma de Puebla, México.

Low-input, non-irrigated agriculture with low yields is predominant in Mexico, with maize and bean as the basic crops. Land for agriculture is very often taken away from forest areas and deforestation rates in Latin-America (due also to timbering, fires and urban use of forest land) are among the highest in the world. 5.3 million forest hectares were lost from 1984 to 1992 in Mexico. It is estimated that at this rate of deforestation, no forests will exist in Mexico by the year 2035. For this reason, we decided to participate in a reforestation program in collaboration with the Ministry of Environment (Secretaría de Desarrollo Ambiental, Morelos).

We proposed the use of native *Leucaena leucocephala*, a fast growing tree (2, 10) with high rates of nitrogen fixation. We germinated 250,000 seeds in our laboratory. This represented 1/8 of the total trees planted for reforestation in Morelos that year. We inoculated the trees with *Rhizobium tropici* (10^5 bacteria/plant) and in some cases plants were also inoculated with *Azospirillum* in hopes of promoting root growth and enhanced plant development. Plants were maintained in greenhouses or in nurseries until they were about 3-5 cm in height. Early plant development was completely dependent upon *Rhizobium* inoculation. Plantlets were transplanted at a low density (1 plant/25 m^2) in eroded areas south of Cuernavaca. These trees had a survival rate of around 80% during the first year but a minimal survival rate after 3 years. For this we have now proposed a different strategy using native leguminous trees. This would have the additional advantage to preserve tree biodiversity. Towards this long-term goal we have started isolating rhizobia from diverse

NATO ASI Series, Vol. G 39
Biological Fixation of Nitrogen
for Ecology and Sustainable Agriculture
Edited by A. Legocki, H. Bothe, A. Pühler
© Springer-Verlag Berlin Heidelberg 1997

leguminous trees. This year we have prepared additional 20,000 Leucaena plantlets inoculated with *Rhizobium* that will be provided to farmers to be used for feeding cattle.

We have also focused our research interest on sugar cane, which is an important crop in Mexico and in 70 other countries, and it is one of the crops with the highest fertilization rates in Mexico (120-320 Kg N/ha) and therefore with high costs.

In contrast to Mexico, in Brazil good sugar cane-yields are obtained with very low amounts of nitrogen (N) fertilizers (12). Different nitrogen-fixing bacteria were isolated from inside sugar cane stems and roots there (1,5). Similarly, Jesús Caballero and his former research team in Puebla succeeded in isolating *Acetobacter diazotrophicus* from sugar cane in Mexico (6). Their isolation frequency was inversely related to levels of crop-fertilization. *A. diazotrophicus* was in low concentration or absent from plants recovered from highly fertilized areas (6,9). Moreover, when the genetic diversity of these populations was analyzed using the methodology established in our laboratory for the analysis of *Rhizobium* populations, it was observed that Mexican populations of *A. diazotrophicus* were significantly less diverse than those from Brazil (3,4). We suppose that these differences might be related to the differences in N-fertilization levels which have been reported to change plant metabolic conditions and even sugar concentrations inside the plant (11). We are presently testing under field conditions the effect of applied N-fertilizer on *A. diazotrophicus* survival and colonization of sugar cane plants and we are also testing if the *A. diazotrophicus* from Mexico are better adapted for plants grown with N-fertilization.

Besides *A. diazotrophicus*, other nitrogen fixers have been identified from sugar cane such as *Herbaspirillum seropedicae* and *Burkholderia cepacia*. We have isolated *Klebsiella* spp. and *Azospirillum* from both inside stems and outside sugar cane plants (rhizosphere). These isolates were analyzed by multilocus enzyme electrophoresis. We found that the endophytic capacity is well distributed over all the different electrophoretic clusters of bacteria. It is worth to pursue the analysis of the genetic differences that determine that one bacteria remains in the rhizosphere while its close relative is capable to penetrate and colonize the sugar cane plant.

It is unknown at present which of all bacteria inside sugar cane contribute to the nitrogen economy of the plant. To test if *A. diazotrophicus nif* genes are expressed *in planta*, an *A. diazotrophicus nif H*-gene promoter-gus A fusion was constructed and introduced into *A. diazotrophicus*. Presence of the bacteria will be monitored by an *A. diazotrophicus* with a constitutively expressed *gusA* gene. With so many nitrogen-fixers inside sugar cane, we suppose that several different bacteria are responsible for fixing nitrogen.

We have isolated *A. diazotrophicus* from mealy bugs associated to sugar cane (3) and recently from coffee plants as well (8). This suggests that nitrogen fixing endophytic bacteria may be more prevalent in plants than was originally expected but that N-fertilization has deleterious effects on these populations. It seems that the highest plant-microbial diversity would be associated to the plants grown in areas where the latter originated and diversified. With this rational, we started to search for endophytic nitrogen fixers of corn (*Zea maiz*) which is native to Mexico. We have isolated different nitrogen fixers from maize recovered from a traditional maize-growing, remote area in the mountains of Oaxaca State that is not N-fertilized and that has presumably high yields (2 ton/ha) (7). In this region, sustainable agriculture has been successfully practiced. The 16S ribosomal sequences from these isolates will be determined to clarify their taxonomic position and their effects on plant will be evaluated. This provides a broad scope of the possibilities towards the application of nitrogen fixation to forestry and agriculture in developing countries.

Acknowledgements:

To Leticia López, Rocio Bustillos, Marco A. Rogel, Laura Candiani, Julio Martínez, Ricardo Méndez and Julian Cabrera (INIFAP) for their technical support, to Michael Dunn for reviewing the manuscript. Research on *A. diazotrophicus* has been partially funded by DGAPA grant IN 209496 (UNAM).

References:
1. Baldani, V. L. D., Baldani, J. I., Olivares, F., and Döbereiner, J. Identification and ecology of *Herbaspirillum seropedicae* and the closely related *Pseudomonas rubrisubalbicans*, *Symbiosis*, 13, 65, 1992.
2. Benge, M. D. *Leucaena* 1: a tree for all purposes, *World Crops*, 1,38, 1981.
3. Caballero-Mellado, J., Fuentes-Ramírez, L. E., Reis, V. M., and Martínez-Romero, E., Genetic structure of *Acetobacter diazotrophicus* populations and identification of a new genetically distant group, *Appl. Environ. Microbiol.*, 61,3008,1995.
4. Caballero-Mellado, J. and Martínez-Romero, E., Limited genetic diversity in the endophytic sugarcane bacterium *Acetobacter diazotrophicus*, *Appl. Environ. Microbiol.*, 60, 1532, 1994.
5. Cavalcante, V. A., and Döbereiner, J. A new acid-tolerant nitrogen-fixing bacterium associated with sugarcane, *Plant Soil*, 108, 23, 1988.
6. Fuentes-Ramírez, L. E., Jiménez-Salgado, T., Abarca-Ocampo, I. R., and Caballero-Mellado, J., *Acetobacter diazotrophicus*, an indoleacetic acid producing bacterium isolated from sugarcane cultivars of Mexico, *Plant Soil*, 154, 145, 1993.
7. Hallberg, T. B., Nitrogen-fixing bacteria associated with maizes native of Oaxaca, in *Nitrogen Fixation: Fundamentals and Applications.*, Tikhonovich, I. A., et al., Eds., Kluwer, the Netherlands, p. 760, 1995.
8. Jiménez-Salgado, T., Fuentes-Ramírez, L. E., Tapia-Hernández, A., Mascarúa-Esparza, M. A., Martínez-Romero, E. and Caballero-Mellado, J. *Acetobacter* endophytic bacteria from *Coffea arabica*, in *International symposium on sustainable agriculture for the tropics. The role of biological nitrogen fixation*, Angra dos Reis-Rio de Janeiro, Brazil, p. 212, 1995.
9. Muthukumarasany, R. Endophytic nitrogen-fixing bacteria from sugarcane in India, in *International symposium on sustainable agriculture for the tropics. The role of biological nitrogen fixation*, Angra dos Reis-Rio de Janeiro, Brazil, p. 216, 1995.
10. National Academy of Sciences, *Tropical Legumes: Resources for the Future*, Washington, D. C., 1979.
11. Pelaez Abellan, I., de Armas Urguiza, R., Valadier, M. H., Campigny, M. L., Short-term effect of nitrate on carbon metabolism of two sugar cane cultivars differing in their biomass production. Phytochem., 36, 819, 1994.
12. Ruschel, A. P., and Vose, P. B., Biological nitrogen fixation in sugar cane, in *Current Development in Biological Nitrogen Fixation*. Subba Rao, N. S. Ed., Edward Arnold (Publishers) Ltd., London, p.219-235, 1984.

Genetic potential of plants for improving the beneficial microbe interactions

Tikhonovich I.A., Kozhemiakov A.P., Provorov N.A. and Kravchenko L.V.

All-Russia Research Institute for Agricultural Microbiology, Sh. Podbelsky 3,
St.-Petersburg, Pushkin 8, 189620, Russia

Abstract. The legume and non-legume crops possess a broad genetic polymorphism for the efficiency of interactions with symbiotic and associative microorganisms. The wild-growing genotypes are characterized by a highest symbiotic potential which should be used in breeding the plants for an improved plant-microbe interactions. New root-associated *Flavobacterium* and *Arthrobacter* strains were isolated which can improve the yield of non-legumes increasing the N_2 fixation, uptake of N and P compounds from soils, protecting the plants from pathogens and environmental stresses. A coordinated selection of plants and microorganisms is suggested to be an important approach for establishing the sustainable agricultural systems.

Keywords. Legumes, rhizobia, root-associated diazotrophs, resistance to pathogens, N_2 fixation, coordinated selection, sustainable agriculture

The beneficial interactions between plants and microbes are of a special importance for the sustainable agriculture. The results of cooperation are genetically controlled by both partners. The role of the plant genotype in determination of efficiency of interaction has been studied extensively for the legume-rhizobia symbiosis and different approaches for breeding the legumes for an improved symbiotic N_2 fixation have been suggested (Phillips, Teuber, 1992). However, for other practically important plant-microbial systems (associations of cereals and root diazotrophs, mycorrhizae, etc.) there are almost no data on the role of plant genes in controlling the mechanisms and an efficiency of interaction. In this paper we shall consider the possibilities for organizing the breeding of cultured plants towards the effective cooperation with different beneficial bacteria.

1. Legume-rhizobia symbiosis

The leguminous crops are characterized by a broad polymorphism for the abilities to nodulate and to fix N_2 (which are dependent on the "major genes" controlling crucial stages of the symbiosis development) and for the quantitative expression of these symbiotic traits (which is polygenically controlled). The wild-growing populations in general exceed the agronomically advanced cultivars for the symbiotic activity in *Pisum sativum*, *Medicago sativa* and *Trigonella foenum-graecum* (Provorov, Tikhonovich, 1996). Therefore, the wild-growing legume populations should be used as the sources of initial material for breeding the plants towards an improved symbiotrophic nitrogen nutrition.

Both major genes and polygenes may be used for improving the symbiotic performance in pea which is one of the most low efficient legume symbionts. For this purpose two breeding strategies have been explored at ARIAM. One of them is based on the spontaneous mutant nodulation restricting allele *sym2* revealed in the

NATO ASI Series, Vol. G 39
Biological Fixation of Nitrogen
for Ecology and Sustainable Agriculture
Edited by A. Legocki, H. Bothe, A. Pühler
© Springer-Verlag Berlin Heidelberg 1997

Afghan peas. This allele permits inoculation of pea with the strains which possess the *nodX* gene only (Firmin et al., 1993). The *Rhizobium leguminosarum* bv. *viceae* A1 strain and its derivatives have been used (Tchetkova, Tikhonovich, 1986) because these strains are highly effective and are not subjected to the competitive nodulation blocking from "European" strains as it was revealed in the previously studied strain TOM. Recessive *sym2* allele has been introduced in an advanced pea cv. Nord (in collaboration with Institute for Grain Legumes and Groats Crops, Orol, Russia). As a result we improved the selectivity of inoculation: *sym2sym2* isolines of cv. Nord have an improved symbiotic efficiency as compared with a parent cultivar (Table 1).

Table 1. Improvement of *Pisum sativum* L. symbiotic efficiency after introduction of the *sym2* allele

plant genotypes	increase after A1 inoculation (%) over control		C_2H_2-reductase activity (mkM C_2H_4/plant/hour)	
	shoot mass	total N in shoots	control	A1
Nord	+70.8	+ 7.3	19.8	27.6
Nord (*sym2sym2*)	+84.4	+45.1	0	24.9
L2150 (Afghan)	+129.0	+21.5	0	20.5

The polygenes which control the efficiency of nitrogen fixation and assimilation may be also involved in the breeding process, e.g., by selection of genotypes with high acetylene reduction activity (ARA). Using ARA we selected a number of pea lines which exceed in this trait the initial cultivars by 30-538% (Tikhonovich et al., 1987). A marked tendency for decreasing the mass of the selected lines grown on nitrate was revealed. Therefore breeding plants for an increased N_2 fixation may improve a balance between the symbiotrophic and combined N nutrition. However, the legume breeding should be aimed at reaching not a maximum possible level of nitrogen fixation, but an optimal one for a defined genotype.

Expression of the legume symbiotic polygenes may depend greatly on the genotypes of rhizobia strains. A two-factor analysis of variance demonstrated that the influence of strain-specific polygenes on the plant growth is most strong for the highly effective symbioses as compared with the low effective ones (Provorov, Tikhonovich, 1996). Thus, for a sufficient improvement of the legume symbiotic efficiency a coordinated selection of plant and bacteria is required to establish the complementary combinations of the partners' genotypes.

2. Associations of non-legumes with root diazotrophs

For non-legumes the application of root-associated bacteria becomes highly urgent. Nowadays a lot of new strains (*Pseudomonas, Agrobacterium, Azospirillum, Acetobacter, Flavobacterium, Arthrobacter*) are available which improve many agronomically important traits of cereals, vegetables, oil producers and other crops. We demonstrated that the associative diazotrophs increase the N_2 fixation on cereal roots by 5-20 times as compared with the uninoculated control (Kozhemiakov et al., 1995) and up to 20% of the total nitrogen of the plants may be fixed biologically. Inoculation also increases the uptake of combined N and P from soils. Therefore, the root-associated bacteria can minimize the application of mineral fertilizers thus decreasing the ecological risk of crop production.

The important property of many root-associated bacteria is the suppression of soil-born diseases. The most effective are the *Arthrobacter* and *Flavobacterium* strains which can protect a wide spectrum of non-legume plants against different pathogens (*Fusarium, Phytophtora, Phoma, Rhizoctonia*). E.g., the bioinocula usually increase the resistance of potatoes to *Rhizoctonia solani* by 2-10 times. From the culture medium of the two most effective strains the antimicrobial peptides have been isolated (Tchebotar et al., 1995). Interestingly, the efficiency of inoculation with these bacteria was highest under the drought conditions. Therefore, the root-associated microflora may protect plants from some environmental stresses.

The results of the non-legume inoculation depend greatly on the plant genotype. We screened a broad collection of wheat genotypes for the susceptibility to inoculation with *Azospirillum, Arthrobacter, Flavobacterium*. About a half of genotypes were capable of increasing significantly the grain yield (by 20-30%), other genotypes did not respond to the introduced bacteria and in some genotypes even a reduction of yield was observed. The similar results were obtained also for sunflowers and tomatoes.

The observed variability may be correlated to some physiological plant traits, e.g., to the composition of root exudates (Kravchenko et al., 1993). We demonstrated that in the wild-growing diploid wheat genotypes the total amount of carbon excreted from the roots is lower than in the agronomically advanced hexaploid cultivars. However, the colonization of root surfaces by *Azospirillum* (per a unit of excreted carbon) is highest in diploids. These data suggest a high potential for establishing the effective associations in the wild-growing non-legume genotypes as it was demonstrated in the legumes. From the results of numerous experiments which have been carried out in the Experimental Network at ARIAM we conclude that application of the conventional breeding methods lead to a decrease in plant symbiotic potential (Fig. 1) possibly due to intensive application of N fertilizers during the breeding process.

The other important similarity between the symbiotic and associative systems concerns the role of strain-specific plant polygenes in controlling the efficiency of the plant-microbe interactions. We demonstrated that these polygenes control up to 47% of variation of the seed mass in sunflower cultivars inoculated with different *Flavobacterium* strains. Therefore, a coordinated plant-bacteria selection will be important for improving the efficiency of microbial interactions in non-legumes.

References

Firmin J.L. et al., 1993. Molec. Microbiol., 10: 351-360.
Kozhemiakov A.P. et al., 1995. In: Nitrogen Fixation: Fundamentals and Applications. Tikhonovich I.A. et al. (Eds). Kluwer Acad. Publ., p. 766.
Kravchenko L.V. et al., 1993. Mikrobiologia (in Russian). 62: 863-868.
Phillips D.A., Teuber L.R., 1992. In: Biological Nitrogen Fixation. Stacey G. et al. (Eds). Chapman and Hall, p. 625-647.

Provorov N.A., Tikhonovich I.A., 1996. Proc. 20-th EUCARPIA Meeting, Poland, October 7-10, 1996 (in press).
Tchebotar V.K. et al. 1995. In: Nitrogen Fixation: Fundamentals and Applications. Tikhonovich I.A. et al. (Eds). Kluwer Acad. Publ. p. 778.
Tikhonovich I.A. et al. 1987. Agricultural Biology (in Russian). 2: 29-34.
Tchetkova S.A., Tikhonovich I.A., 1986. Mikrobiologia (in Russian). 55:143-147.

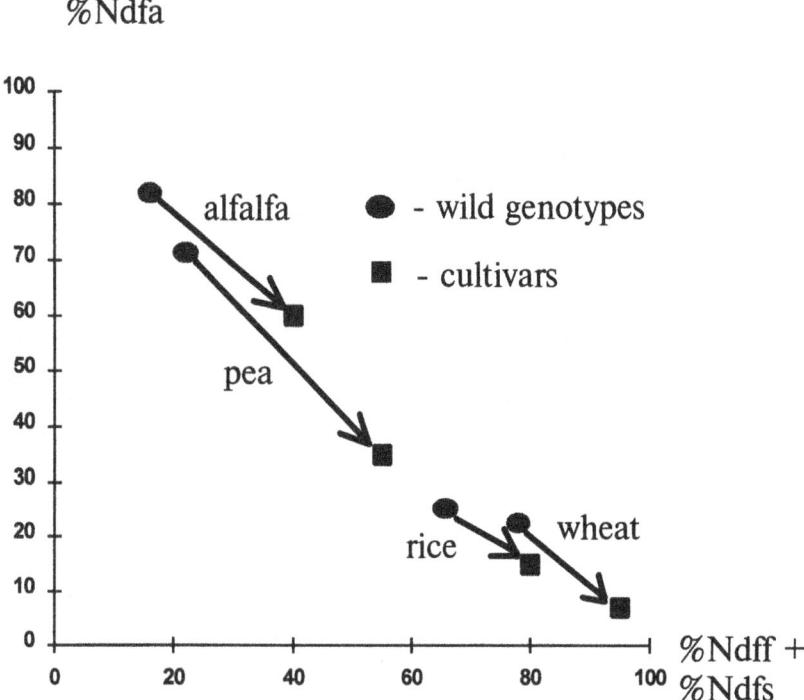

Fig. 1. Effect of breeding on the balance between symbiotrophic
and combined N nutrition in plants

DNA-probing as a tool to monitor the distribution of N₂-fixing, denitrifying and nitrifying bacteria in soils

H. Bothe, K. Kloos, K. Kaiser, U. Hüsgen, S. Sonnwald, B. Schmitz

Botanisches Institut, Universität zu Köln, Gyrhofstr. 15,
D-50923 Köln, Germany

Microorganisms and plants are believed to significantly contribute to the annual changes in the global N-cycle, but their relative contributions in N₂-fixation, denitrification and nitrification are poorly understood as yet. Molecular biological techniques have rarely been applied to identify the percentage of bacteria with such traits in the N-cycle in water or soil ecosystems (e. g. 1,2,3). Recently, this laboratory developed gene probes for all steps of denitrification (with the exception of NO-reduction), nitrification (*amoA* gene) and N₂-fixation (4). To obtain such probes, published sequences of the genes were screened for conserved regions by computer analysis, and oligonucleotide primers were developed of such regions which allowed to amplify gene segments of 400 - 850 bp by PCR. The amplicates obtained were cloned and sequenced and verified by sequence comparison. The following probes were used:

a) denitrification:

nitrate reductase: a 414 bp segment of the *narG* gene coding for one subunit of the enzyme from *Escherichia coli* (5).

nitrite reductase (cytochrome c,d containing): a 596 bp segment of *nirS* coding for the apoprotein of the enzyme from *Pseudomonas aeruginosa* DSM 6195, showing 100% homology to the amino acid sequence published for the protein from *Pseudomonas aeruginosa* NCTC 6750 (6).

nitrite reductase (Cu containing): a 576 bp segment of *nirK* coding for the enzyme from *Alcaligenes xylosoxidans* NCIMB 11015 showing 67 % sequence homology on the amino acid level to the enzyme from *Alcaligenes faecalis* S-6 (7).

nitrous oxide reductase: a 598 bp segment of *nosZ* coding for the enzyme from *Pseudomonas stutzeri* ZoBell, showing 91% homology to the published sequence of the protein from the same organism (8).

b) nitrification:

an 820 bp segment of the *amoA* gene coding for one of the subunits of ammonium monooxygenase from *Nitrosomonas europaea*, that shows 98 % homology on the amino acid level to the published sequence from the same organism (9).

c) nitrogen fixation:

a 450 bp segment of the *nifH* gene (coding for the nitrogenase reductase) of *Azospirillum brasilense* Sp7, showing 100 % homology to the published sequence of the gene from the same organism (4).

d) an universal probe 5´- GCTGCCTCCCGTAGGAGT-3´ of the 16S-rRNA, which is used to control the release of DNA from the bacterium to be investigated.

NATO ASI Series, Vol. G 39
Biological Fixation of Nitrogen
for Ecology and Sustainable Agriculture
Edited by A. Legocki, H. Bothe, A. Pühler
© Springer-Verlag Berlin Heidelberg 1997

These probes were used to screen for the occurrence of such genes in a lot of bacteria of diverse systematic groups (α,β,γ subdivison of proteobacteria, Gram-positive bacteria with either a high or a low GC content, cyanobacteria). On the whole, the results from the DNA-DNA hybridizations with these probes coincided with the known physiological traits (10). The probes *nirK, nosZ* and *nifH* recognized the genes practically from all organisms. The *nirS* probe (gene of cytochrome c,d containing NO_2^- -reductase) from *Pseudomonas aeruginosa* gave distinct hybridization bands with DNA from only a restricted number of bacteria possessing this gene. In the case of *Azospirillum*, e. g., the probe hybridized only with DNA from *A. brasilense* Sp7 and Cd, but not that from other isolates. The *narG* probe hybridized with DNA from a lot of systematically unrelated bacteria, but unexpectedly gave negative results in the case of some bacteria known to perform dissimilatory NO_3^--reduction. The informations about the capability of the *amoA* probe are limited as yet.

These probes were used to screen for the occurrence of genes coding for denitrification and N_2-fixation in *Hyphomicrobium* isolates (4). Bacteria of this genus abundantly occur in lakes, but can be grown only slowly under laboratory culture conditions. Some isolates could be cultivated in quantities sufficient for DNA-hybridization and activity measurements. The data showed that some *Hyphomicrobium* isolates possess the *nifH* gene and therefore likely can perform N_2-fixation, and the DNA of almost half of the isolates hydrizied with at least two different probes for the denitrification steps. Denitrifying hyphomicrobia preferentially occurred in the influent, activated sludge and effluent of the sewage disposal plants of the city of Plön in Northern Germany, and the limited body of information indicated that N_2-fixing hyphomicrobia may preferentially live in N-poorer water. Thus, it could be shown by means of this DNA probing that hyphomicrobia strains apparently occupy different ecological niches in waters (4).

The gene probes have now been used to monitor the distribution of bacteria in different soils. Two different sets of experiments were performed:

i) characterization of the microorganism flora in four different soils in the Cologne area a) acid forest soil, pH 4.3, Chorbusch near Pulheim, b) neutral, nutrient rich soil close to the Rhine river (Worringer Bruch) c) and d) soil of chalk meadows in the Eifel mountains, Biesberg near Muldenau and Hühlesberg near Iversberg, pH ~ 8.

ii) characterization of the fluctuations in the bacterial populations during the vegetation period in an acid Norway spruce forest near Villingen/Black forest, Southern Germany.

For the investigations, soil samples were taken from each site, in all cases (whenever possible) from the three layers < 5 cm, 10 cm and 25 cm. The bacteria were extracted from the soils by stirring with water, and the bacteria separated from the soil particles were diluted, plated on media in agar, isolated on plates, used by replica plating or after growth in liquid media for colony hybridizations. Details of this protocol and also of the data are in the course of being published, and an example of a typical pattern is given in Fig. 1. The major results can be summarized here as follows:

(1) The absolute number of bacteria is always highest in the upper layer (< 5 cm) and decreases with the depth of the soil.

(2) The number of denitrifying and N_2-fixing amounts to ~ 5 % and never exceeds 20 % of the total bacterial population.

absolute number of isolates

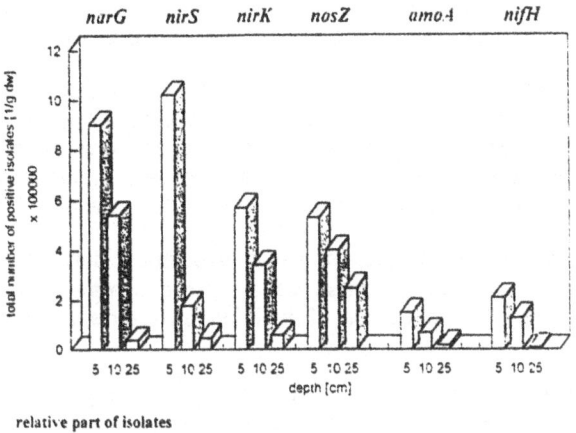

relative part of isolates

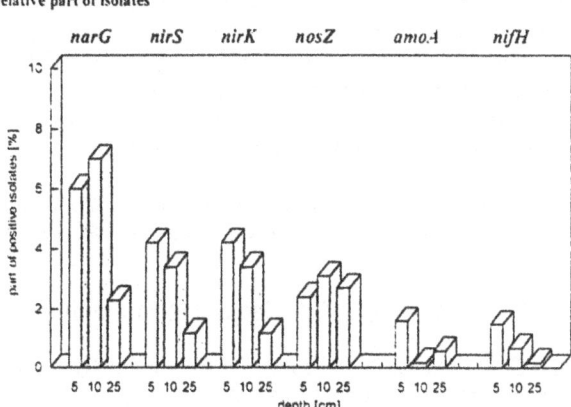

Fig. 1. Distribution of denitrifying, nitrifying, N$_2$-fixing bacteria in an acid soil near Cologne (Chorbusch). The Bacteria were isolated from soil samples of 5, 10 and 25 cm depth, and the number of denitrifying, nitrifying and N$_2$-fixing bacteria was determined by DNA-probing (see text).

(3) The number of denitrifying and N$_2$-fixing bacteria is highest in the upper layer (< 5 cm). This is consistently observed with the absolute number. In addition, the percentage of denitrifying bacteria relative to the total amount of bacteria is very often increased in the upper layer. This is somewhat surprising, since the concentration of nitrate is only 1/3 lower in 25 cm depth than in 5 cm. Nitrate apparently does not determine the distribution of either denitrifying and nitrifying bacteria in these soil layers.

(4) The gene probe amoA (for ammonium monooxygenase) hybridizes with DNA from only few bacteria. Autotrophic nitrifying bacteria seem to play at best a minor role in such soils. The situation may be different with heterotrophic nitrifiers (11).

(5) The absolute number of bacteria is always considerably higher in the vicinity of roots than in the bulk, plant free soil, probably due to the better supply with organic carbon from the roots. A relative enrichment of denitrifying bacteria in the vicinity of plant roots is often, but not always observed in the different soils.

(6) In parallel measurements (testing two samples from the same location), the data vary by a factor of 2 to 3 due to the complexity of soils. Despite of this, the experiments with the gene probes give reliable data about the distribution of bacteria in soils.

(7) The total number of bacteria and the number of denitrifying and N_2-fixing bacteria can drastically vary within a year. In the spruce forests trial of Villingen, the number was highest in spring, and next in autumn, whereas the values were low in summer. It remains to be elucidated whether such fluctuations represent annual rhythms, are consequences of longer drought in summer or are due to other, unresolved factors.

(8) Soil samples were assayed in the lab. for maximal potential denitrification (N_2O-formation $\pm$ C_2H_2) activities. Although these assays were somewhat artificial, the results also indicated the highest population of denitrifying bacteria in the upper layer (< 5 cm) of soils.

In situ hybridizations have not yet been performed due to the complexity of soils. For the DNA-probing, the bacteria had always to be enriched to get enough cells for DNA-extractions. Thus all the data refer to those soil bacteria which can be enriched from soils. Nevertheless, the data obtained from the DNA-probing experiments matched with those from activity measurements. All the results obtained thus far indicate that the upper layer (probably partial anaerobic microsites) and the vicinity of roots are the areas which govern the occurrence of N_2-fixing and denitrifying bacteria.

Acknowledgement. This work was kindly supported by a grant from the Deutsche Forschungsgemeinschaft within the Schwerpunktprogramm "Mikrobielle Ökologie"

References:
(1) Holben, W.E., Jansson, J.,K., Chelm, B.K. & Tiedje, J.M. (1988) Appl. Environm. Microbiol. 54,703-711
(2) Norris, P.R., Murrel, J.C. & Hinson, D. (1995) Arch. Microbiol. 164,294-300
(3) Linne von Berg, K.-H. & Bothe, H. (1992) FEMS Microbiol. Ecol. 86,331-340 (1992)
(4) Kloos, K., Fesefeldt, A., Gliesche, C.G. & Bothe, H. (1995) FEMS Microbiol. Ecol. 18,205-213
(5) McPherson, M.J., Baron, A.J., Pappin, D.J.C. & Wootton, J.C. (1984) FEBS Lett. 177,260-264
(6) Silvestrini, M.C., Galeotti, C.L., Gervais, M., Schinina, E., Barra, D., Bossa, F. & Brunori, M. (1989) FEBS Lett. 254,33-38
(7) Masuko, M., Iwasaki, H., Sakurao, T., Suzuki, S., Nakahara, A. (1984) J. Biochem. 96,447-453
(8) Viebrock, A., Zumft, W.G. (1988) J. Bacteriol. 170,4658-4668
(9) McTavish, H., Fuchs, J.A., & Hooper, A.B. (1993) J. Bacteriol. 175,2436-2444
(10) Kloos, K. (1996) thesis, The University of Cologne, Germany
(11) Papen, H., von Berg, R., Hinkle, I., Thoene, B. & Rennenberg, H. (1989) Appl. Environm. Microbiol. 55,2068-2072

Part 8

Carbon-Nitrogen Metabolism
in Symbiotic Systems

Carbon-Nitrogen Metabolism in Symbiotic Systems: Integration and Overall Regulation

Nicholas J. Brewin and Preeti Dahiya

Department of Genetics, John Innes Centre, Norwich NR4 7UH, Great Britain

Keywords. Carbon, nitrogen, bacteroids, symbiosomes, metabolism

1 Biological Importance of Legume Nodules

Symbiotic nitrogen fixation is important in the development of sustainable agriculture because it reduces the need for application of fertiliser nitrogen to crops. As a result, both the economic and environmental costs of agriculture are reduced. A heightened awareness of the mechanism by which available nitrogen controls legume nodule activity is a key element in designing strategies to enhance the productivity of legume crops by genetic engineering.

2 The Nodule as an Organ for Carbon-Nitrogen Exchange

The symbiotic legume root nodule is, in effect, a chemical engine converting carbon compounds derived from photosynthesis into nitrogenous compounds that are made available for plant growth. The overall function of legume nodules requires very large amounts of reducing equivalents which are delivered to the nodule in the form of sucrose, the product of photosynthesis in the aerial parts of the plant. The energy costs for biological nitrogen fixation have been summarised by Streeter (1991). For soybean (*Glycine max*), there is a requirement for about 12.2g of carbohydrate for each gram of nitrogen fixed. This includes a direct "cost" of 7.3g for the biochemical operation of nitrogenase; 1.9g for the assimilation of ammonia and the export of fixed nitrogen; and 0.26g for growth. Translated into a field crop, this means that a hectare of soybeans capable of fixing approximately 75kg of nitrogen during the growing season would consume about 1.5 tonnes of sucrose in the process. Similar estimates of the energy cost of nitrogen fixation have been derived for other legumes, both grain legumes and forage crops.

In nitrogen fixing plants, nodule growth and activity is consistently matched to whole plant requirements for nitrogen. This has led to the concept of an "N-stat", but the underlying mechanisms for this regulatory system have not been elucidated. The sensing of plant nitrogen may involve the translocation of N in the phloem sap into nodules and it has been suggested (Parsons et al., 1993) that the concentration of certain key amino compounds may signal plant N status to the nodule.

NATO ASI Series, Vol. G 39
Biological Fixation of Nitrogen
for Ecology and Sustainable Agriculture
Edited by A. Legocki, H. Bothe, A. Pühler
© Springer-Verlag Berlin Heidelberg 1997

3 The Symbiosome: an Organelle for Carbon-Nitrogen Exchange

The legume nodule is formed by two interacting genomes and operates through two interacting sets of metabolic capabilities (Brewin, 1991). Superimposed on this metabolic complexity is a system of specialised tissues, cells and compartments within cells (Werner, 1992). Of particular importance to the functioning of the nitrogen-fixing symbiosis in host nodule cells is the symbiosome compartment. This organelle-like structure comprises the differentiated, bacteroid, form of *Rhizobium*, enclosed by a plant-derived membrane, the peribacteroid membrane (Verma & Hong, 1996). Transport across the peribacteroid membrane controls carbon-nitrogen exchanges between the energy-consuming, NH_3-excreting bacteroid and the surrounding plant cytoplasm which assimilates the fixed nitrogen and provides the bacteroid with substrates for metabolism and growth (Whitehead et al., 1995).

While there is abundant evidence that dicarboxylic acids, e.g. malate are transported across the peribacteroid membrane and may serve as respiratory substrates for bacteroids, the nitrogen status of bacteroids is not well understood (Whitehead et al., 1995). How do bacteroids get their nitrogen? This is an important question because more than half the nitrogen content of nodules is ascribable to endophytic bacteria. Many of these rhizobia do not "fix" nitrogen and even the nitrogen-fixing cells excrete NH_3 to host plant cells without assimilating it for their own growth. Because there is strong evidence demonstrating that the peribacteroid membrane lacks transport systems for amino acids and sugars, it has been assumed that organic acids are the only major metabolic substrate for bacteroids. However, this hypothesis ignores the observation that glycoproteins (e.g. pea nodule lectin) are targetted by vesicle fusion into the symbiosome compartment where they might subsequently be degraded by proteases and glycosidases. (Similar arguments apply to the possible nutritional status of rhizobia in infection threads.)

We therefore propose that endophytic *Rhizobium* bacteria may utilise plant-derived protein hydrolysate as a major nutritional source. Our recent studies with nodule-specific lectin (Kardailsky et al., 1996) and nodule thiolproteases (Kardailsky & Brewin) provide preliminary evidence supporting this model for protein turnover. The model, which could have important implications for the regulation of nitrogen-fixing activity in bacteroids, is outlined in Fig. 1. A partial version of this new model was proposed by Kahn et al. (1985), who presented the issue as follows:- *"How can the plant manipulate the [symbiotic] interaction to maximise the fixed nitrogen that it receives? We [Kahn et al.] propose that, by feeding the bacteroids a nitrogen (N)-containing compound, the plant can obtain fixed nitrogen from the interaction with less risk of being exploited. We also propose that, if an N-containing compound is used to feed the bacteroid, the unusual pattern of fixing nitrogen and exporting ammonia could be part of a bacterial strategy to increase the carbon and energy supply to it by the host plant."*

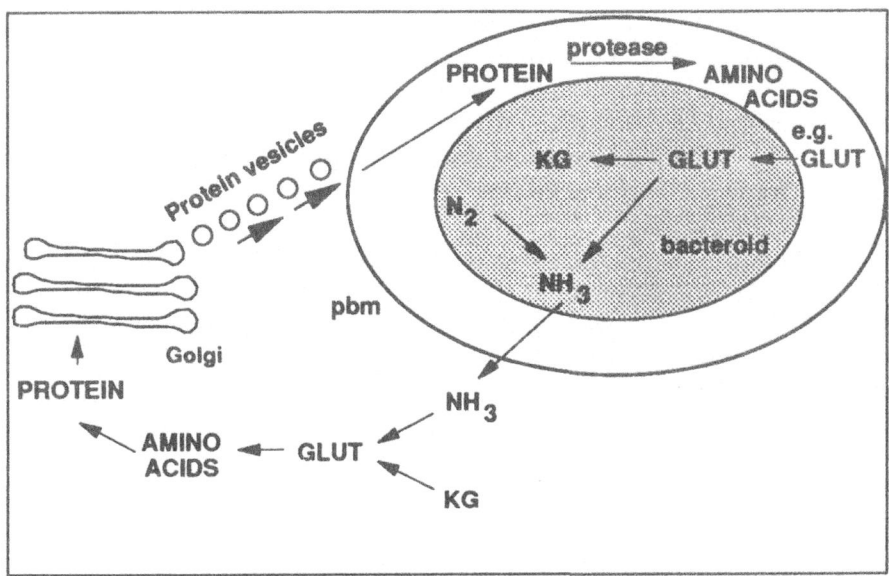

Fig. 1. Model for protein hydrolysis and the nutrition of bacteroids

Kahn then went on to propose that amino acids (e.g. glutamate) are supplied through the peribacteroid membrane as a major nutritional source for bacteroids. However, this proposal has subsequently been shown to be unsustainable because of the weight of evidence demonstrating that the peribacteroid membrane (pbm) is impermeable to amino acids and sugars, and the conventional wisdom (Whitehead et al., 1995) currently considers organic acids to be the only major carbon source for bacteroids. In our model for nutrient exchange (Fig. 1) the rate of protein synthesis, and hence of protein vesicle targetting to the symbiosome compartment, is controlled by the nitrogen status of the host plant cell. This in turn depends on the rate of biological nitrogen fixation by bacteroids within the host cell. The testable prediction of this model is that proteins targetted to the symbiosome compartment turn over very quickly, and will disappear rapidly when biological nitrogen fixation is experimentally interrupted.

4 References

Brewin, N.J. (1991) Development of the legume root nodule. Annu. Rev. Cell Biol. 7: 191-226

Kardailsky, I.V., Sherrier, D.J. and Brewin, N.J. (1996) Identification of a new pea gene, *PsNlec1*, encoding a lectin-like glycoprotein isolated from the symbiosomes of root nodules. Plant Physiol. 111: 49-60.

Kardailsky, I.V. and Brewin, N.J. (1996) Expression of cysteine protease genes in pea nodule development. MPMI 9:*In Press.*

Parsons, R., Stanforth, A., Raven, J.A., and Sprent, J.I. (1993) Nodule growth and activity may be regulated by a feedback mechanism involving phloem nitrogen. Plant Cell Environ. 16: 125-136

Streeter, J.G. (1991) Transport and metabolism of carbon and nitrogen in legume nodules. Adv. Bot. Res. 18: 129-187

Verma, D.P.S. (1996) Biogenesis of the peribacteroid membrane in root nodules. Trends in Microbiol. 364-368

Werner, D. (1992) Physiology of nitrogen-fixing legume nodules: compartments and functions. In Biological Nitrogen Fixation (Eds Stacey, G., Burris, R.H. and Evans, H.J.) Chapman & Hall, New York. Chapter 10, pp399-431

Kahn, M.L., Kraus, J. and Somerville, J.E. (1985) In Nitrogen Fixn. Res. Prog. (Ed. Evans, H.J.) Njhoff, 193-199.

Whitehead, L.F. Tyerman, S.D., Salom, C.L. and Day, D.A. (1995) Transport of fixed nitrogen across symbiotic membranes of legume nodules. Symbiosis 19, 141-154.

ASSIMILATION OF REDUCED NITROGEN IN TROPICAL LEGUME NODULES: REGULATION OF *DE NOVO* PURINE BIOSYNTHESIS AND PEROXISOME PROLIFERATION

D.P.S. Verma, J. H. Kim and T. Wu

Department of Molecular Genetics and Plant Biotechnology Center, The Ohio State University, Columbus, OH 43210 USA

Abstract. The symbiotically reduced nitrogen is assimilated in the host cytoplasm with the help of cytosolic glutamine synthetase. We have shown that this enzyme is induced directly in response to ammonia. The glutamine formed in the cytoplasm is funneled to the plastids where it is converted to purines. The latter are synthesized *via* *de novo* biosynthesis pathway. The purines are then oxidized into ureides inside the uninfected cells of root nodules. We have demonstrated that the *de novo* purine biosynthesis pathway is induced prior to the onset of nitrogen fixation, apparently due to high demand for purines for DNA endoreduplication in the infected cells. We have also demonstrated that glutamine enhances the expression of PRAT gene, encoding the first enzyme of the pathway. Because the level of uricase increases prior to that of PRAT, it suggests that purine catabolism may begin prior to the increase in *de novo* purine biosynthesis and the commencement of nitrogen fixation. The access of purines, exported to the uninfected cells, apparently induce peroxisome proliferation for the oxidation of uric acid to ureides. We have demonstrated that purines are able to induce peroxisome proliferation in yeast. Using urate as a source of nitrogen, we have been able to isolate several yeast mutants defective in peroxisome proliferation/assembly. We have complemented one of these mutants with a soybean gene encoding a putative peroxisome proliferation transcriptional factor with similarity to G-box binding proteins. In addition, we have determined two independent pathways for the conversion of uric acid to ureides in root nodules and have isolated corresponding genes involved in these pathways.

1. Introduction

Tropical legumes (e.g. soybean) transport reduced nitrogen, particularly symbiotically-fixed nitrogen, to the shoot as ureides (allantoin and allantoic acid). Ureides are the most efficient form of nitrogen transport compounds because they contain a higher molar ratio of nitrogen per carbon atom as compared to amides (Schubert, 1986). The production of ureides requires large quantities of purines. Accordingly, the rate of *de novo* purine biosynthesis in root nodules is increased. It has long been observed that the enzymes of the *de novo* purine biosynthesis pathway are induced with the onset of N_2-fixation (Reynolds *et al.*, 1982). However, our recent observations (Kim *et al.*, 1995) suggested that this pathway may be activated in young developing nodules due to a high demand for purines for DNA endoreduplication in the infected cells, the level of which reaches up to 64N (Mitchell, 1965). With the commencement of nitrogen fixation, the DNA synthesis stops while the expression of the genes of purine pathway is further enhanced to meet the increasing demand of purines for ureide production (Fig 1). This occurs 10-12 days after infection when nitrogen fixation commences. We have demonstrated that glutamine induces the expression of the gene encoding the first enzyme of the

NATO ASI Series, Vol. G 39
Biological Fixation of Nitrogen
for Ecology and Sustainable Agriculture
Edited by A. Legocki, H. Bothe, A. Pühler
© Springer-Verlag Berlin Heidelberg 1997

pathway, phosphoribosylpyrophosphate (PRPP) amidotransferase (PRAT; Kim *et al.*, 1995). Glutamine is the N donor for purine biosynthesis that takes place in the plastids of the infected cells (Schubert, 1986; Reynolds *et al.*, 1982). The purines are then transferred to the uninfected cells of root nodules where they are oxidized to ureides (Nguyen *et al.*, 1985). The fact that uricase is induced prior to the induction of PRAT suggests that purine oxidation begins prior to the onset of nitrogen fixation and *de novo* purine biosynthesis..

2. Isolation of purine biosynthesis genes from soybean and *Vigna*

By functional complementation of *E. coli* mutants defective in purine biosynthesis, we succeeded in isolating cDNA clones of *purF*, *purC* and *purE* homologs from a *Vigna* (mothbean) nodule cDNA expression library. They encode PRPP aminotransferase (PRAT), 5-aminoimidazole ribonucleotide (AIR) carboxylase and 5-aminoimidazole-4-*N*-succinocarboxamide ribonucleotide (SAICAR) synthetase, respectively (Chapman *et al.*, 1994; Kim *et al.*, 1995). The mothbean PurC and PurE enzymes are distinct proteins while in animals, both activities are associated with a single bifunctional polypeptide (Chapman *et al.*, 1994). A soybean *purF* homolog cDNA was isolated by PCR approach (Kim *et al.*, 1995). It encodes a peptide with an N-terminal plastid-targeting sequence, confirming the plastidial location of this enzyme (Reynolds *et al.*, 1984). Two soybean PRAT genomic clones were isolated in our lab and found to share almost 90% homology in the coding regions but differ with each other in their promoter and 5'-noncoding regions. The soybean PRAT promoter sequence was found to contain a putative Pur-Box that is present in most of the prokaryotic *pur* gene promoters. In bacteria and some insects, *pur* genes are regulated by a repressor mechanism. It is possible that such a mechanism also exists in the *pur* genes of plants.

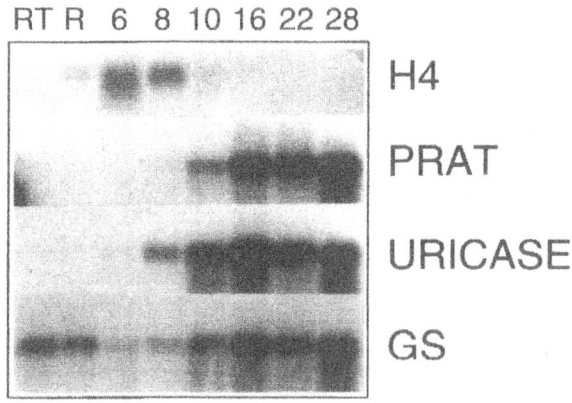

Figure 1: Expression of glutamine synthetase, PAPP-AT and uricase genes in relation to DNA synthesis and the onset of nitrogen fixation in soybean root nodules. H4, histone; PRAT, PRPP amidotransferase; GS, glutamine synthetase. Numbers are days after *Rhizobium* infection. Day 10-12 represents commencement of nitrogen fixation.

3.   Isolation of yeast peroxisome biogenesis mutants and their complementation by plant genes

We found that the catalase and uricase activities of *Saccharomyces cerevisiae* were enhanced when xanthine is used as a sole N-source. This indicated that xanthine could induce the proliferation of peroxisomes. Similarly, when uric acid is provided as a sole N-source, the peroxisome proliferation is induced, a situation likely to occur in soybean root nodules. We have recently isolated several yeast mutants that failed to grow on media containing oleic acid as the sole carbon source or uric acid as a nitrogen source. Using urate utilization ability, we were able to identify 4 groups of mutants likely defective in peroxisome assembly/biogenesis (T. Wu and Verma, unpublished data). One of these mutants (*ppb1*, for putative peroxisome biogenesis) was used for complementation using a soybean nodule cDNA library constructed in pYEUra3 yeast expression vector. A soybean cDNA (pS-PPB1) was able to restore the ability of this mutant to grow on oleic acid or uric acid-containing media. This demonstrated that it would be possible to isolate plant *pas/peb* genes by the complementation approach. Sequence analysis of pS-PPB1 indicated that it contains a basic helix-loop-helix motif near the N terminus (Fig 2) and shares sequence homology with *Phaseolus* G-box binding protein and rice transcriptional factor R gene (Purugganan and Wessler, 1994) as well as myc-like regulatory genes (PePinho *et al.*, 1987). SPPB1 may regulate the transcription of genes involved in peroxisome biogenesis. In addition, we were able to complement a known *peb* mutant (*peb4*) provided by Dr. P. Lazarow. Analysis of this gene indicated that it may encode an ion channel protein. Active ion channels are required for the assembly of peroxisomal proteins (Bellion and Goodman, 1987). Thus, we have succeeded in isolating two putative *pas/peb* genes of soybean.

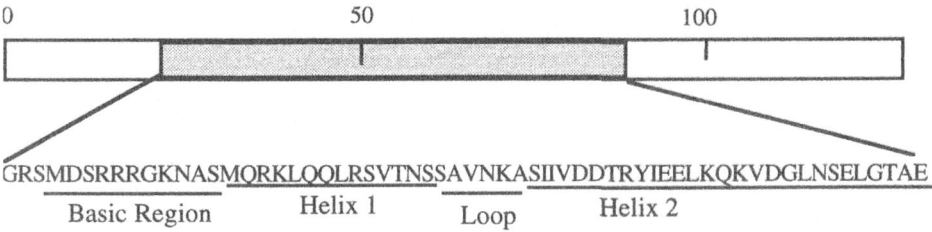

Fig. 2. Structure of the putative soybean transcription factor S-PPB1 that restores peroxisome proliferation in the yeast mutant *ppb1*. Amino acid sequence (20-82) of the bHLH region is presented.

4.   Cellular compartmentation and the control of ureide production

The compartmentation of the ureide biosynthetic pathway between different organelles and cell types in root nodules is important for maintaining metabolite channeling and levels of key intermediates and effectors (Schubert and Boland, 1990). Oxygen appears to play an important role in the inter-cellular compartmentation of the enzymes of the ureide biosynthesis pathway. For example, uricase requires oxygen and accordingly is located in the uninfected cells, whereas PRAT is oxygen sensitive and is located in the infected cells that are very low in oxygen; thus, different steps of this pathways are divided between the infected and uninfected cells of the nodules (see Schubert, 1986; Verma, 1989). Moreover, purines inhibit PRAT and uricase activities and purine catabolism is facilitated in tissues with low oxygen tension. Transcriptions of both uricase and PRAT genes are induced prior to the commencement of N_2-fixation (Nguyen *et al.*, 1985; Kim *et al.*, 1995), indicating a developmental control on the induction of these pathways. The expression of genes of these pathways is further enhanced following the onset of N_2-fixation and the reduced nitrogen is assimilated *via* this pathway in tropical legumes.

5. Catabolism of uric acid through an alternative pathway

The uric acid is generally oxidized in peroxisomes in root nodules (Equation 1), however, an urate degrading enzyme system has been found in radicals. Diamine oxidase and peroxidase are able to catalyze urate oxidation in the presence of a cofactor, cadaverin (Tajima *et al.*, 1985). Therefore, expression of one of these enzymes may confer the ability of the cells to grow on urate (Equations 2-3), but not on oleate, even in the absence of functional peroxisomes. This would allow rescue of one of the uric acid non utilizing mutant. Thus we have cloned both diamine oxidase and cytochrome P-450 cDNAs from soybean and confirmed the existance of this alternative pathway for urate oxidation in root nodules. Both of these enzymes apparently exist in peroxisomes since they contain SKL peroxisomal targeting sequence near the C terminus (Miura, 1992) but they function in cytosol as was shown for uricase (Suzuki and Verma, 1991).

$$\text{Urate} + O_2 + 2H_2O \xrightarrow{\text{Uricase}} \text{Allantoin} + H_2O_2 + CO_2 \quad \cdots (1)$$

$$\text{Urate} + O_2 + 2H_2O \xrightarrow{\text{Diamine Oxidase}} \text{Allantoin} + H_2O_2 + CO_2 \quad \cdots (2)$$

$$\text{Urate} + H_2O_2 \xrightarrow{\text{P450-Peroxidase}} \text{Allantoin} + CO_2 \quad \cdots (3)$$

These results suggest that nitrogen assimilation pathways are primarily induced in response to an increase in substrate concentration, NH_4 induces GS and glutamine induces PRAT. These pathways may be further regulated by the supply of carbon the flux of which is drastically altered as the nodule shifts to C_4 type metabolism. Thus, the nitrogen assimilation pathway in tropical legume nodules is adapted with the carbon metabolism which in turn is adapted to deal with the low level of oxygen in the nodule tissue.

6 Bibliography

Bellion, Z. and Goodman, J.M. (1987) Proton inonophores prevent assembly of peroxisomal protein. Cell 48:165-73.

Chapman, K, Delauney, A. J., Kim, J. and Verma, D.P.S. (1994) Isolation and expression of *purE* and *purC* genes of *de novo* purine biosynthesis pathway in soybean by functional complementation. Plant Mol Biol. 24: 389-395.

Gould, S. J., Keller, G.-A., Hosken, N., Wilkinson, J. and Subramani, S. (1989) A conserved tripeptide sorts proteins to peroxisomes. J Cell Biol. 108:1657-1664.

Purugganan, M. D. and Wessler S. R.(1994) Molecular evolution of the plant R regulatory gene family. Genetics 138: 849-854.

Kim, J., Delauney, A. and Verma, D. P. S. (1995) Control of *de novo* purine biosynthesis genes in ureide-producing legumes: induction of glutamine phosphoribosylpyrophosphate amidotransferase gene and characterization of its cDNA from soybean and *Vigna*. Plant J. 7: 77-86.

Mitchell, J. P. (1965) The DNA content of nuclei in pea root nodules. Annals of Bot. 29: 371-376.

Nguyen, T., Zelechowska, M., Foster, V., Bergmann, H. and Verma, D.P.S. (1985) Primary structure of the soybean nodulin-35 gene encoding uricase II localized in the peroxisomes of uninfected cells of nodules. Proc. Natl. Acad. Sci. USA 82: 5040-5044.

PePinho, R., Hatton, K., Tesfaye, A., Yancopoulos, G., and F. Alt (1987) The human myc gene family: structure and activity of L-myc and L-myc pseudogene. Genes and Development. 1, 1311-1326.

Reynolds, P.H.S., Belvins, D.G. and Randall, D.D. (1984) 5-Phosphoribosyl-pyrophosphate amidotransferase from soybean root nodules: kinetic and regulatory properties. Arch. Biochem. Biophys. 229: 623-631.

Reynolds, P.H.S., Boland, M.J., Blevins, D.G., Schubert, K.R. and Randall, D.D. (1982) Enzymes of amide and ureide biogenesis in developing soybean nodules. Plant Physiol. 69: 1334-1338.

Schubert, K., Boland, M. (1990) The ureides. In: The Biochemistry of Plants, pp. 197-283, Miflin, B. and P. Lea, eds., Academic Press, New York

Schubert, K.R. (1986) Products of biological nitrogen fixation in higher plants: synthesis, transport and metabolism. Ann. Rev. Plant Physiol. 37: 539-574.

Suzuki, H. and Verma, D. P. S. (1991) Soybean nodule-specific uricase (nodulin-35) is expressed and assembled in to functional tetrameric holoenzyme in *Escherichia coli*. Plant Physiol 95, 384-389.

Tajima, S., Kanazawa, T., Takeuch, Z., Yamamoto, Y. (1985) Characteristics of a urate-degrading diamine oxidase-peroxidase enzyme system in soybean radicals. Plant and Cell Physiol. 26, 787-795.

Verma, D.P.S. (1989) Plant genes involved in carbon and nitrogen assimilation in root nodules. In Plant nitrogen metabolism, (Poulton, J.E., Romeo, J.T. and Conn, Z.Z., eds). New York: Plenum Pub. Corp., pp. 43-63.

Contrasting C Supply, N Assimilation and N Transport Across a Range of Symbiotic Plants

Richard Parsons

Biological Sciences, University of Dundee, Dundee, DD1 4HN, UK

Keywords. Nitrogen fixation, GC-MS, metabolism, amino acid, translocation.

1. General

Nitrogen fixing symbioses can be broadly divided by the type of symbiotic bacteria involved: rhizobia, *Frankia* or cyanobacteria. With the exception of the aquatic plant *Azolla* (and the specialist adaptation that stem nodules represent) all the symbioses are located within the soil or at the soil surface. Each symbiosis involves one plant and one bacteria and, with the exception of the cyanobacterial symbioses, rbcL sequence analysis (Soltis *et al.*, 1996) groups all the other symbiotic nitrogen-fixing plants within a single broad clade within the dicotyledons. All the symbiotic bacteria are eubacteria. Studies on the physiology of these symbioses have detailed many aspects of the C and N relationships in different examples from these groups. Armed with this knowledge, it is worthwhile to contrast the different evolutionary solutions that provide the mechanisms that supply C to bacteria, assimilate the N fixed and translocate this N within the plant.

2. Carbon supply to bacteria within plant tissues

In rhizobia/legume systems evidence that organic acids, particularly malate, are supplied by the plant to the bacteria has been obtained (eg. Day and Udvardi, 1989; Werner, 1992) and it is likely that the rhizobia within *Parasponia* receive C in a similar form. There has been little recent work investigating the transfer of C compounds in actinorhizal nodules, but the metabolic capabilities of *Frankia* (see Bensen and Schultz, 1990) and the carbon metabolism of actinorhizal plants also suggests that organic acids are also transferred in these symbioses (Akkermans, 1981). Cyanobacterial symbioses are likely to utilise very different mechanisms of C transfer from the plant to bacteria, as cyanobacteria lack a key enzyme of the TCA cycle and use the alternative oxidative pentose phosphate (OPP) pathway to generate ATP and reductant for nitrogenase activity. In a series of careful experiments using clusters of *Nostoc* isolated from *Gunnera* nodules, we have evidence that the sugars glucose, sucrose and fructose directly support cyanobacterial N_2 fixation. These sugars are present in high concentrations within the plant tissues (Parsons and Silvester, unpublished)

The supply of reductant to nitrogenase is achieved using oxidative metabolism within the bacteria of all symbioses (except *Azolla*, where a contribution may be made via photosynthesis). As discussed above, this is the TCA cycle in rhizobia and probably *Frankia* and the OPP pathway in cyanobacteria.

3. N_2 fixation

The fixation of nitrogen in symbiotic systems appears to be via Mo-Fe dinitrogenase, perhaps in part because plants have other enzymes dependent upon Mo (nitrate reductase) and no enzymes requiring vanadium.

4. N release

Fixed N is released from nitrogenase as NH_3 and equilibrates with NH_4^+ within the bacterial cell. The first step in the transfer of NH_3/NH_4^+ to the plant involves transfer across the bacterial membrane and then uptake across the plant membrane into the plant cell. Recent results have demonstrated the presence of an ammonium transporter on the peribacteroid membrane in soybean (Tyerman *et al.* 1995) and this probably forms a good model for other symbiotic systems. We have evidence that NH_3 transfer and assimilation is very rapid in a range of plants (Fig. 1). In each of these plants, which represent the legumes (*Lupinus*), *Parasponia*, actinorhizal plants (*Myrica*) and

NATO ASI Series, Vol. G 39
Biological Fixation of Nitrogen
for Ecology and Sustainable Agriculture
Edited by A. Legocki, H. Bothe, A. Pühler
© Springer-Verlag Berlin Heidelberg 1997

cyanobacterial symbioses (*Gunnera*) label appears in the amino acids rapidly and significant amounts are detectable within typically 30 to 60 seconds of $^{15}N_2$ addition. These experiments demonstrate that the diffusion of N_2 into the nodules, its reduction by nitrogenase, the release of NH_3 by the bacteria and the subsequent uptake into the plant cells and assimilation by cytoplasmic glutamine synthetase all occurs within 30 seconds in many plants. This provides evidence that there is no large pool of ammonia within the symbioses, and that the levels of free ammonia are maintained at a minimum. These experiments provide evidence that NH_3/NH_4^+ are transferred between the bacteria and plant and I am not aware of evidence for the transfer of substantial amounts of N as amino compounds in any symbioses.

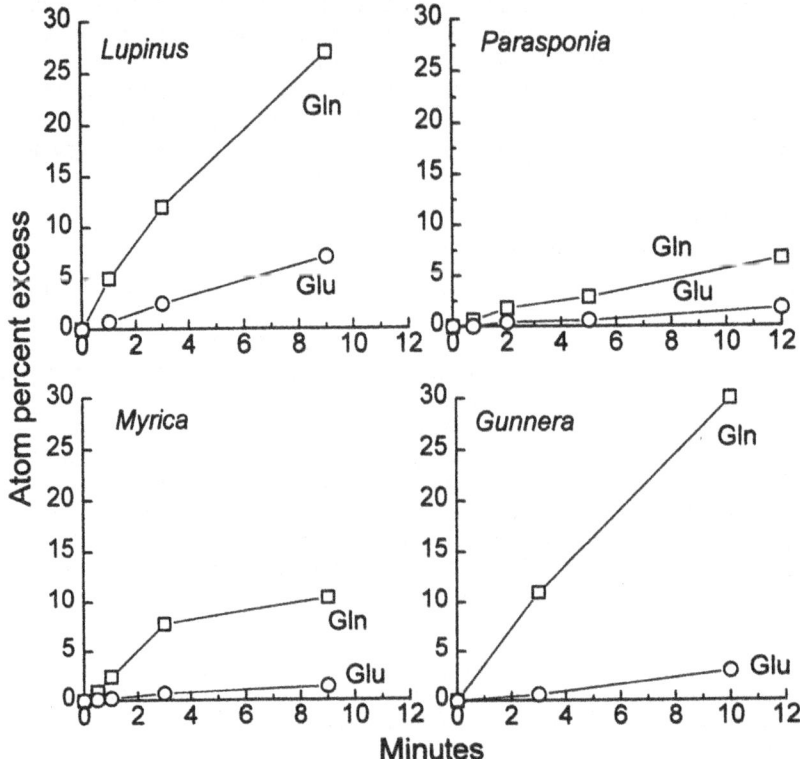

Figure 1. The rapid incorporation of ^{15}N into glutamine and glutamate within nodules from a range of plants representing different kinds of symbioses. Nodules were detached from the plants and enclosed within an atmosphere containing 80% $^{15}N_2$ and 20% O_2 at time 0 and the enrichment determined by GC-MS SIM analysis of MTBSTFA derivatised amino acids. Each point is the mean of six samples.

5. N assimilation

Our experiments following $^{15}N_2$ addition and incorporation into amino acids in a range of symbioses have always shown label appearing first in glutamine with subsequent labelling of glutamate and other amino acids. This work supports many studies that have measured the activity of the GS/GOGAT enzyme cycle within symbiotic tissues. In all the symbioses we have examined, where the final translocated product may be amides, ureides, citrulline or other compounds, the first amino compounds labelled are glutamine and glutamate. We have not observed glutamate becoming labelled before glutamine and so, although GDH is found in many systems, we have no evidence of it acting in an assimilatory role.

6. N transport

In symbiotic plants the initial translocation of fixed N is via the xylem, passing from the nodulated roots (most kinds of symbioses), or stem tissues (*Sesbania* and *Gunnera*), to the transpiring leaves within the transpiration stream (Fig. 2). Subsequent translocation will involve phloem recirculation requiring the loading of the phloem with N compounds. Distribution of N to distal regions such as root tips and growing buds or fruits will, in general, be via the phloem. However, while the mechanisms of N transport within plants are similar, there is considerable diversity in amino compounds used in different plants. Legumes are well known to fall broadly into two groups, those that synthesise predominantly ureides in nodules (eg. cowpea and soybean) and those that synthesise predominantly the amides, asparagine and glutamine (eg. *Lupinus* and *Medicago*) (Atkins, 1991). Beyond the legumes we find many other plants also favour asparagine as a translocation compound; examples include *Gunnera*, *Myrica*, *Hippophae* and *Parasponia*. Citrulline, a compound that is considered as both a ureide and an amino acid, is common in trees, and predominates in the xylem sap of *Alnus* and *Casuarina*. In *Alnus* we have carried out a detailed study using ^{15}N to trace both nitrogen fixation (Fig. 2) and nitrate uptake and we find in both cases that citrulline is the dominant compound translocated (Baker and Parsons, unpublished). Our recognition of the diversity of compounds translocated by plants increases as we continue our investigations. We recently discovered that *Parasponia* translocates the unusual amino acid 4-methylglutamate in its xylem sap (along with other amino acids including Asn and Gln), and provided evidence for the synthesis of this compound in *Parasponia* nodules (Baker, Dodd and Parsons, 1996).

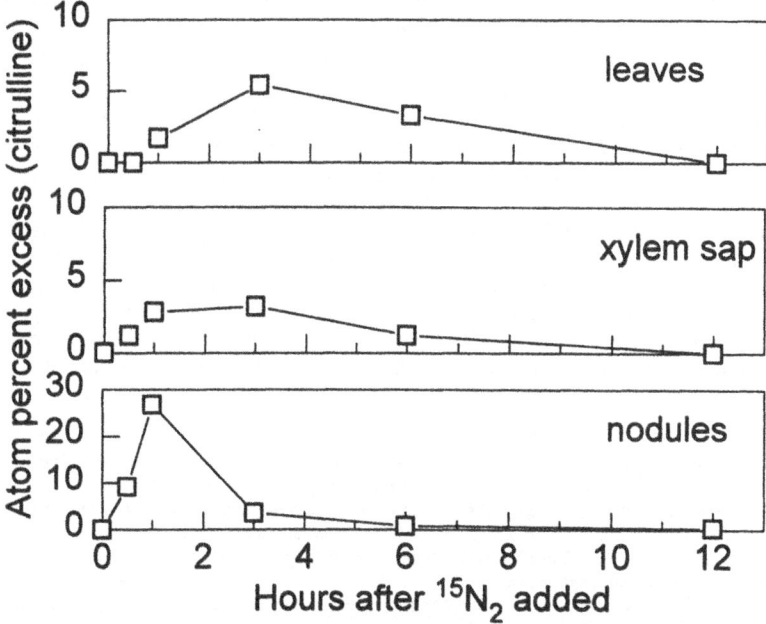

Figure 2. Time course of ^{15}N labelling in citrulline in the nodules, xylem sap and leaves of *Alnus glutinosa* plants following the addition of [15]N$_2$ to the nodulated roots for one hour at time 0.

Although diverse, all of the N compounds translocated (except 4-methylglutamate in *Parasponia*) represent compounds present within the primary metabolism of plants. In different plants, different aspects of this primary metabolism has clearly been enhanced to allow excess synthesis of ureides, amides or citrulline. Similarly the degradation and subsequent reassimilation of the N translocated in these compounds has also coevolved in different plants.

Summary of aspects of C and N metabolism of N$_2$ fixing symbioses

Plant/bacteria/ location	Carbon transfer	Nitrogen transfer	Nitrogen assimilation	Nitrogen transport in plant	Resistance to O$_2$ diffusion
soybean/rhizobia root nodule	malate (succinate)	NH$_3$/NH$_4^+$	GS/GOGAT then ureide production	ureides	Plant cortex + haemoglobin
lupin/rhizobia root nodule	malate	NH$_3$/NH$_4^+$	GS/GOGAT then asparagine synthesis	gln, asn, asp, glu	Plant cortex + haemoglobin
Parasponia/ rhizobia root nodule	malate ?	NH$_3$/NH$_4^+$	GS/GOGAT then asparagine, 4-methylglutamate	asn, gln, glu, asp, 4-methylglutamate	Plant cortex + haemoglobin
Alnus/Frankia root nodule	unknown, organic acids proposed	NH$_3$/NH$_4^+$	GS/GOGAT then citrulline	citrulline, glu	*Frankia* vesicles
Myrica/Frankia root nodule	unknown	NH$_3$/NH$_4^+$	GS/GOGAT then asparagine synthesis	asn, gln, glu, asp	*Frankia* vesicles, plant cell structure
Azolla/Anabaena leaf cavity	possibly sucrose	NH$_3$/NH$_4^+$	GS/GOGAT then alanine synthesis	gln, glu, ala	cyanobacteria heterocyst
Gunnera/Nostoc stem nodule	glucose (sucrose)	NH$_3$/NH$_4^+$	GS/GOGAT then asparagine synthesis	asn, gln, glu, asp	cyanobacteria heterocyst, plant cell structure

7. Plant N status and the regulation of N$_2$ fixation.

Almost all N$_2$ fixing plants are able to reduce N$_2$ fixation when combined N is available from the soil. Furthermore, because N$_2$ fixation represents an almost unlimited supply of N, feedback regulation of N$_2$ fixation must involve the N status of the plant being sensed and signals returned to the nodules (probably via the phloem sap) to stimulate or inhibit further N$_2$ fixation. These mechanisms remain to be elucidated, but given the diversity of N transport compounds, we might expect a similar diversity of signal molecules. In some types of symbioses there is evidence for the involvement of the control of O$_2$ diffusion in the overall regulation of N$_2$ fixation. However, this mechanism of control is unlikely to extend to all symbioses, as in cyanobacterial and many *Frankia* systems the bulk of the resistance to O$_2$ diffusion is provided by bacterial structures (heterocyst and vesicle wall respectively) and these would appear unlikely to be involved in the plant mediated regulation of activity.

8. References:

Akkermans ADL, Huss-Danell K, Roelofsen W. 1981. Enzymes of the tricarboxylic acid cycle and the malate-aspartate shuttle in the N$_2$ fixing endophyte of *Alnus glutinosa*. *Physiologia Plantarum*, 53, 289-294.

Atkins CA. 1991. Ammonia assimilation and export of nitrogen from the legume nodule. In *Biology and Biochemistry of Nitrogen Fixation* (eds. M.J. Dilworth & A.R. Glenn) pp. 293-311. Elsevier, Amsterdam.

Baker A, Dodd CD, and Parsons R. 1996. Identification of amino compounds synthesised and translocated in symbiotic *Parasponia*. *Plant Cell and Environment*. In press.

Bensen, DR and Schultz, NA. 1990. Physiology and Biochemistry of *Frankia* in culture. In: *The Biology of Frankia and Actinorhizal Plants*. Eds: CR Schwintzer and JD Tjepkema. Academic Press, London.

Day DA and Udvardi, MK. 1993. Metabolite exchange across symbiosome membranes. *Symbiosis*, 14, 175-189.

Soltis, DE, Soltis, PS, Morgan, DR, Swensen, SM, Mullin, BC, David, JM & Martin, PG. 1996. Chloroplast gene sequence data suggest a single origin of the predisposition for symbiotic nitrogen fixation in angiosperms. *Proc. Natl. Acad. Sci. USA* 92, 2647-2651.

Tyerman SD, Whitehead LF & Day DA. 1995. A channel like transporter for NH$_4^+$ on the symbiotic interface of N$_2$-fixing plants. *Nature* 378, 629-632.

Werner, D. 1992. Physiology of nitrogen fixing legume nodules: compartments and functions. In *Biological Nitrogen Fixation* (Eds G Stacey, RH Burris, HJ Evans) pp 399-431. Chapman and Hall, New York.

Robinia pseudoacacia, a model tree legume

Astrid Wetzel, Patrick von Berswordt-Wallrabe, Marie-Luise Meinhold, Mechthild Röhm, Petra Scheidemann, Wolfgang Streit and Dietrich Werner

Fachbereich Biologie, FG Angewandte Botanik und Zellbiologie, Philipps-Universität, Karl-von-Frischstraße, D-35032 Marburg, Germany

1.   Physiological and genetical characters of *Robinia pseudoacacia*

Robinia pseudoacacia has a number of attributes as a model tree legume, due to one of the highest net photosynthetic rates amoung woody plants (up to 36 μM CO_2 x m^{-2} x s^{-1}), resistance to a number of stresses, early flowering, production of abundant seeds already after three years and a significant genetic variation (Hanover, 1990). With these characters it has a number of advantages compared to the list of 50 nitrogen fixing trees proposed by the Nitrogen Fixing Tree Association considered for their economical or ecological importance (Brewbaker, 1990). All these specific characters can be studied in relation to nodulation and nitrogen fixation by *Rhizobium loti* (Werner et al., 1996), to VA-mycorrhiza infection (Werner, 1992) and also to other ecological important aspects such as growth on degraded soils due to the very plastic and efficient root system.

2.   Signalling compounds in the root exudate

In the root exudate of *Robinia peusodoacacia* seedlings 8 different components had nod gene-inducing activity (Scheidemann and Wetzel, 1996, with *Rhizobium* NGR235 pM P22, Lewin et al., 1990). Five of these compounds were identified by two-dimensional HPTLC, HPLC and GC-MS methods as naringenin, apigenin, chrysoeriol, isoliquiritigenin and 4',7-dihydroxyflavone. Three other flavonoids with nod gene-iducing activity were also present in significant quantities with a final identification still pending. These flavonoids are different from the flavonoids found in the hardwood of *Robinia pseudoacacia* which have been identified as robinetin, dihydrorobinetin, dihydrofisetin, fisetin, robtin, butin, robtein and leucorobinetinidin (Table 1). However these hardwood flavonoids may also play a role in the rhizosphere in soils, released from decaying hardwood affecting a complex interaction in the rhizosphere of *Robinia* roots with *Rhizobium* strains and arbuscular mycorrhiza strains. Altough it is established that flavonoids can stimulate to a certain extent spore germination and also the first phases of hyphal growth of AM fungi it has been found that they are not necessary plant signals in this symbiosis (Bécard et al. 1995). So the complexity of signalling in a tree rhizosphere may be larger than in agricultural annual legumes with the molecular basis of specificity in the centre of interest over the last year (Fellay et al. 1990). On the other side, the nutrient uptake rate per area in forest trees in general is significantly lower than in agricultural plants, as found for nitrogen, phosphorus, potassium, calcium and magnesium (George and Marschner, 1996).

NATO ASI Series, Vol. G 39
Biological Fixation of Nitrogen
for Ecology and Sustainable Agriculture
Edited by A. Legocki, H. Bothe, A. Pühler
© Springer-Verlag Berlin Heidelberg 1997

Table 1. Flavonoids in root exudate and hardwood from *Robinia pseudoacacia*

Flavonoids in root exudates of seedlings	Flavonoids from hardwood
naringenin	robinetin
apigenin	dihydrorobinetin
chrysoeriol	dihydrofisetin
isoliquiritigenin	fisetin
4',7-dihydroxyflavone	robtin
	butin
three non-identified nod	robtein
gene-inducing flavonoids	leucorobinetinidin

Table 2. Parameter of glucose uptake kinetics in bean nodulating rhizobia (CIAT 611, CIAT 899, KIM5s) in comparison to robinia nodulating NGR234. The uptake follows a Michaelis-Menten-Kinetik of the type $v = V_{max} \times s/(K_m + S)$. V_{max} gives nmol/min/mg protein, K_m the µM glucose (data from M. Meinhold and D. Werner)

strain	pH	parameter	glucose-uptake	with addition of 10 µM apigenin
CIAT 611	6,8	V_{max}	36,2	15,9
	6,8	K_m	11,5	11,5
	5,2	V_{max}	39,5	28,2
	5,2	K_m	15,0	16,8
CIAT 899	6,8	V_{max}	26,4	12,9
	6,8	K_m	13,7	15,4
	5,2	V_{max}	23,1	12,6
	5,2	K_m	11,4	13,4
KIM 5s	6,8	V_{max}	13,9	8,4
	6,8	K_m	10,2	11,0
	5,2	V_{max}	11,1	8,9
	5,2	K_m	13,2	10,9
NGR234	6,8	V_{max}	1,62	1,16
	6,8	K_m	12,8	14,5

3.   Nitrogen/Hydrogen/Carbon metabolism

Robinia pseudoacacia nodulating rhizobia isolated from soils in North America and Asia are very heterogenous with generation times between 3 and 9 hours and more than 30 different groups on the basis of the protein profile (McCray Batzli et al., 1992). By direct isolation of strains from rediffentiated bacteroids of black locust nodules, 8 different strains with different efficiencies in nitrogen fixation in infected plants were isolated. One strain (*Rhizobium loti* R1) was most efficient in terms of nodulation and N_2-fixation (Röhm and Werner, 1992; Schäfer and Werner, 1993). In most legumes between 30 and 40% of the nitrogen electron flask is lost as H_2 (Evans and Burris, 1992). All black locust nodulating *Rhizobium* strains tested so far can take up molecular hydrogen. Hydrogen uptake by nodule homogenates from *Robinia pseudoacacia* was markedly lower than the activities found in soybean nodules, however in the same order of magnitude as H_2 uptake in *Phaseolus vulgaris* nodule homogenates (Röhm et al., 1993). Another specific character of *Robinia pseudoacacia* nodulation is the pronounced sensitivity towards low concentrations of nitrate, where already 1 mM concentration can reduce nitrogenase activity to very low levels (Röhm and Werner, 1991).

An interesting relationship between nodulating signals and carbon metabolism is given in the results of Table 2. A 10μM concentration of apigenin significantly affect glucose uptake kinetics in the *Rhizobium* strains CIAT 611, CIAT 899 and KIM5s (bean nodulating rhizobia), reducing V_{max} by 40 to 60%. The V_{max} data are in the order of 14 to 36 nM glucose x min x mg protein without additional apigenin. Compared to these activities, the activity in the black locust nodulating strains NGR234 is significantly lower whereas K_m is unchanged. New results in non linear sugar transport kinetics have been used (Fuhrmann and Völker, 1993). The data so far received seem to indicate that there is no relationship between the flavonoid effect on glucose uptake and the competitiveness of the strains. But this must be confirmed with a large number of strains tested.

The work was partially supported through SFB 395 - Project A6.

4. References

Bécard, G., L.P. Taylor, D.D. Douds, Jr., P.E. Pfeffer and L.W. Doner (1995) Flavonoids are not necessary plant signal compounds in arbuscular mycorrhizal symbioses. MPMI 8, 252-258

Brewbaker, J.L. (1990) Nitrogen fixing trees. In: Fast Growing Trees and Nitrogen Fixing Trees (eds. D. Werner, P. Müller), 253-262, G. Fischer Verlag, Stuttgart, New York

Evans, H.J. and R.H. Burris (1992) Highlights in biological nitrogen fixation during the last 50 years. In: Biological Nitrogen Fixation (eds. G. Stacey, R.H. Burris, H.J. Evans), 1-42, Chapman & Hall, New York

Fellay, R., P. Rochepeau, B. Relic and W.J. Broughton (1995) Signals to and emanating from Rhizobium largely control symbiotic specificity. In: Pathogenesis and Host Specificity in Plant Diseases. Vol. I. Prokaryotes (eds. U.S. Singh, R.P. Singh, K. Kohmoto), 199-220, Pergamon/Elsevier Sci. Ltd., Oxford

Fuhrmann, G.F. and B. Völker (1993) Misuse of graphical analysis in nonlinear sugar transport kinetics by Eadie-Hofstee plots. Bioch. Biophys. Acta 1145, 180-182

George, E. and H. Marschner (1996) Nutrient and water uptake by roots of forest trees. Z. Pflanzenernähr. Bodenk. 159, 11-21

Hanover, J.W. (1990) Physiological genetics of black locust (*Robinia pseudoacacia* L.): A model multipurpose tree species. In: Fast Growing Trees and Nitrogen Fixing Trees (eds. D. Werner, P. Müller), 253-262, G. Fischer Verlag, Stuttgart, New York

Lewin, A., E. Cervantes, W. Chee-Hoong and W.J. Broughton (1990) NodSU, two new nod genes of the broad host range *Rhizobium* strain NGR234 encode host-specific nodulation of the tropical tree *Leucaena leucocephala*. MPMI 5, 517-326

McCray Batzli, J., W.R. Graves and P. van Berkum (1992) Diversity among rhizobia effective with Robinia pseudoacacia L. Appl. Environ. Microbiol. 58, 2137-2143

Röhm, M. and D. Werner (1991) Nitrate levels affect the development of the black locust-*Rhizobium* symbiosis. Trees 5, 227-231

Röhm, M. and D. Werner (1992) *Robinia pseudoacacia - Rhizobium* symbiosis: Isolation and characterization of a fast nodulating and efficiently nitrogen fixing *Rhizobium* strain. Nitrogen Fixing Tree Res. Reports 10, 193-197

Röhm, M., W. Streit, H.J. Evans and D. Werner (1993) Hydrogen uptake by *Robinia pseudoacacia* nodules. Trees 8, 99-103

Schäfers, B. and D. Werner (1993) Nodulation of *Robinia pseudoacacia* by two *Rhizobium* strains. Nitrogen Fixing Tree Res. Reports 11, 121-126

Scheidemann, P. and A. Wetzel (1996) Identification and characterization of flavonoids in the root exudate of *Robinia pseudoacacia*. Trees (*in press*)

Werner, D. (1992) Symbiosis of Plants and Microbes. Chapman & Hall, London, New York

Werner, D., M. Röhm, B. Schäfers, P. Scheidemann and A. Wetzel (1996) Signalling in the *Robinia-Rhizobium* Symbiosis. In Trees - Contributions to Modern Tree Physiology (eds. H. Renneberg, W. Eschrich, H. Ziegler), SPB Academic Publ., The Hague (*in press*)

Unsolved Mysteries in Carbon Metabolism in Legume Nodules

John Streeter

Department of Horticulture and Crop Science, Ohio State University/OARDC, 1680 Madison Ave, Wooster, OH 44691 USA

1 Poly-ß-hydroxybutyric acid (PHB)

The accumulation of massive quantities of PHB in bacteroids of many rhizobia has been known for many decades. Based on a small sample of genera and species, it appears that the fast-growing rhizobia do not accumulate PHB whereas the slow-growing rhizobia do (see reviews by McDermott et al., 1989; Streeter, 1995). A role for the polymer, which can amount to 50% of bacteroid dry weight in *Bradyrhizobium japonicum* (Wong & Evans, 1971), has never been established. Recent genetic evidence suggests that PHB metabolism does not play a major role in *Rhizobium meliloti*; however, the possibility of some role in slow-growing rhizobia remains an open and potentially important question.

1.1 *R. meliloti*

Recently, the genes coding for ß-ketothiolase, acetoacetyl-CoA reductase and PHB synthase (the three enzymes required to convert acetyl-CoA to PHB) have been cloned from *R. meliloti* (Tombolini, et al. 1995). Earlier, the same group reported a transposon mutant of *R. meliloti* that lacked PHB synthase and was unable to synthesize PHB (Povolo et al., 1994). This mutant did not differ significantly from wild type in the nodulation of alfalfa (*Medicago sativa*) or in the fixation of nitrogen.

Very recently, Aneja and Charles (1996) have reported a transposon mutant of *R. meliloti* lacking ß-hydroxybutyrate (BHB) dehydrogenase. This mutant was also Nod^+ and, although N_2 fixation was slightly delayed, the mutant must be considered Fix^+. For *R. meliloti*, the formation and breakdown of PHB in bacteroids must be considered to be of little consequence to the fixation of nitrogen.

1.2 Slow-growing rhizobia

Part of the reason for the renewal of interest in PHB in bacteroids stems from a series of papers by Bergersen et al. in the early 1990's; only two of these papers are cited here (1990, 1991). In these papers, *B. japonicum* bacteroids were carefully isolated anaerobically and studied *in vitro* for consumption of dicarboxylic acids, PHB turnover, CO_2 and NH_3 release, etc. under a range of O_2 concentrations and other conditions. The results of these very detailed studies indicates that PHB does turnover under the conditions studied. Bergersen concludes that PHB probably serves as an internal, ancillary source of reduced carbon and may be important to the sustained operation of nitrogenase under carbon "stress" conditions such as darkness.

NATO ASI Series, Vol. G 39
Biological Fixation of Nitrogen
for Ecology and Sustainable Agriculture
Edited by A. Legocki, H. Bothe, A. Pühler
© Springer-Verlag Berlin Heidelberg 1997

Cevallos et al. (1996) have recently succeeded in the construction of a *R. etli* mutant lacking PHB synthase. The mutant has interesting metabolic characteristics, and bacteroids accumulate glycogen instead of PHB. Nodules formed by the mutant on *Phaseolus vulgaris* had slightly higher acetylene reduction rates and this was especially pronounced at later plant growth stages. The most important conclusion is that bacteroids lacking the ability to form PHB can sustain higher rates of N_2 fixation.

My interest in PHB stems from a little-known review by David Emerich (1985) in which he describes UV mutagenesis of *B. japonicum* (USDA 143) and selection of mutants unable to grow on BHB. One mutant was obtained which possessed all of the enzymes of the PHB shunt except BHB dehydrogenase. This mutant was Fix[-] and a revertant was obtained that had wild type BHB dehydrogenase activity and acetylene reduction activity. The work was not pursued. A few years ago, we attempted to rescue this mutant from a freezer culture (courtesy of D. Emerich) but failed.

Recently, I have screened about 7500 Tn5 insertion mutants of *B. japonicum* USDA 110 for ability to grow on BHB. A few mutants with the desired phenotype were found but all had wild type or lowered but positive BHB dehydrogenase activity. One mutant which had zero growth on BHB probably lacked acetoacetyl-succinyl CoA transferase (not tested); this mutant has recently been found to be Fix[+] on soybeans (*Glycine max*).

My conclusion is that PHB metabolism in slow-growing rhizobia should be pursued. Although the evidence for *R. etli* indicates that PHB synthesis is not useful, a transposon mutant lacking BHB dehydrogenase should be found and studied. Assuming adequate genetic homology, the recent work with *R. meliloti* should be helpful in finding such a mutant.

2 Trehalose

Trehalose is a simple glucose-glucose disaccharide but with an unusual linkage, namely α1-1. It is the blood sugar of insects, a common (universal?) constituent of fungi, and is often, but not always, found in bacteria. Trehalose has not been conclusively demonstrated in higher plants; in fact it is toxic to plants lacking sufficient trehalase - the most common enzyme of trehalose breakdown (Veluthambi et al., 1981). Because trehalose is a major constituent of soybean nodules and because all of our early evidence indicated considerable trehalose synthesis by bacteroids, we surveyed most of the genera and species of rhizobia and found active trehalose synthesis in all strains tested (Streeter, 1985).

Müller et al. (1994) have recently reported that they could not detect trehalose in nodules from half of the 12 legume species tested. This may relate to the fact that trehalose synthesis is highly variable among rhizobial strains and that much of the trehalose synthesized by bacteroids is released to the peribacteroid space and to the host cytoplasm where there is substantial trehalase activity (Streeter, 1985). The apparently futile cycling of glucose through trehalose and back to glucose in nodules is curious.

Trehalose has many unusual biochemical properties; two that may be relevant to nodules are the inhibition of cell wall synthesis (Veluthambi et al., 1982) and the stabilization of membranes and proteins (Crowe et al., 1987). Several recent reports are consistent with the idea that trehalose synthesis in rhizobia is stimulated by salt and/or osmotic stress (Breedveld et al., 1991; Pfeffer et al., 1994; Ghittoni et al., 1995). Unfortunately, the osmotic potential of the peribacteroid space in intact nodules is very difficult to determine.

An obvious way to approach the question of the role of trehalose in symbiotic systems is to obtain mutants of rhizobia unable to produce the compound. We first studied the enzymes of trehalose synthesis in *B. japonicum* (Salminen & Streeter, 1986) and obtained evidence for the pathway previously reported in *Escherichia coli*:

UDP-glucose + glucose-6-P --> trehalose-6-P (synthetase)

trehalose-6-P --> trehalose + phosphate (phosphatase)

Activity of the synthetase was very low and could not account for the calculated accumulation of trehalose in nodules.

In spite of questions about the presence of the *"E. coli* mechanism" for trehalose synthesis in rhizobia, we attempted to clone the genes and to obtain mutants (I. Hoezle, S. Salminen, J. Streeter; unpublished data). Using a probe for the *E. coli* synthetase gene we have screened libraries for *R. etli* (USDA 2667) and *B. japonicum* (USDA 110) but were unsuccessful in identifying clones in these rhizobia. Attempts to complement *E. coli* mutants (supplied by A.R. Strøm; see 1993 review) were also unsuccessful.

Recently, a totally new mechanism for the synthesis of trehalose in bacteria has been discovered (Maruta et al., 1995). The substrates for the first enzyme are maltooligosaccharides (MOS); i.e. ......glucose-α1→4-glucose-α1→4-glucose...... The synthase enzyme flips the terminal glucose (reducing end) to give: ......glucose-α1→4-glucose-α1→1-glucose (maltooligosyl-trehalose). The second enzyme hydrolyzes the "rightmost" two glucose residues to give trehalose plus a shortened MOS.

Maruta et al. (1995) suggested that the mechanism is present in rhizobia, but the ability of the presumed rhizobial strain to form nodules was not established. Recent unpublished evidence from my lab indicates that the MOS mechanism is present in *B. japonicum*. Trehalose synthesis and accumulation are stimulated by high sugar concentrations (50 to 100 mM) as reported by Maruta et al. (1995), and this may be simply a response to high osmotic potential. We are presently extending the survey for the "MOS mechanism" to other genera and species of rhizobia and are actively seeking mutants of *B. japonicum* using a variety of molecular genetic approaches. The sequences of the new genes for trehalose synthesis have recently been published (Maruta et al., 1996), making it possible to generate oligonucleotide probes for these new genes.

Is trehalose important? It is hard to say. In addition to the anecdotal observations already mentioned, it should be noted that trehalose is also synthesized by *Frankia* and cyanobacteria. The implication is that this plant toxin plays some role in all nitrogen-fixing associations. The opportunities for solving this dilemma are best in rhizobia where genetic approaches are most feasible.

3 Acknowledgement

I thank Arvind Bhagwat, David Emerich, Kent Peters, and Vassily Romanov for stimulating discussions on these topics.

4 References

Aneja, P. & T.C. Charles. Abstract H-2 from the Eighth International Symposium on Molecular Plant-Microbe Interactions, Knoxville, TN, 1996.

Bergersen, F.J. & G.L. Turner. Proc. R. Soc. Lond. B 240(1990)39.

Bergersen, F.J., M.B. Peoples, & G.L. Turner. Proc. R. Soc. Lond. B 245(1991)59.

Breedveld, M.W., L.P.T.M. Zevenhuizen, & A.J.B. Zehnder. Arch. Microbiol. 156(1991)501.

Cevallos, M.A., S. Encarnacion, A. Leija, Y. Mora, & J. Mora. J. Bacteriol. 178(1996)1646.

Crowe, J.H., L.M. Crowe, J.F. Carpenter, & C.A. Wistrom. Biochem. J. 242(1987)1.

Emerich, D.W. In: Ludden, P.W & J.E. Burris. Nitrogen Fixation and CO_2 Metabolism. 1985. Elsevier, New York., pp. 21-30.

Ghittoni, N.E. & M.A. Bueno. Can. J. Microbiol. 41(1995)1021.

Maruta, K., T. Nakada, M, Kubota, H. Chaen, T. Sugimoto, M. Kurimoto, and Y. Tsujisaka. Biosci. Biotech. Biochem. 59(1995)1829.

Maruta, K., K. Hattori, T. Nakada, M. Kubota, T. Suginoto, & M. Kurimoto. Biochim. Biophys. Acta. 1289(1996)10.

McDermott, T.R., S.M. Griffith, C.P. Vance, & P.H. Graham. FEMS Microbiol. Rev. 63(1989)327.

Müller, J., Z.-P. Xie, C. Staehelin, R.B. Mellor, T. Boller, & A. Wiemken. Physiol. Plant. 90(1994)86.

Pfeffer, P.E., G. Becard, D.B. Rolin, J. Uknalis, P. Cooke, and S.-I. Tu. Appl. Environ. Microbiol. 60(1994)2137.

Povolo, S., R. Tombolini, A. Morea, A.J. Anderson, S. Casella, & M.P. Nuti. Can. J. Microbiol. 40(1994)823.

Salminen, S.O. & J.G. Streeter. Plant Physiol. 81(1986)538.

Streeter, J.G. J. Bacteriol. 164(1985)78.

Streeter. J.G. Symbiosis 19(1995)175.

Strøm, A.R. & I. Kaasen. Mol. Microbiol. 8(1993)205.

Tombolini, R., S. Povolo, A. Buson, A. Squartini, & M.P. Nuti. Microbiol. 141(1995)2553.

Veluthambi, K., S. Mahadevan, & R. Maheshwari. Plant Physiol. 68(1981)1369.

Veluthambi, K., S. Mahadevan, & R. Maheshwari. Plant Physiol. 70(1982)686.

Wong, P.P. & H.J. Evans. Plant Physiol. 47(1971)750.

Phosphorous Deficiency Increases the Respiratory Cost of Symbiotic N2 Fixation

Jean-Jacques Drevon[1], Hélène Payré[1], Jérome Ribet[1][2] and Vincent Vadez[1]

[1] Laboratoire de Recherche sur les Symbiotes des Racines, Institut National de la Recherche Agronomique, Place Viala, Montpellier, F-34060 cedex, France.
[2] Bean Nutrition Program, CIAT, AA 56, Cali, Colombia

Abstract. Phosphorus deficiency is a major environmental factor limiting plant growth in more than 60 % of tropical soils. This affects particularly symbiotic legumes which usually have higher P requirements than non-symbiotic plants (Israel 1987). As partners in a multidisciplinary project to improve legume nitrogen fixation and yield in low-P soils, we investigated the effects of P deficiency on nodule respiratory (CO_2 evolution and O_2 uptake) and nitrogenase activities (C_2H_2 reduction). Nodulated soybean (*Glycine max* L. Merr.) and common bean (*Phaseolus vulgaris*) were grown in intensely aerated liquid nutrient solutions. This hydroponic culture optimized (i) the control of P supply (ii) the nodulation and nitrogen fixation - dependent growth, and (iii) the *in situ* measurement of the nodule gas-exchanges. During measurements of the nitrogenase acetylene reduction activity (ARA), there was an acetylene-induced decline (C_2H_2-ID) in nitrogenase activity which was significantly increased by P deficiency. By plotting CO_2 evolution as a function of ARA during the decline, we found that nodules of P-deficient soybean plants evolved significantly more CO_2 per C_2H_4 than nodules of P-sufficient plants. However, both types of nodules had similar nitrogenase activity, measured as pre-decline ARA. P deficiency also increased the nitrogenase-linked O_2 uptake, which was the variable component of the response of nodulated-root O_2 uptake to changes in rhizosphere pO_2 between 15 and 21 kPa O_2. These effects of P deficiency were generally associated with an increase in nodule permeability to O_2 diffusion, except for the more tolerant common bean genotype, G19839. It is concluded on P deficiency effects on nodule alternative oxidase activity, cell shape variation and gene expression in the nodule cortex where the variable nodule permeability is thought to operate.

Keywords. Acetylene reduction, biodiversity, *Glycine*, nitrogenase, nodule respiration, oxygen permeability, *Phaseolus*, *Rhizobium*

1 *In situ* measurements of nodulated-root gas exchanges

Soybean and common bean were inoculated and grown in hydroponic culture in a glass house or a phytotron as described in details by Drevon et al. (1988) and Hernandez et al (1991). P supplies varied between 50 and 75 versus 250 and 500 μmol P plant^{-1} week^{-1}, for P deficiency vs. P sufficiciency, respectively. For

NATO ASI Series, Vol. G 39
Biological Fixation of Nitrogen
for Ecology and Sustainable Agriculture
Edited by A. Legocki, H. Bothe, A. Pühler
© Springer-Verlag Berlin Heidelberg 1997

soybean (cv Kinsoy) or common bean (genotype G3010) shoot, root and nodule dry weights were in the ranges of 7 to 13, 2 to 4 and 0.5 to 1 g plant^{-1} for P sufficicient plants at the early flowering stage, between 40 and 45 days after sowing (DAS) (Ribet and Drevon, 1995a; Vadez et al., 1996). P deficiency decreased these parameters to ca. 30, 70 and 27 % of the above values, respectively. However, for the common bean genotype G19839 these parameters were decreased to only 50, 85 and 36 % of the above values. The mean individual nodule mass and size were generally decreased by P deficiency.

Measurements of the acetylene reduction activity (ARA) were performed between 35 and 60 DAS using the flow–system described by Drevon et al. (1988). Repeating measurements every 6 days had no effect on plant growth and N_2 fixation when exposures to C_2H_2 did not exceed 2 h (Vidal et al., 1995). Values of ARA (peak ARA before the C_2H_2-ID) were about 70 vs. 10 and 180 vs. 30 μmol C_2H_4 h^{-1} plant^{-1} for P-sufficient vs. P-deficient soybean plants at 42 and 56 DAS, respectively (Ribet and Drevon, 1995b).

The nodulated-root CO_2 production was measured simultaneously using an IRGA. The same flow system was also used to measure the nodulated-root O_2 uptake either by flushing the system with a flowing gas stream (1) or by successive confinments in a circulating gas stream (2) (Roy, 1993). In (1) the value of O_2 uptake was equal to (Oin - Oex) D, with Oin and Oex as the O_2 concentrations ([O_2] in mol ml^{-1}) of the in- and out- flow and D as the flow rate (ml min^{-1}). In (2) O_2 uptake was equal to (∂Oex / ∂t) V, with V as the volume of the nodulated-root cuvette and ∂Oex / ∂t as the rate of decrease in the out-flow [O_2] during the confinment period, the gas-flow being driven by a peristaltic pump. After less than a 2 % decrease in O_{ex}, the cuvette was reconnected to the gas-flow system. Then the gas phase was renewed with either the same Oin to perform a replicate measurement, or a different Oin to study the effects of changing rhizosphere pO_2 on the nodulated-root O_2 uptake.

An increase from cultivation pO_2, ie. air containing 21 KPaO_2, to 27 kPaO_2 induced an immediate increase in O_2 uptake (Drevon et al., 1995). However, this increase was only transient. After ca. 20 min, the O_2 uptake started to decrease and returned to about its initial level within ca. 60 min. This corresponded to a decrease in the nodule O_2 permeability (g$_{no}$) and was associated with a contraction of the nodule inner-cortex cells in soybean (Serraj et al., 1995).

A decrease from 21 to 15 kPaO_2 induced an immediate decrease in O_2 uptake (Drevon et al., 1995) which returned to the initial rate within a 6h period, ie. there was no rapid change in g$_{no}$. This made it possible to measure the nodule permeability by measuring the O_2 uptake in successive confinments as pO_2 was decreased from 21 to 15 kPa O_2. In this range of pO_2, the nodulated-root O_2 uptake was a linear function of pO_2 (Roy, 1993). The fixed component, ie. the intercept of the regression with the Y axis, was interpreted as the root respiration plus the nodule respiration for growth and maintenance whereas the variable component corresponded to the nodule O_2 uptake linked to N_2 fixation. The latter was divided by the nodule surface (S = $\Sigma n_i \pi d_i^2$, with n_i = the number of nodules in the class of mean diameter of d_i, and d_i is divided in 5 classes between 1.5 and 5.5 mm) to calculate the flux of O_2 diffusion into the nodule (Jon). According to Fick's law the nodule O_2 permeability was calculated as g$_{no}$ = Jon (Oex - Oi)$^{-1}$ which can be shortened to Jon Oex^{-1} since Oi (nodule infected zone [O_2]) is 10^4 times lower than Oex (rhizosphere [O_2]) (Hunt & Layzell, 1993).

2 Effects of P deficiency on nodule gas exchanges

The above methodology was used to compare nodulated-root respiration and ARA for P-deficient vs. P-sufficient soybean plants. The O_2 uptake for growth and maintenance of the root and nodule tissue for P-deficient plants was 35 % of the value for P-sufficient plants. By contrast, the nodule nitrogenase-linked O_2 uptake for P-deficient plants was 210 % of the value for P-sufficient plants at ambient pO_2, and the nodule O_2 permeability was twice as high for P-deficient plants (Table 1). This was also observed for common bean genotype G3010. By contrast the g_{no} was not affected in genotype G19839 in which the N_2 fixation is more tolerant to P deficiency (Vadez et al., 1996a).

During ARA assays, nitrogenase activity declined a few minutes after exposure of the nodulated roots to C_2H_2. This acetylene-induced decline (C_2H_2-ID) was less than 25% of the peak ARA value for P-sufficient plants, which was contrastingly lower than the C_2H_2-ID described by Minchin et al. (1986) for vermiculite-grown soybean. However, the C_2H_2-ID was significantly increased by P-deficiency (Ribet & Drevon, 1995b) and a curvilinear correlation was found between C_2H_2-ID and P supply (Fig. 1).

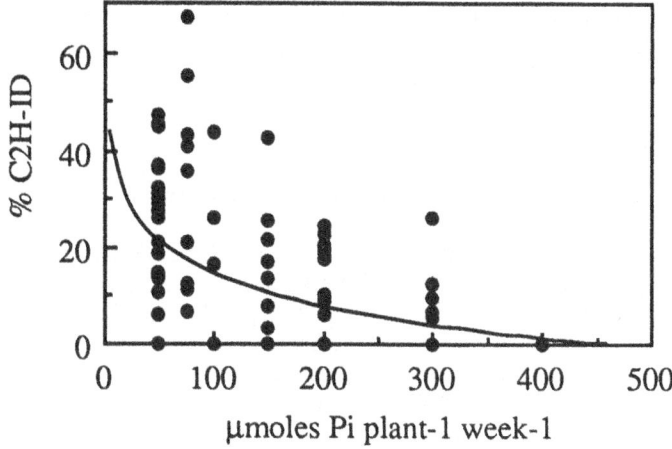

Figure 1. Influence of P nutrition on the acetylene-induced decline of nodule nitrogenase activity in soybean . % C2H2-ID = 100 (1 - ARAmin/ARAmax). Shown individual measurements adjust to a non-linear regression ($y = 81 - 30$ Logx R=0.62)

There was a transient increase in nodulated-root CO_2 production following the exposure to C_2H_2. This increase was twice as high under P deficiency. Following this increase, the nodulated-root CO_2 production decreased concomitantly with nodule ARA and a linear regression was found between CO_2 and C_2H_4 productions during the C_2H_2-ID (Ribet & Drevon, 1995b). The nodule nitrogenase-linked CO_2 production, calculated as the variable component of this regression, was twice as high under P deficiency (4.5 mol CO_2 mol^{-1} C_2H_4). P deficiency did not affect the ARA (104 mol C_2H_4 h^{-1} g^{-1} DW nod) nor the ratio of whole plant N fixed to nodule mass at 75 DAS (Ribet & Drevon, 1995b). This agrees with data of Pongsakul & Jensen (1991) using ^{15}N. However it may vary with species and genotypes (Vadez et al. 1996b).

Conclusions and perspectives

Taken altogether, our data lead us to hypothesize that P deficiency increases the nodule O_2 uptake per electron transfered by nitrogenase which is associated with an increase in nodule O_2 permeability. Whether this is a consequence of decreased size of nodule or increased alternative respiratory pathways in nodule like in roots (Rychter et al., 1992) needs further elucidation. It is also required to analyse how P deficiency affects the distribution of P in the nodule. A major conclusion is that the relative magnitude of the C_2H_2-ID depends on the level of nodule permeability prior the assay. Consequently, a large C_2H_2-ID may occur whenever an environmental factor limits nodule development and growth more than that of other plant organs, therefore causing a saturation of the capacity of the symbiosis to satisfy plant N demand. More comparative studies are needed to address how growth, anatomy and permeability of nodule interact with the C_2H_2-ID and the environment, including P deficiency and solid versus liquid substrates for growth of legume-symbioses.

References
Drevon JJ, Kalia VC, Heckmann MO & Pédelahore P. 1988. *Plant Physiol. Biochem.* 26, 73-8

Drevon JJ, Deransart C, Irekti H, Payré H, Roy G & Serraj R. 1995. *In* Drevon JJ (ed). *Facteurs Limitant la Fixation Symbiotique de l'Azote dans le Bassin Méditerrranéen.* INRA Editions (Les Colloques N° 77), Paris, pp. 73-84

Hernandez G, Vasquez H, Toscano V, Sanchez T, Franchi-Alfaro A, Mendez N & Drevon JJ. 1993 *Trop. Agric. (Trinidad)* 70, 230-34.

Hunt S & Layzell DB. 1993. *Annu. Rev. Plant Physiol. Mol. Biol.* 44, 483-511

Israel DW. 1987. *Plant Physiol.* 69, 835-840

Minchin FR, Sheehy JE, Witty JF. 1986. *J. Exp. Bot.* 37, 1581-91

Pongsakul P & Jensen ES. 1991. *J. Plant Nutr.* 14, 809-23

Ribet J & Drevon JJ. 1995a. *Physiol. Plant.* 94, 298-304

Ribet J & Drevon JJ. 1995b. *J. Exp. Bot.* 46, 1479-86

Roy, 1993. Respiration et Diffusion de l'Oxygène dans la Symbiose Fixatrice d'Azote : Soja - *Bradhyryzobium japonicum.* ENSA Montpellier. 98 p

Rychter AM, Chauveau M, Bomsel JL & Lance C. 1992. *Physiol. Plant.* 84, 80-6

Serraj R, Fleurat-Lessard P, Jaillard B & Drevon JJ. 1995. *Plant Cell Environm.* 18, 455-62

Vadez V, Rodier F, Payré H, Beck D & Drevon JJ. 1996a. *Plant Physiol. Biochem.* (in press)

Vadez V, Beck D, Lasso JH & Drevon JJ. 1996b. *Physiol. Plant.* (in press)

Vidal R, Gerbaud A., Vidal D. & Drevon JJ. 1995. *J. Plant Physiol.* 145, 201-8

Carbon and Nitrogen Metabolism in Plant-Derived Ineffective Nodules of Pea (Pisum sativum L.)

Vassily I. Romanov[1], Anthony J. Gordon[2], Frank R. Minchin[2], John F. Witty[2], Leif Skøt[2], Caron L. James[2], Alexej Y. Borisov[3] and Igor A. Tikhonovich[3]

[1]A.N.Bach Institute of Biochemistry, Leninsky Prospect 33, Moscow, 117071, Russia.
[2]Institute of Grassland and Environmental Research, Aberystwyth, SY23 3EB, UK.
[3]All-Russian Institute of Agricultural Microbiology, Podbelsky 3, St Peterburg-Pushkin, 188620, Russia.

Keywords. Fix⁻ pea mutants, nodules, metabolism

1 Introduction

Organogenesis of legume root nodules and the construction of the nitrogen fixation system require exchanges of molecular signals between rhizobia and the host plant to activate the expression of all the necessary genes. Plant mutants with ineffective nodules are potentially useful for studies of the host-plant control and regulation of symbiotic nitrogen fixation. In this paper the nodules of three different Nod+ Fix- pea mutants (Sym 13, Sym 31 and FN1) as well as their parent lines (Sparkle, Sprint 2 and Rondo) formed on plants by the same rhizobial strain and grown in the same controlled environment cabinet, were compared for selected physiological and biochemical parameters. All mutants are monogenic, recessive and non-allelic. The Sprint 2 Fix- (Sym 31) mutant is characterized by a block in bacteroid differentiation and abnormal symbiosome structure (Borisov et al., 1993). By contrast, mutants E135 (Sym 13) and FN1 were characterized by early senescence of the symbiosomes and nodules as a whole (Kneen et al., 1990; Postma et al., 1990).

The purpose of this investigation was to determine: (1) nodule respiration and nitrogen fixation in relation to the appearance of nitrogenase proteins, (2) nodule structure and O_2 concentration profiles in relation to leghemoglobin (Lb) and Lb apoprotein content, (3) the concentration of key carbon reserve compounds and (4) the activity of key enzymes involved in carbon and nitrogen metabolism and immunodetection of enzyme proteins.

2 Results and Discussion

2.1 Nitrogen Fixation and the presence of Lb and nitrogenase proteins

The plant-derived Sym 31 and Sym 13 mutants of pea, inoculated with *Rhizobium leguminosarum* bv. *viciae*, RCR 1045, produced large (Sym 31) or small (Sym 13) white nodules which had no detectable nitrogenase (acetylene reduction) activity, whilst it was readily detectable in wild-type nodules on parental cultivars (Table 1). Unexpectedly, the FN1 mutant formed partially pink nodules which showed between 5 and 16% of the acetylene reduction activity of the Rondo parent line. This activity was enhanced by raising the external O_2 concentration (with a maximum at 40 - 50% O_2), confirming the nitrogenase origin of this acetylene reduction.

Table 1. Nitrogen fixation (μmol C_2H_2 g^{-1} DW min^{-1}) and Lb content (mg g^{-1} FW) in normal (Sprint 2 Fix$^+$, Sparkle and Rondo) and mutant (Sym 31, Sym 13 and FN1) nodules of pea. Data are means of at least 3 replicates. Standard errors were less than 25% of the mean for N_2 fixation and less than 10% for Lb determination.

	Sprint Fix$^+$	Sym 31	Sparkle	Sym 13	Rondo[*]	FN1[*]	Rondo	FN1
N_2 Fixation	2.01	nd	1.39	nd	1.53	0.24	1.70	0.09
Lb	4.68	nd	4.14	0.61	4.17	1.65	4.18	0.99

[*] Nodules from 3-week-old plants; all others were from 4-week-old plants.
nd = not detectable.

Western blotting with antiserum against nitrogenase revealed that the two components of nitrogenase were clearly visible in FN1 nodules and supports the observation of effective nitrogenase activity (Table 1). However, nitrogenase was hardly detectable in Sym 13 and completely absent in Sym 31 nodules. This data, obtained with antibodies from three different sources, did not confirm our previous results where we were able to detect nitrogenase component 1 in the bacteroids from Sym 31 nodules (Romanov et al., 1995). Lb was easily detectable in the nodules of FN1 and Sym 13 by immunoblotting (data not shown) and colourimetric assay (Table 1), but Lb was not detectable in the Sym 31 nodules. The levels of Lb appear to be correlated with the presence and activity of nitrogenase.

2.2 Nodule Structure and O_2 Gradients
The light micrographs of thin sections of all ineffective (or partially effective) pea nodules show a similar cortical structure to that of wild-type nodules with no obvious abnormalities when compared with nodules from parent plants (data not shown). Micro-electrode measurements of oxygen gradients across the nodule cortex showed that within all wild-types and FN1 mutant nodules, O_2 concentration decreased rapidly and was close to zero within about 350 μm of the nodule surface. In nodules of the Fix$^-$ mutants, the O_2 concentration also decreased rapidly. However, in Sym 13 nodules the O_2 gradient was not as steep so that O_2 became undetectable at a greater distance from the nodule surface (1400 μm). In contrast, in nodules of the Sym 31 mutant the O_2 concentration did not fall below 0.5%. There seems to be a correlation between Lb content (Table 1) and O_2 concentration in the central region of the mutant nodules.

2.3 Nodule Carbohydrates
Gas chromatograms of trimethylsilylated derivatives of soluble carbohydrates showed two major peaks, identified as sucrose and ononitol, from wild-type nodules (Table

2). Sucrose concentrations in the mutant nodules was either similar (for FN1) or 50-60% higher than that in nodules of the parent lines. On the contrary, ononitol concentration in all mutant nodules was lower than in the corresponding parental line. Hexose levels were much lower (data not shown).

Table 2. Carbohydrate content in normal and mutant nodules of pea (mg g^{-1} FW) Data are means of at least 3 replicates. Standard errors were less than 15% of the means.

	Sprint 2 Fix+	Sym 31	Sparkle	Sym 13	Rondo	FN1
Ononitol	3.79	0.27	2.65	1.89	2.76	1.17
Sucrose	5.34	7.73	2.87	4.71	2.66	2.29
Starch	30.16	33.08	9.14	20.66	10.62	23.08

2.4 Nodule Protein Content and Enzyme Activities

The low host plant protein levels in nodules of Sym 31 and Sym 13, in comparison with the wild-type nodule protein levels, suggest that these mutations affected the sequence of development events which normally lead to an approximately ten-fold increase in protein content. In the FN1 nodules, however, the higher protein content correlates with a partially effective symbiosis and the presence of both nitrogenase components and Lb (although in reduced amounts).

Table 3. Activity of host plant enzymes of carbon and nitrogen metabolism in normal and mutant nodules of pea (expressed as μmol min^{-1} mg^{-1} protein). Data are means of at least 3 replicates. Standard errors were less than 10% of the means.

	Sprint 2 Fix[+]	Sym 31	Sparkle[*]	Sym 13[*]	Rondo	FN1
Protein (mg g^{-1} FW)	25.01	4.40	16.04	7.59	25.43	18.24
SS	0.304	0.133	0.161	0.052	0.570	0.169
PEPC	0.408	0.225	0.153	0.017	0.550	0.174
ADH	0.107	0.627	0.089	0.367	0.197	0.213
GS	0.196	0.011	0.127	0.003	0.120	0.080
APAT	0.323	0.186	0.065	0.018	0.358	0.176

[*]Data of Sugunuma and La Rue (1993).

Of the 15 enzymes of carbon and nitrogen metabolism investigated, significant differences between normal and mutant nodules for all three pairs were found for the 5 enzymes listed in Table 3. The lower activity of sucrose synthase (SS), phosphoenolpyruvate carboxylase (PEPC) and alanine pyruvate aminotransferase (APAT) in the ineffective (or partly effective) nodules confirm that these enzymes play an important role in the nitrogenase-linked metabolism of normal nodules. Interestingly, the data of Suganuma and LaRue (1993) for the Sym 13 mutant showed exceptionally low activities of PEPC, APAT and GS in comparison with wild-type nodules and also with our data for both the less effective (Sym 31) and more effective (FN1) mutants. This may indicate a lesion which directly affects carbon metabolism in the Sym 13 mutant. The high activity of alcohol dehydrogenase (ADH) in Sym 31 and Sym 13 nodules, together with low internal O_2 concentrations, suggests that part of the carbon coming from shoots may be metabolized through anaerobic pathways. Glutamine synthetase (GS) activity was very low in Fix⁻ (Sym 31 and Sym 13) nodules, but not in the FN1 mutant, which was able to fix small amounts of nitrogen. Western immunoblotting confirmed reduced levels of SS in all mutants and the almost complete absence of GS in Sym 31 and Sym 13.

3 Conclusions

(i) In spite of information available from the literature, in our experiment nodules on the FN1 mutant (formed from *R. leguminosarum* RCR 1045) were able to fix some nitrogen. However, nitrogenase activity and Lb levels were apparently reduced between 21 and 28 days, suggesting that the plant was unable to sustain effective nodules for more than a few weeks.

(ii) In Sym 31 Fix⁻ nodules (blocked in bacteroid differentiation and with an abnormal symbiosome structure) we were unable to find any traces of nitrogenase proteins or Lb (Lb-apoprotein hardly detectable) and the O_2 concentration in the central region was 0.5-3.0%.

(iii) In Sym 13 Fix⁻ nodules (with early senescence of the symbiosomes) nitrogenase components were found in small amounts, Lb content was 15-18% of that in the parent line (Lb-apoprotein was easily detectable using antibodies) and the O_2 concentration in the central region was very low. Enzyme data also suggests that carbon metabolism may be impaired.

(iv) The high level of starch content in the nodules of all 3 mutants suggests there was no block in carbon supply to nodules from the shoots.

(v) The very low ononitol content in Sym 31 (but not in Sym 13) nodules suggest that this component may play an important role in symbiosome formation.

4 Acknowledgments

Dr. Romanov's visit to IGER was funded by The Royal Society and an IGER Fellowship. The work was also supported by The Russian Fund for Fundamental Science (A.Borisov and I.Tikhonovich).

5 References

Borisov A.Y. et. al., (1992) Symbiosis 14, 293-313.
Kneen B.E. et. al., (1990) Plant Physiology 94, 899-905.
Postma J.G. et. al., (1990) Plant Science 68, 151-161.
Romanov V.I. et. al., (1995) J.Exp.Botany 46, 1809-1816.
Suganuma N. and La Rue T.A. (1993), Plant Cell Physiol. 34, 761-765.

Part 9

Oxygen Regulation in Nitrogen Fixation

Genetic Regulation and Bioenergetics of Symbiotic Nitrogen Fixation in *Bradyrhizobium japonicum*

H. Hennecke, M. Babst, H.M. Fischer, T. Kaspar, I. Kullik, D. Nellen-Anthamatten, O. Preisig, P. Rossi, K. Schneider, L. Thöny-Meyer and R. Zufferey

Mikrobiologisches Institut, Eidgenössische Technische Hochschule, Schmelzbergstrasse 7, CH-8092 Zürich, Switzerland

The low free oxygen concentration (< 25 nM) within soybean root nodules leads to two consequences that are relevant for this report: (i) it induces the synthesis of a high-affinity terminal oxidase that enables endosymbiotic bacteroids to respire at these extremely microaerobic conditions; (ii) it triggers the derepression of genes involved in the formation of the nitrogenase complex. We summarize here how both events are genetically controlled in *Bradyrhizobium japonicum*, and report on some recently discovered functions of the regulated genes.

1. The FixLJ-FixK₂ Regulatory Cascade

B. japonicum possesses an oxygen-controlled two-component regulatory system (FixL, FixJ) similar as that present in *Rhizobium meliloti* (1). Like in many other rhizobia, FixJ controls another regulatory gene, *fixK*, whose product activates a number of different target genes and operons (2). We had previously described a *B. japonicum* *fixK*-like gene (now called *fixK₁*) which did not play a role in symbiotic nitrogen fixation even though it was subject to control by FixJ (3). Intriguingly, when *fixK₁* was expressed from a constitutive promoter, it partially complemented the defective phenotype of a *fixJ* mutant (3). The enigma was solved by the discovery of a second *fixK*-like gene (termed *fixK₂*) which was shown to be regulated by FixJ and whose product, the FixK₂ protein, was an activator for the expression of the *fixK₁* gene (4). As expected, a *fixK₂* mutant had a Fix⁻ phenotype, suggesting that *fixK₂*, but not *fixK₁*, was the regulator of symbiotically relevant target genes (see below). What role FixK₁ plays under microaerobic or anaerobic conditions it not known.

2. Genes Controlled by the FixLJ-FixK₂ System

Apart from *fixK₁* we have identified at least three further target genes or operons that are under the control of the FixLJ-FixK₂ regulatory cascade: (i) the *fixNOQP* operon (5, 6); (ii) the *fixGHIS* operon (5, 7); and the *rpoN₁* gene, one of two homologous σ^{54} genes present in *B. japonicum* (8, 9). The promoter regions of all of these genes and operons share a characteristic FixK binding site centered around nucleotide position −40/−41 relative to the transcription start site (2), and the dependency of transcription from these promoters on microaerobiosis and an intact FixLJ-FixK₂ cascade has been documented by various means (e.g. primer extension analysis of mRNA, expression of reporter gene fusions). In addition, we have evidence that suggests a FixLJ-FixK₂-dependent regulation of certain genes involved in anaerobic respiration of *B. japonicum* with nitrate as the terminal electron acceptor (1, 3, 4); however, no attempts were made to identify those genes.

NATO ASI Series, Vol. G 39
Biological Fixation of Nitrogen
for Ecology and Sustainable Agriculture
Edited by A. Legocki, H. Bothe, A. Pühler
© Springer-Verlag Berlin Heidelberg 1997

3. Functions of the Proteins Encoded by the *fixNOQP* and *fixGHIS* Operons

Mutants with lesions in *fixN*, *fixO* and *fixP*, but not *fixQ*, show a severe defect in symbiotic nitrogen fixation (Fix⁻) and a strong decrease in cytochrome oxidase activity (6, 10). Likewise, a deletion affecting the *fixGHIS* operon displayed similar phenotypes, suggesting a functional connection between both operons (7). The FixN, FixO and FixP proteins are the subunits of a so-called *cbb₃*-type cytochrome oxidase in which subunit I (FixN) binds two B-type hemes, and a copper ion is probably associated with one of the B hemes in a binuclear heme-Cu$_B$ metal center, an arrangement typically found in all heme-copper oxidases known to date (10-12). FixO and FixP are membrane-anchored mono- and di-heme *c*-type cytochromes (6, 10). The *B. japonicum cbb₃* oxidase was solubilized from the cytoplasmic membrane and partially purified; it contained FixN, FixO and FixP as shown by N-terminal sequence analysis and immunoblotting. The oxidase had a K_M for O$_2$ as low as 7 nM which is six- to eightfold lower than that of the aerobic *aa₃*-type cytochrome *c* oxidase (13). This suggests strongly that the *cbb₃*-type cytochrome oxidase terminates a symbiosis-specific respiratory chain supporting microaerobic respiration in endosymbiotic bacteroids.

Only traces, if any, of cytochrome *cbb₃* subunits were present in membranes isolated from a Δ*fixGHI* mutant (7). As this was not due to lack of *fixNOQP* transcription and translation, the results suggested a critical involvement of the *fixGHIS* gene products in the assembly and/or stability of the *cbb₃*-type heme-copper oxidase (7). On the basis of sequence similarities between the FixI protein and a Cu-transporting P-type ATPase (CopA) of *Enterococcus hirae*, and between FixG and a membrane-bound oxidoreductase (RdxA) of *Rhodobacter sphaeroides* (7), we postulate that a membrane-bound FixGHIS complex might play a role in uptake and metabolism of copper required for the *cbb₃*-type heme-copper oxidase.

In conclusion the FixLJ-FixK$_2$ regulatory cascade primarily controls genes that are involved in microaerobic or anaerobic respiratory energy conservation in *B. japonicum*.

4. The NifA Cascade

A second set of *B. japonicum* genes induced in response to microaerobiosis includes those that are concerned with the formation of an active nitrogenase enzyme complex. Apart from the nitrogenase structural genes (*nifH*, *nifDK*) and genes for metal cofactor synthesis (e.g. *nifEN*, *nifB*, *nifS*) this group also contains a special copy of the chaperonin genes, *groESL₃* (see section 6). All of these genes have in common that they are transcribed by the σ⁵⁴-RNA polymerase and activated by the oxygen-responsive NifA protein. Two overlapping promoters are responsible for expression of the *nifA* gene, and transcription of both is positively regulated in a complex way (see section 5) so that NifA, like FixK$_2$ (see above), is also part of a regulatory cascade. There is a connection between the NifA cascade and the FixLJ-FixK$_2$ cascade in that the latter activates one of the two σ⁵⁴ genes (*rpoN₁*) needed in the former (8, 9). While the structural features of NifA and the possible mechanisms by which this protein might be modulated by oxygen have been reviewed extensively (2, 14) we focus in the following two sections on two other aspects of the NifA cascade.

5. Regulation of the *fixRnifA* Operon

The *B. japonicum fixRnifA* operon is already transcribed during free-living, aerobic growth, albeit at low level. Synthesis of the corresponding transcript (T_2) is probably driven by a −35/−10-type promoter (15). Under low-oxygen conditions, transcription initiates predominantly at a −24/−12-type promoter leading to transcript T_1 whose start site is just two nucleotides upstream of the T_2 start site (15). While transcription from the −24/−12 promoter requires the σ^{54}-RNA polymerase and (auto)activation by NifA, the −35/−10 promoter is probably recognized by the housekeeping σ factor and activated by a postulated activator binding further upstream of the promoter (15, 16). Some progress has been made recently towards the characterization of the protein that binds to the *fixR*-upstream region around position −68. Using a DNA mobility shift as a specific assay, this protein was purified 6'000-fold from aerobically grown *B. japonicum* cells and shown to have an apparent molecular mass of 21.5 kDa as determined by SDS-gel electrophoresis (17). The N-terminal amino acid sequence of this protein was then determined, which should now pave the way to identify its gene.

6. What Is the Role of the NifA-regulated *groESL₃* Genes?

Five highly conserved, but differently regulated *groESL* operons are present in *B. japonicum* (18, 19). Most strikingly, expression of *groESL₃* is co-regulated with symbiotic nitrogen fixation genes via σ^{54} and NifA (18). Knock-out mutation of individual *groEL* genes does not impair symbiotic nitrogen fixation activity of the respective mutant strains. By contrast, the *groEL₃₄* double mutant strain D4, which is mutated in those *groEL* genes that contribute most to the GroEL pool under symbiotic conditions, exhibits less than 5% Fix activity as compared with the wild type (20). Expression of *lacZ* fusions to six representative *nif* and *fix* genes is hardly affected in mutant D4. However, the level of NifH protein was found to be drastically reduced in extracts prepared from D4 bacteroids or from free-living cells grown anaerobically. These results indicate that efficient synthesis of NifH requires GroEL chaperonin at a post-transcriptional level. Transcriptional fusions of the *groESL₃* promoter (P3) to all five *B. japonicum groESL* operons as well as to *groESL* from *E. coli* were integrated into the chromosome of mutant D4 and the resulting strains were tested for their ability to restore the symbiotic defect of D4. Partial complementation was observed in strains harboring P3 fused to *groESL₁*, *groESL₂*, *groESL₅* or *E. coli groESL* whereas the full wild-type phenotype was restored in strains complemented with P3 fused to *groESL₃* (control) or *groESL₄* (20). Thus, the function of GroEL₃ and GroEL₄ in symbiosis can be substituted partially by forced expression of alternative *B. japonicum* or *E. coli groESL* operons. In conclusion, the complementation analyses were not indicative of a strict specificity of any of the *B. japonicum groESL* gene products, which is in good agreement with their high degree of sequence conservation.

Acknowledgements. Research in our laboratory has been supported by grants from the Swiss National Foundation for Scientific Research and from the Federal Institute of Technology, Zürich.

References

1. Anthamatten, D. and Hennecke, H. (1991) Mol. Gen. Genet. **225**, 38-48.
2. Fischer, H.M. (1994) Microbiol. Rev. **58**, 352-386.
3. Anthamatten, D., Scherb, B. and Hennecke, H. (1992) J. Bacteriol. **174**, 2111-2120.
4. Nellen-Anthamatten, D. and Rossi, P., unpublished results.
5. Preisig, O. (1995) Ph.D. Thesis no. 11306, Federal Institute of Technology Zürich.
6. Preisig, O., Anthamatten, D. and Hennecke, H. (1993) Proc. Natl. Acad. Sci. USA **90**, 3309-3313.
7. Preisig, O., Zufferey, R. and Hennecke, H. (1996) Arch. Microbiol. **165**, 297-305.
8. Kullik, I., Fritsche, S., Knobel, H., Sanjuan, J., Hennecke, H. and Fischer, H.M. (1991) J. Bacteriol. **173**, 1125-1138.
9. Kullik, I. (1992) Ph.D. Thesis no. 9895, Federal Institute of Technology Zürich.
10. Zufferey, R., Preisig, O., Hennecke, H. and Thöny-Meyer, L. (1996) J. Biol. Chem. **271**, 9114-9119.
11. Iwata, S., Ostermeier, C., Ludwig, B. and Michel, H. (1995) Nature **376**, 660-669.
12. Tsukihara, T., Aoyama, H., Yamashita, E., Tomizaki, T., Yamaguchi, H., Shinzawa-Itoh, K., Nakashima, R., Yaono, R. and Yoshikawa, S. (1996) Science **272**, 1136-1144.
13. Preisig, O., Zufferey, R., Thöny-Meyer, L., Appleby, C.A. and Hennecke, H. (1996) J. Bacteriol. **178**, 1532-1538.
14. Morett, E. and Segovia, L. (1993) J. Bacteriol. **175**, 6067-6074.
15. Barrios, H., Fischer, H.M., Hennecke, H. and Morett, E. (1995) J. Bacteriol. **177**, 1760-1765.
16. Thöny, B., Anthamatten, D. and Hennecke, H. (1989) J. Bacteriol. **171**, 4162-4169.
17. Kaspar, T., unpublished results.
18. Fischer, H.M., Babst, M., Kaspar, T., Acuña, G., Arigoni, F. and Hennecke, H. (1993) EMBO J. **12**, 3373-3383.
19. Babst, M., Hennecke, H. and Fischer, H.M. (1996) Mol. Microbiol. **19**, 827-839.
20. Fischer, H.M., Babst, M. and Schneider, K., unpublished results.

Dynamic Control of Oxygen Diffusion Resistance in Nodules

John F. Witty and Frank R. Minchin
Institute of Grassland and Environmental Research, Aberystwyth, SY23 3EB, UK.

Keywords. Legume nodules, oxygen, hydrogen, control of fixation

1 Introduction

Excess inputs of O_2 to infected cells of legume nodules will lead to loss of nitrogenase activity (Witty et al., 1986), whilst insufficient O_2 initially curtails N_2 fixation and eventually causes oxidant-induced nodule senescence (Escuredo et al., 1996). There is now a considerable weight of evidence showing that the O_2 supply to N_2-fixing bacteroids is controlled by a variable physical barrier to diffusion which balances O_2 influx with its respiratory consumption. However, some workers have suggested biochemical/metabolic alternatives which might substitute for a physical barrier.

Evidence for a physical barrier is supplied by (a) micro-electrode studies which show decreases in pO_2 across the nodule cortex (Witty et al., 1987), (b) only small increases in nodular O_2 uptake when external O_2 is increased by a factor of 3 (Minchin and Witty, 1990), (c) H_2 micro-electrode studies which show that intercellular spaces within the infected zone are interconnected (Witty, 1991) and (d) He/O_2 studies which show open pores traversing the cortex which close under stress (Witty and Minchin, 1994). Arguments against a physical barrier are largely based upon a lack of evidence for rapid physical changes within the nodule cortex (Van Cauwenberge et al., 1994) but also include speculation about changes in carbohydrate biochemistry within the infected cells (Streeter, 1995).

Within Hup^{-ve} nodules H_2 from nitrogenase cannot be further metabolized and the relationship between the internal concentration of this gas and H_2 efflux from the nodule and can only be determined by the resistance of a physical barrier to diffusion. This paper presents data on H_2 concentrations and efflux which, we believe, provides incontrovertible evidence for such a variable barrier.

Continuous measurements of CO_2 and H_2 production by intact undisturbed nodulated roots of Soyabean (cv. Clarke inoculated with Hup^{-ve} USDA 16) were made using an open, flow-through system (Minchin et al., 1983). Plants were grown in cabinets under conditions described by Minchin and Witty (1990). H_2 concentrations inside attached root nodules were measured using H_2 specific micro-electrodes (Witty, 1991).

2 Results and Discussion

2.1 Effects of argon stepping

Hydrogen production increased progressively as N_2 in the gas stream was replaced with Ar (Fig 1) because of both the saturation kinetics of nitrogenase with respect to N_2 and inhibition of N_2 reduction by H_2. Respiratory CO_2 production remained constant despite the diversion of e^- to H^+ until almost all the N_2 had been displaced. Internal pH_2 increased in parallel with H_2 production until the rapid increase and subsequent decline in production when pN_2 reached zero. Relative diffusion resistance to H_2 ($R[H_2]$), calculated by applying Fick's Law to rates of H_2 efflux and internal pH_2, remained constant up to this point and then increased by a factor of 4.4 (Table 1).

Figure 1. Effect of displacing gas phase N_2 with Ar on H_2 production and respiration. The O_2 concentration was maintained at 20.9 %. Numbers above arrows show pN_2 (At).

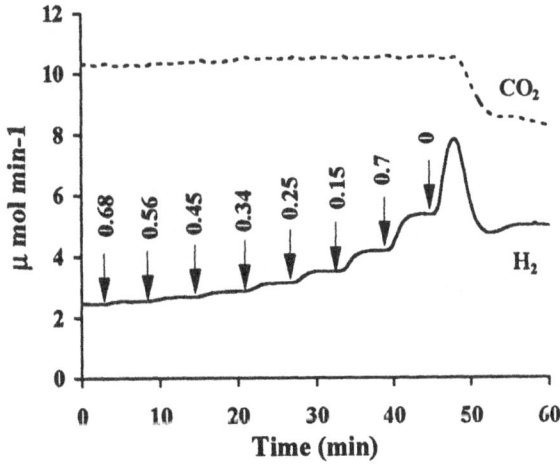

Table 1. Effect of N_2 partial pressure on H_2 production, pH_2 within the nodule and on calculated relative diffusion resistance to H_2 ($R[H_2]$). SE < 10% of the mean.

PN_2	H_2 Production (μ mol min^{-1})	H_2 Concentration (μ mol m^{-3})	$R[H_2]$ (Ratio)
0.69	1.23	0.40	1.00
0.45	1.33	0.44	1.02
0.25	1.55	0.52	1.02
0.07	2.09	0.66	0.97
0.00[a]	3.90	1.77	1.39
0.00[b]	2.20	3.12	4.43

[a]From maximum rate of H_2 production (50 min, Fig 1)
[b]Following maximum rate of H_2 production (55 min, Fig 1)

2.2 Effects of O_2 stepping

Initial increases in pO_2 from 20.9 to 30% and from 30 to 40% resulted in a transient decrease in H_2 production and associated respiration but these decreases disappeared with higher O_2 steps to 50 and 60%, suggesting that they do not relate to O_2 damage of nitrogenase (Fig. 2).

Internal pH_2 increased to a greater extent than H_2 production and $R[H_2]$ increased steadily to 4.5 at 60% O_2 (Table 2). The observed increase in H_2 production from 0.8 to 1.2 μ mol min^{-1} as O_2 increased to 60% does not appear to be associated with an increase in overall enzyme activity because, following perturbations after each O_2 step, respiration returned to its initial rate (Fig. 2). Thus, this increase appears to be due to the same causes as the increase in the Ar experiment.

Minchin and Witty (1990) demonstrated that RQ values for respiration of soyabean nodules remained close to unity as pO_2 increased and it can be inferred from rates of CO_2 production in these experiments that the rate of O_2 influx remained constant. Thus, values for relative diffusion resistance with respect to O_2 $R[H_2]$ can be calculated from external pO_2 on the assumption that internal concentrations are negligible. The 60% O_2 value for $R[O_2]$ of 2.87 is considerable lower that the $R[H_2]$ value (Table 2)

Figure 2. Effect of stepped increases in pO_2 on H_2 production and respiration by attached, undisturbed soyabean nodules.

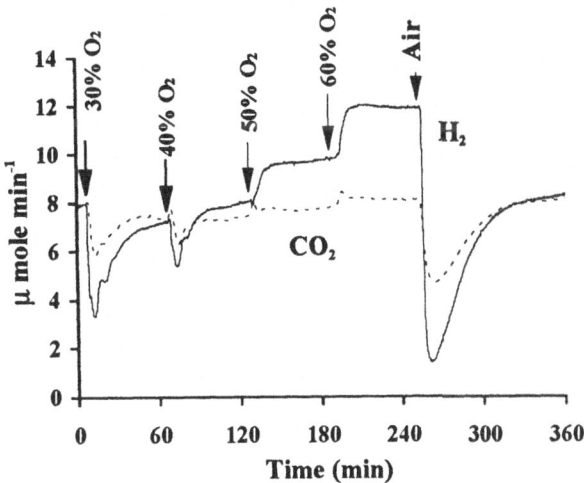

Table 2. Effect of stepped O_2 increases on H_2 production, H_2 concentration and calculated diffusion resistance to H_2 ($R[H_2]$) and O_2 ($R[O_2]$). SE<10% of mean

pO_2 (%)	H_2 Production (μ mol min^{-1})	H_2 Concentration (μ mol m^{-3})	$R[H_2]$ (Ratio)	$R[O_2]$ (Ratio)
20.9	0.83	0.41	1.00	1.00
30	0.73	0.58	1.57	1.43
40	0.87	1.22	2.81	1.91
50	1.08	1.64	3.24	2.39
60	1.24	2.79	4.52	2.87

2.3 Effects of de-topping

Both H_2 production and respiration declined progressively during a 120 min period following de-topping; to about one half and one third of the initial values, respectively (Table 3), whilst internal H_2 concentration remained constant and $R[H_2]$ increased by a factor of 4. In the same experiment, values for $R[O_2]$; based on the intercept value for CO_2 upon H_2 (2.23 μ mol CO_2 min^{-1}, r^2= 0.998), an RQ of unity and negligible internal O_2, increased by a factor of 3 (Table 3).

Table 3. Effect of de-topping on H_2 production, concentration and calculated diffusion resistance to H_2 ($R[H_2]$) and O_2 ($R[O_2]$). SE < 10% of the mean.

Time (min)	H_2 Production (μ mol min^{-1})	H_2 Concentration (μ mol m^{-3})	$R[H_2]$ (Ratio)	$R[O_2]$ (Ratio)
0	1.22	0.41	1.00	1.00
30	1.14	0.41	1.07	0.98
60	0.72	0.46	1.88	1.45
90	0.45	0.48	3.11	2.19
120	0.33	0.44	3.99	3.03

3 General Discussion
3.1 Existence and position of a physical barrier

Rates of H_2 production from nitrogenase can be affected by a plethora of biochemical processes. However, the only factor which can relate changes in internal pH_2 with rates of efflux from the nodules is diffusion resistance. Thus, measured changes in the relationship between these two parameters, as presented above, can only be interpreted by accepting the existence of a variable physical barrier.

At the termination of all the above experiments micro-electrode measurements showed that H_2 gradients across the infected zone of the nodules were almost flat, indicating that the spaces between infected cells remained open and interconnected. Thus, the impediment to H_2 efflux is outside this zone and corresponds with the rapid decrease in H_2 concentration observed across the inner cortex (Witty 1991).

3.2 Operating Mechanism - differences between $R[H_2]$ and $R[O_2]$

In experiments to determine the nature of the diffusion barrier Witty and Minchin (1994) exploited the fact that O_2 will diffuse through a He atmosphere 3.7 times faster than through air, but that O_2 flux through a liquid phase is unaffected by changes in the mixing gas. We concluded that with unstressed soyabean nodules about half of the O_2 flux to the infected zone passed through air filled pores which closed to produce a liquid filled barrier in response to stress. One consequence of this change arises from the fact that the relative rates of diffusion of H_2 and O_2 through gas and liquid phases are different. Denison et al. (1992) calculated that with an "air pore" barrier internal pH_2 would be 0.87 Kp while with a liquid barrier, having the same resistance to O_2, pH_2 would increase to 2.5 Kp. These predictions represent a 2.87 fold increase in the ratio of $R[H_2]:R[O_2]$, whilst the increase when going from a system where half the O_2 passes through open pores to a completely liquid filled barrier would be 1.44.

In the experiments reported above the ratio $R[H_2]:R[O_2]$ increased to 1.6 at 60% pO_2 (Table 2) and to about 1.3 after de-topping (Table 3). This suggests that all (O_2 stepping) or almost all (de-topping) of the open pores were lost as a result of these treatments. Such pore closures would be consistent with the operation of osmocontractile cells within the inner cortex or the production of intercellular occlusions within the mid-cortex (Minchin, 1996).

The H_2 data above confirms the presence of a variable physical barrier to diffusion in the nodule cortex. This does not preclude the existence of other metabolic 'fine tuning' mechanisms at the infected cell level (Bergersen 1994, Thumfort et al 1994).

4 References

Bergersen (1994) Protoplasma 183, 49-61

Denison R.F. et. al., (1992) Plant Physiology 100, 1863-1868

Escuredo P.R. et. al., (1996) Plant Physiology 110, 1187-1195.

Minchin F.R. et. al., (1983) J. Exp. Botany 34, 641-649.

Witty J.F. et. al., (1986) Oxford Surveys of Plant Molecular and Cell Biology 3, 275-314.

Minchin F.R. and Witty J.F. (1990) J. Exp. Botany 41, 1271-1277

Minchin F.R. (1996) Soil Biol. Biochem. (in press).

Witty J.F. et. al., (1987) J. Exp. Botany 38, 1129-1140.

Streeter J.G. (1995) J. Theor. Biol. 174, 441-452.

Thumfort P.P. et al, (1994) Plant Physiology 105, 1321-1333

Van Cauwenberghe O.R. et. al, (1994) Physiol. Plant. 91, 477-487.

Witty J.F. (1991) J. Exp. Botany, 42, 765-771.

Witty J.F. and Minchin F.R. (1994) J. Exp. Botany 45, 967-978.

Is the Variable Oxygen Permeability in Nodules a Physical or a Physiological Phenomenon?

Ueli A. Hartwig[1], Jan Trommler[1], Carina Weisbach[1], Paul Walther[2],
Paola Curioni[1], Kathryn A. Schuller[3] and Josef Nösberger[1]

[1] Institute of Plant Sciences, ETH-Zürich, 8092 Zürich, Switzerland
[2] Laboratory for Electron Microscopy 1, Institute of Cell Biology, ETH-Zürich, 8092 Zürich, Switzerland
[3] School of Biological Sciences, Flinders University, Adelaide S.A. 5001, Australia

1. Introduction

Beside persistent adaptations (e.g. nodule morphology or nodule number) of N_2 fixing symbiotic systems to distinct environmental conditions, there exist highly flexible mechanisms to quickly and reversibly change N_2 fixing- (nitrogenase) activity in the root nodules. Such changes usually take place within less than an hour to up to three to four hours (Witty et al. 1984; Hartwig et al. 1987). Due to the fact that quickly depressed nitrogenase activity can be reversed by supplying an elevated rhizospheric pO_2 this mechanism is usually referred to as "changing O_2 permeability", "changing O_2 diffusion resistance" or "changing O_2 barrier", however without really understanding the underlying mechanisms.

Many attempts have been made to understand the nature of this variable O_2 permeability. Nevertheless, no convincing model is available yet. If one presumes that a common mechanism exists, the following phenomena must be explainable by a model: i) response within less than one to a few hours, ii) depression of nitrogenase activity induced by environmental perturbations is immediately reversible by gradually increasing rhizospheric pO_2, iii) responsive to changing rhizospheric pO_2 and iv) inducible through preventing current N_2 fixation by replacing rhizospheric N_2 with Ar or by supplying 10% C_2H_2.

2. Can morphological changes in the nodules account for the rapidly changing O_2 permeability?

The association of persistent morphological features with differences in O_2 permeability in nodules implies that short term and reversible changes in gas permeability could also be regulated by morphological adaptations. In such models, reduced O_2 permeability could be associated with changes in the volume or percentage of liquid- or gas- filled intercellular spaces in the nodules (Van Cauwenberghe et al. 1994; Serraj et al. 1995). Another hypothesis, based on investigations with soybean and lupin, states that O_2 permeability is regulated by an occluding glycoprotein in the intercellular spaces of the nodule cortex (Iannetta et al. 1995). This study

NATO ASI Series, Vol. G 39
Biological Fixation of Nitrogen
for Ecology and Sustainable Agriculture
Edited by A. Legocki, H. Bothe, A. Pühler
© Springer-Verlag Berlin Heidelberg 1997

provides in fact the only evidence so far for a morphological response within less than one hour.

In an attempt to test these hypotheses on typical indeterminate lucerne nodules, the cortex region was investigated in a cry- scanning-electron microscope (cryo-SEM) after using liquid ethane to rapidly freeze the nodules. Subsequent investigations of freeze fractured nodules showed no changes in the water content of intercellular spaces in the nodule cortex after decreasing the nodule O_2 permeability by either detopping the plants or after carefully adapting the nodules to a rhizospheric pO_2 of 80 kPa (Weisbach et al. unpublished data). An antibody against the sugar epitope of a glycoprotein (generously provided by Dr. N. J. Brewin, Norwich, UK) cross-reacted with an antigen in the intercellular spaces of lucerne nodules. However, changes in nodule O_2 permeability had no effect on the abundance of label in the intercellular spacs, as investigated on immuno-gold labelled thin sections in the transmission electron microscope. Likewise, the percentage of intercellular spaces labelled with the antibody remained unchanged. These data do not support the hypothesis that fast and reversible changes in the O_2 permeability of indeterminate nodules like the ones in lucerne are associated with morphological changes.

3. The fast and reversible changes in nitrogenase activity are associated with an apparent change in nodule oxygen permeability

The only unquestioned knowledge about the fast and reversible changes in nitrogenase activity we have so far is that we can at least partially reverse them by supplying extra O_2 in the rhizosphere. In order to explain this phenomenon further, an understanding of the physiological processes at the plant level that are associated with the appropriate changes in nitrogenase activity is a prerequisite. In the view that nitrogen fixation is a fine-tuned process between two organisms, such an understanding will finally link the physiological changes in plants and nodules to the changes in nitrogenase activity and O_2 permeability.

As one attempt to gain such a fundamental understanding, the possible role of carbon metabolites either from storage or from current photosynthesis was investigated. In these studies, Hartwig et al. (1990) demonstrated that a 2-fold increase in the amount of carbon reserves cannot prevent a fast and dramatic decrease in nitrogenase activity after a complete defoliation of white clover. In further studies, Denison et al. (1992) and Weisbach et al. (1996) provided evidence that nitrogenase activity is always O_2 and not carbon limited after defoliation in lucerne, birdsfoot trefoil and white clover. The hypothesis that carbon sources from current photosynthesis are governing nitrogenase activity must also be rejected since Culvenor and Simpson (1990) were not able to restore nitrogenase activity after partial defoliation of *Trifolium subterraneum* by increasing net photosynthesis to the pre-defoliation value by elevated atmospheric pCO_2 and supplemental light. The same conclusion was drawn by Hartwig et al. (1994) after they demonstrated that even when current photosynthesis was prevented for six hours by exposing the shoot to a CO_2 partial pressure of less than 10 µl l^{-1}, nitrogenase activity remained unaffected.

As another attempt to understand nodule activity in relation to plant physiology, amino acid concentrations in the xylem of white clover were measured before and after a complete defoliation. In these experiments we observed a fast and significant increase in the concentrations of several amino acids (Table 1).

Table 1. Asparagine and glutamine concentrations in the xylem sap of white clover (*Trifolium repens* L.) before and one hour after complete defoliation. Means of 5 replicates ± SE are shown (Trommler et al. unpublished data).

Amino acid	before defoliation	30 minutes after defoliation
	(μmol ml^{-1})	
Asparagine	3.01 ± 0.25	6.03 ± 0.75
Glutamine	0.77 ± 0.05	2.05 ± 0.17

The increase in the concentrations of key amino acids clearly indicates that after a complete defoliation the supply of amino compounds from the nodules is in excess compared to the demand in the sink organs. This is consistent with the view that after defoliation the N sink strength is temporarily reduced (Hartwig et al. 1994). Assuming a xylem-phloem continuum (Parsons and Baker 1996), it is thus feasible that a N-feedback mechanism is involved in the regulation of the apparent O_2 permeability. Evidence for the operation of such a mechanism on a short-term basis comes from work of Heim et al. (1993). In this study it was shown that the decline in nitrogenase activity in white clover nodules, within six hours of defoliation, was greatly reduced when N_2 in the rhizosphere was replaced by Ar. It was concluded that under such conditions where N_2 fixation was prevented, a possible N feedback mechanism could not became effective.

. A new concept

To date the focus has been on a physical rather than a physiological mechanism which could explain the response of nitrogenase activity to environmental changes such as defoliation, altered rhizosphere pO_2 or prevention of N_2 fixation through Ar or C_2H_2 exposure. We suggest that too little attention has been paid to a possible physiological mechanism. The N-feedback hypothesis may in this respect draw the path towards a physiological mechanism which integrates whole plant mechanism with nodule N metabolism. In such a concept, the nodules would sense the plant N status which could act on an energy dependant biochemical step. This concept would then explain why increasing rhizospheric pO_2 would stimulate nitrogenase activity after nodule perturbation.

5. References

Culvenor RA, Simpson RJ (1990) Studies on the relation between net photosynthesis and nitrogenase-linked respiration in subterranean clover. J Exp Bot 41:933-939

Denison RF, Hunt S, Layzell DB (1992) Nitrogenase activity, nodule respiration, and O_2 permeability following detopping of alfalfa and birdsfoot trefoil. Plant Physiol 98:894-900

Hartwig UA, Boller BC, Baur-Höch B, Nösberger J (1987) The influence of carbohydrate reserves on the response of nodulated white clover to defoliation. Ann Bot 65:97-105

Hartwig UA, Boller BC, Nösberger J (1987) Oxygen supply limits nitrogenase activity of clover nodules after defoliation. Ann Bot 59:285-291

Hartwig UA, Heim I, Lüscher A, Nösberger J (1994) The nitrogen-sink is involved in the regulation of nitrogenase activity in white clover after defoliation. Physiol Plant 92:97-105

Heim I, Hartwig UA, Nösberger J (1993) Current nitrogen fixation is involved in the regulation of nitrogenase activity in white clover (*Trifolium repens* L.). Plant Physiol 103:1009-1014

Iannetta PPM, James EK, Sprent Ji, Minchin FR (1995) Time-course of changes involved in the operation of the oxygen diffusion barrier in white lupin nodules. J Exp Bot 46:565-575

Parsons R, Baker A (1996) Cycling amino compounds in symbiotic lupin. J Exp Bot 47:421-429

Serraj R, Fleurat-Lessard P, Jaillard B, Drevon JJ (1995) Sructural changes in the inner-cortex cells of soybean root nodules are induced by short-term exposure to high salt or oxygen concentrations. Plant Cell Environ 18:455-462

Van Cauwenberghe OR, Hunt S, Newcomb W, Canny MJ, Layzell DB (1994) Evidence that short-term regulation of nodule permeability does not occur in the inner cortex. Physiol Plant 91:477-487

Weisbach C, Hartwig UA, Heim I, Nösberger J (1996) Whole nodule carbon metabolites are not involved in the regulation of the oxygen permeability and nitrogenase activity in white clover nodules. Plant Physiol 110:539-545

Witty JF, Minchin FR, Sheehy JE, Minguez MI (1984) Acetylene induced changes in the oxygen diffusion resistance and nitrogenase activity of legume root nodules. Ann Bot 53:13-20

Regulation of Nitrogen Fixation Genes by the NIFA and NIFL Regulatory Proteins

Ray Dixon[1], Sara Austin[1], Trevor Eydmann[1], Tamera Jones[1], Peter Macheroux[1], Erik Söderbäck[2] and Susan Hill[3]

[1]Nitrogen Fixation Laboratory, John Innes Centre, Colney Lane, Norwich NR4 7UH, UK
[2]Department of Botany, University of Stockholm, Stockholm, Sweden.
[3]School of Biological Sciences, University of Sussex, Brighton BN1 9QJ UK

1. Introduction

The extreme oxygen sensitivity of the nitrogenase component proteins and the energetic demands of nitrogen fixation impose stringent physiological constraints on diazotrophs and hence *nif* gene transcription is tightly regulated in response to fixed nitrogen and the external oxygen concentration (1). Over the past ten years it has become apparent that mechanisms for regulating expression of nitrogen fixation (*nif*) genes are surprisingly novel in a number of different respects. Firstly transcription from most *nif* promoters requires a novel form of RNA polymerase holoenzyme containing a unique sigma factor, σ^N (σ^{54}), which is competent to bind its target promoters but is inactive in the absence of a specific transcriptional activator. Secondly, the activator binds to upstream enhancer-like sequences and contacts the downstream bound σ^N-containing RNA polymerase via DNA looping. The principal role of the activator is to catalyse the isomerisation step in transcription initiation by interacting with the σ^N-holoenzyme to promote the transition of the closed promoter complex to the open complex. An unusual feature of this reaction with respect to prokaryotic transcription systems is that it requires the hydrolysis of a nucleoside triphosphate, principally ATP or GTP (2).

The enhancer binding protein, NIFA is ubiquitous among diazotrophic proteobacteria as an activator of *nif* gene transcription but the activity of this protein as well as the expression of *nifA* is regulated in a variety of different ways in nitrogen fixing bacteria. In diazotrophic representatives of the gamma subdivision of the proteobacteria, namely *Klebsiella pneumoniae*, *Enterobacter agglomerans* and *Azotobacter vinelandii*, NIFA activity is modulated by the product of the *nifL* gene which is co-transcribed with *nifA*. However searches for *nifL* in other diazotrophic purple bacteria have been uniformly unsuccessful and in these organisms, the NIFA protein itself appears to be intrinsically oxygen sensitive and in some examples may also be responsive to the level of fixed nitrogen. These differences in the regulation of NIFA activity in different diazotrophs are intriguing although it is not possible at present to speculate on the potential advantages which each distinct mode of regulation may confer. In this review we will focus on the characteristics of the NIFA and NIFL proteins in the obligately aerobic diazotroph, *A. vinelandii*.

NATO ASI Series, Vol. G 39
Biological Fixation of Nitrogen
for Ecology and Sustainable Agriculture
Edited by A. Legocki, H. Bothe, A. Pühler
© Springer-Verlag Berlin Heidelberg 1997

2. Domain Structure of NIFL and NIFA

NIFL and NIFA comprise an atypical two-component regulatory system in which sensory communication between the partners is apparently relayed by protein: protein communication rather than by phosphotransfer. *A.vinelandii* NIFA has the characteristic three-domain structure of σ^{54} (σ^N)-dependent transcriptional activators (Fig 1), comprising (i) an amino-terminal domain with potential regulatory properties, which is not homologous to the canonical phosphoacceptor domain of the response regulator family, (ii) a central domain possessing nucleoside triphosphatase activity, which interacts with the σ^{54}-RNA polymerase holoenzyme, and (iii) a C-terminal domain which recognises specific DNA target sites (3). Sequence analysis indicates that

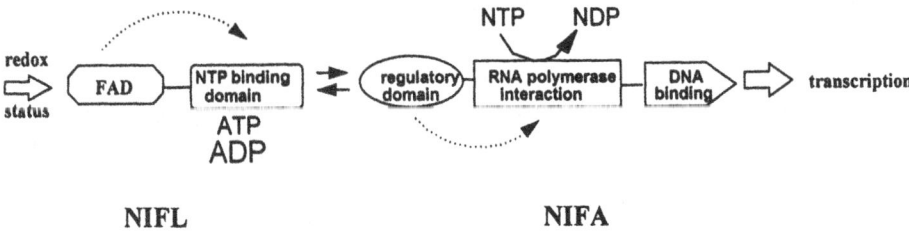

Fig1 Domain Structure of the *A.vinelandii* NIFL and NIFA proteins

A.vinelandii NIFL is comprised of two domains and we have confirmed this by limited proteolysis. The amino terminal domain appears to be sensory and shows homology to other oxygen responsive regulatory proteins including FixL and Bat whereas the C-terminus resembles the conserved transmitter domain found in members of the histidine protein kinase family of two-component regulatory proteins (4) (Fig1). Although this domain possesses all five of the conserved regions found in other transmitters, including the region containing the conserved histidine residue which is the site of autophosphorylation, several substitutions of this residue do not impair NIFL function (5), and neither autophosphorylation of NIFL nor phosphotransfer to NIFA has been detected *in vitro* (6, 7). Signalling between NIFL and NIFA does not require ATP hydrolysis but apparently involves stoichiometric interaction between both proteins. It therefore seems likely that sensory communication in the NIFL-NIFA system does not involve phosphorylation.

3. NIFL is a flavoprotein which modulates NIFA activity via a redox switch

Purified NIFL is yellow in colour indicative of a bound chromophore. The visible absorbance spectrum reveals an absorbance maximum of 446 nm with shoulders at 420 and 470 nm indicative of a bound flavin moiety (8). After unfolding of the protein in guanidium hydrochloride, FAD is released from the protein. The FAD appears to be quite tightly bound to the protein since it is not released on precipitation with trichloroacetic acid; CD spectra indicate that the FAD is bound in an asymmetric environment and that interactions between the ribityl side chain and the apoprotein occur. The FAD moiety is slowly reduced by dithionite and is difficult to reduce photochemically, suggesting that NIFL may interact with a specific physiological reductant *in vivo*. The flavin is located in the amino terminal domain of the protein, since the purified amino terminus has similar spectral features to intact NIFL.

When NIFL is overexpressed aerobically in nitrogen-rich medium and purified under aerobic conditions it is competent to inhibit NIFA activity *in vitro* (9). Since NIFL itself does not have any known catalytic function it is necessary to measure its ability to inhibit trancriptional activation by NIFA in order to assay its activity. The effect of NIFL on the formation of open promoter complexes by NIFA at the *nifH* promoter was measured in the presence or absence of sodium dithionite. To ensure that NIFL was maintained in the reduced form these experiments were carried out in a glove box under anaerobic conditions. Reactions were perfomed on linear DNA templates in the presence of σ^{54}-RNA polymerase holoenzyme, integration host factor and GTP. Promoter complexes which have undergone the transition from the closed to the open complex are resistant to heparin challenge and were quantitated using a gel shift assay (8). In the absence of NIFL, sodium dithionite had little effect on the activity of NIFA indicating that it did not strongly effect the ability of NIFA to catalyse open complex formation. However when NIFL was present in the oxidised form, the formation of open promoter complexes was strongly inhibited (Table 1). In contrast when NIFL was reduced with sodium dithionite, it failed to inhibit NIFA activity. Control experiments indicated that NIFL did not influence the activity of the related σ^{N}-dependent activator protein NTRC, either in the presence or absence of sodium dithionite. NIFL therefore specifically modulates the activity of NIFA and a major switch in activity is observed when NIFL is converted from the oxidised to the reduced form.

Table 1 Influence of redox status and ADP on NIFL activity

	% radioactivity in open complex [a]	
Additions [b]	GTP only [c]	GTP + ADP [d]
NIFA only	15.2	14.5
NIFA + dithionite	17.4	13.9
NIFA + NIFL	2.1	0.4
NIFA +NIFL + dithionite	25.2	0.4

[a] Reactions were carried out in an anaerobic glove box and quantitated as the percentage of total radioactivity in heparin resistant open complexes separated on native gels.
[b] Final concentration of NIFA and NIFL-6his was 200nM and 400nM respectively (calculated as the dimer forms).
[c] Final concentration 4mM.
[d] Final concentration 3.95 mM GTP plus 0.05mM ADP

4. NIFL activity is also responsive to adenosine nucleotides.

In addition to the switch in activity observed under reducing conditions, the activity of NIFL is also influenced by the presence of adenosine nucleotides., particularly ADP. When ADP was added to anaerobic reactions containing NIFA and GTP, NIFL strongly inhibited open complex formation by NIFA irrespective of whether sodium dithionite was present (Table 1). Although ATP also stimulated inhibition by NIFL, this inhibition is presumably a consequence of the formation of ADP by the catalytic activity of NIFA, since we find that this inhibition can be prevented by the addition of an ATP-regenerating system to the reaction mixture (9). Limited proteolysis experiments indicate that the conformation of the C-terminal domain of NIFL is altered by the binding of adenosine nucleotides, and that this domain has a higher affinity for ADP compared with ATP. The response of NIFL to ADP is apparently independent of its ability to sense the redox status, since, as expected, refolded NIFL apoprotein lacking FAD is not activated upon oxidation, but is still responsive to the presence of ADP.

5. Conclusions

As indicated in Fig 2 , NIFL appears to be activated independently by redox status and ADP and may exist in three distinct conformational states. This may allow

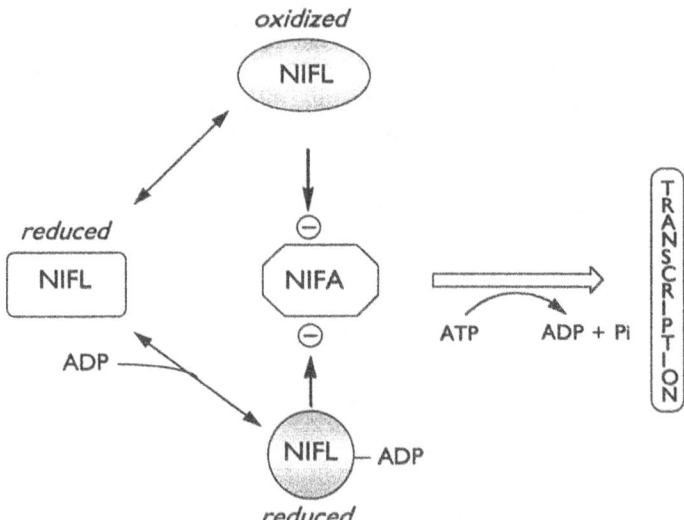

Fig 2 Model for Regulation of NIFA activity by NIFL

regulation in response to energy status *in vivo* in addition to   redox responsive modulation of NIFA activity.   A further sensing pathway which remains to be elucidated concerns the mechanism whereby NIFL responds to the nitrogen status.

6. References

1. Hill, S. (1992) in *Biological Nitrogen Fixation*, eds. Stacey, G., Burris, R. H. & Evans, H. J. (Chapman and Hall, New York), pp. 87-134.
2. Kustu, S., Santero, E., Keener, J., Popham, D. & Weiss, D. (1989) *Microbiol.Revs.* **53,** 367-376.
3. Morett, E. & Segovia, L. (1993) *J.Bacteriol.* **175,** 6067-6074.
4. Blanco, G., Drummond, M., Woodley, P. & Kennedy, C. (1993) *Mol.Microbiol.* **9,** 869-880.
5. Woodley, P. & Drummond, M. (1994) *Mol.Microbiol.* **13,** 619-626.
6. Lee, H.-S., Narberhaus, F. & Kustu, S. (1993) *J.Bacteriol.* **175,** 7683-7688.
7. Austin, S., Buck, M., Cannon, W., Eydmann, T. & Dixon, R. (1994) *J.Bacteriol.* **1 76,** 3460-3465.
8. Hill, S., Austin, S., Eydmann, T., Jones, T. & Dixon, R. (1996) *Proceedings of the National Academy of Sciences (USA)* **93,** 2143-2148.
9. Eydmann, T., Söderbäck, E., Jones, T., Hill, S., Austin, S. & Dixon, R. (1995) *J.Bacteriol.* **177,** 1186-1195.

Part 10

Model Plants for Nitrogen Fixation and Legume Genetics

MODEL PLANTS FOR NITROGEN FIXATION:
Genetic and Molecular Tools for Analysing the Infection and Infestation of Clovers

B G Rolfe, M A Djordjevic, J J Weinman, J W Redmond, S Natera, U Mathesius, N Guerreiro, C Pittock, A Morris, L Roddam, J de Majnik, E Gärtner and J McIver

Plant Microbe Interaction Group, Research School of Biological Sciences, Australian National University, PO Box 475, Canberra, ACT 2601, AUSTRALIA, Fax: (61)(6) 249-0754; Email: rolfe@rsbs-central.anu.edu.au

ABSTRACT.

Trifolium repens (white clover) and *Trifolium subterraneum* (subterranean clover) have proved to be good model systems for the investigation of *Rhizobium* symbiosis and plant defence systems. Considerable research has been done on the interaction of these clovers with *Rhizobium leguminosarum* bv. *trifolii* at the practical agricultural level. This work has now been supplemented with the application of many recent molecular techniques. Our research has investigated the interactions involved in development, defence and symbiosis of clovers. We have taken the dual approach of constructing transgenic plant tools and Proteome analysis to examine the activity of phytohormones and flavonoids as well as the roles of pathogenesis-related (PR) protein genes in these biological processes.

KEYWORDS. Transgenic plants, GUS reporter gene, proteome, legume root morphogenesis

INTRODUCTION.

Our investigation into the symbiotic relationship between *R. l.* bv. *trifolii* and white and subterranean clovers (1) and our molecular analysis of defence related genes (2) has enabled us to develop a range of ways to perturb the normal patterns of clover root development and to report on physiological and genetic processes occuring during plant microbe interactions. Our techniques range from bacterial molecular genetics, plant bioassays, plant tissue culture, cloning of plant genes and the construction of various transgenic plants, to the extensive use of 2-Dimensional Gel Electrophoresis (2-DE). We summarised a number of these studies and advantages of the *Trifolium* species in earlier publications (3, 4); this is shown in Fig. 1.

ROOT AND NODULE DEVELOPMENT:
PHYTOHORMONE LEVELS.

Spot inoculation using rooted leaf bioassays (5) were used to introduce auxin, cytokinins, flavonoids, NPA (N-(1-naphthyl) phthalamic acid), *Rhizobium* and *Rhizobium* Nod factors to precise sites on the roots of transgenic white clovers (6). The results suggest that the application of *Rhizobium*, NPA, flavonoids or Nod factors result in alterations of auxin levels, via blockage of polar auxin transport, in the roots within 24 h of their application. In addition, rapid and localised increases in chalcone synthase expression occur after the application of *Rhizobium* and Nod factors.

CHITINASE EXPRESSION.

A tobacco basic-chitinase promoter-GUS gene construct was found to show comparable developmental expression patterns in tobacco and clovers: high expression in root tips and very low expression in healthy aerial tissues. Wounding in all tissues tested induced expression to high levels. Experiments using transgenic clovers to study *Rhizobium*-clover interactions found no significant expression in nodules over a period of 1 day to 14 days post-infection. However, *Rhizobium* infection induces a temporal expression in the cortical cells in the zone of infection between 1 and 4 hours post-infection which is absent thereafter. In contrast, expression was observed during the early stages of development and emergence of lateral root formation.

PROTEOME ANALYSIS.

Proteome analysis has been used to monitor global changes in gene expression: in *R.l.* bv. *trifolii* strain ANU843 in response to exposure to plant signals; in infected and non-infected roots and in subclover suspension cultures; and in roots under attack by clover pests. In strain ANU843 over 1000 well resolved spots were detected using 2-D gel electrophoresis (2-DE) with a number of proteins being induced and the expression levels of some others being up- or down-regulated. The identity of 88 constitutively expressed proteins was

NATO ASI Series, Vol. G 39
Biological Fixation of Nitrogen
for Ecology and Sustainable Agriculture
Edited by A. Legocki, H. Bothe, A. Pühler
© Springer-Verlag Berlin Heidelberg 1997

examined using amino acid composition analysis. In addition, 2-DE was used to study the global changes of *Trifolium* gene expression in response to the attack of predatory mites. Redlegged Earth Mite (RLEM) resistant cultivars of subterranean clover were examined and up to 15 individual proteins were shown to be either elevated or strongly induced during predation by RLEM. Thus, we have shown that 2-DE can be used to analyse genomic structure and global changes of gene expression in *Rhizobium* under different biological conditions and also in plants under various situations.

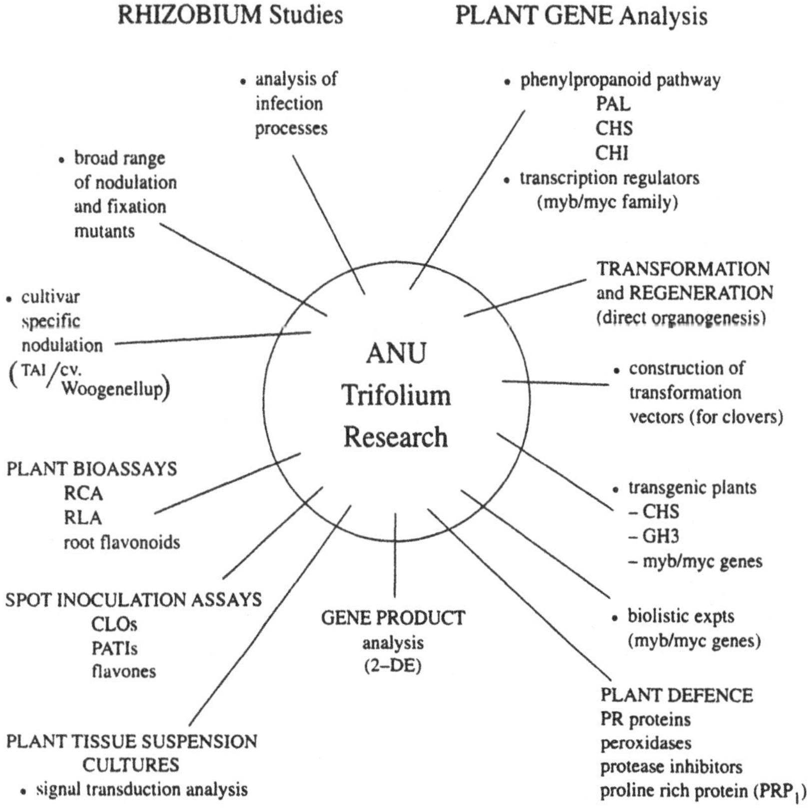

Figure 1. Representation of ANU *Trifolium* research. PAL/CHS/CHI, phenylpropanoid pathway genes; GH3, soybean GH3 auxin responsive promoter; CLO, chito-lipo-oligosaccharides; PATI, polar auxin transport inhibitor; RCA, rooted cotyledon assay; RLA, rooted leaf assay.

SELECTED REFERENCES

(1) Djordjevic MA and JJ Weinman (1991) Factors determining host recognition in the clover-*Rhizobium* symbiosis. Aust. J. Plant Physiol. **18:** 543-557.

(2) Arioli T, Howles PA, Weinman JJ and BG Rolfe (1994) In *Trifolium subterraneum*, chalcone synthase is encoded by a multigene family. Gene **138:** 79-86.

(3) Weinman JJ, Djordjevic MA, Howles PA, Arioli T, Lewis-Henderson W, McIver J, Oakes M, Creaser EH and BG Rolfe (1991) The use of the genus *Trifolium* for the study of plant-microbe interactions. Hennecke H and DPS Verma (eds.) In: Advances in Molecular Genetics of Plant-Microbe Interactions, Vol 1, 168-173, Kluwer Academic Publishers, Dordrecht.

(4) Weinman JJ, Djordjevic MA, Creaser EH, Lawson CGR, Broderick K, Mathesius U, Gärtner E, Pittock C, de Majnik J and BG Rolfe Preparing subterranean clovers for future biotechnology: molecular analysis of genes and proteins involved in stress and defence reactions and the construction of transgenic plants. (1995) Plant Protection Quarterly **10:** 47-49.

(5) Rolfe BG and J McIver (1996) Single-leaf bioassays for the study of root morphogenesis and *Rhizobium*-legume nodulation. Aust. J. Plant Physiol. **23:** 271-283.

(6) Larkin P, Gibson JM, Mathesius U, Weinman JJ, Gärtner E, Hall E, Tanner G, Rolfe BG and MA Djordjevic (1996) Transgenic white clover. Studies with the auxin-responsive promoter, GH3, in root gravitropism and lateral root development. Transgenic Research **5:** 1-11.

Recent Advances in the Molecular Genetics of The Model Legume *Lotus japonicus*

Leif Schauser, Leszek Boron, Eloisa Pajuelo, Thomas Thykjær, Dorthe Danielsen, Jørgen Finneman, Knud Larsen, Niels Sandal and Jens Stougaard

Laboratory of Gene Expression, Department of Molecular and Structural Biology, University of Aarhus, DK-8000 Aarhus C, Denmark

In an attempt to accelerate the progress in understanding the plant contribution to symbiotic nitrogen fixation, the model legume *Lotus japonicus* was suggested as experimental organism for a focused international venture (Handberg and Stougaard, 1992). At the Laboratory of Gene Expression the main emphasis has been concentrated on the molecular genetics of root nodule organogenesis, together with the signal transduction and the transcriptional gene regulation occurring during root nodule development. Different approaches aimed at developing the molecular genetic techniques applicable to *L. japonicus* has been initiated and some of the progress will be reported here. The approaches taken to establish transposon tagging using the maize elements *Ac* and *Ds* were presented recently (Thykjær et al. 1995, Pajuelo et al. 1996). This proceeding will therefore concentrate on presenting some of the other methodologies under investigation or in use.

1. T-DNA insertion mutagenesis

The apparently arbitrary integration of the *Agrobacterium* T-DNA segment into the plant genome opens the possibility for random insertion mutagenesis using the T-DNA as tag. With a genetic approach it is possible to identify important genes by investigating the phenotypes of mutants appearing after a mutagenesis program. The phenotype will give an indication of the function of the gene and it should additionally be possible to identify genes expressed at low levels or only transiently in a few specialised cell types. The functional hierarchy of the genes may also be determined by future genetic analysis of progeny from mutant crosses. One of the mutagenesis programmes started on *L. japonicus* is insertion mutagenesis using the *Agrobacterium tumefaciens* T-DNA as insertion mutagen. Progeny from approximately 1000 independent transgenic plants selected for Hygromycin or Geneticin resistance was screened for symbiotic phenotypes after inoculation with *Rhizobium loti* NZP2235 and growth in nitrogen free nutrient solution. Under these conditions the characteristic nitrogen deficiency symptoms of *L. japonicus* are easily observable by visual inspection. Removal of the inert growth support allows observation on the root system and scoring of the nodule appearance. Plant mutants with root nodules impaired in nodule function can on this basis be distinguished from mutants where root nodule development is changed. As an example the phenotypic appearance of a mutant line unable to develop root nodules is shown in Figure 1. At the primary screening several criteria are used for selecting putative mutant lines for further investigation. Nitrogen starvation observed as the characteristic light green to white leaves, bright red stem and poor growth is the first indication for an ineffective symbiosis and points towards

NATO ASI Series, Vol. G 39
Biological Fixation of Nitrogen
for Ecology and Sustainable Agriculture
Edited by A. Legocki, H. Bothe, A. Pühler
© Springer-Verlag Berlin Heidelberg 1997

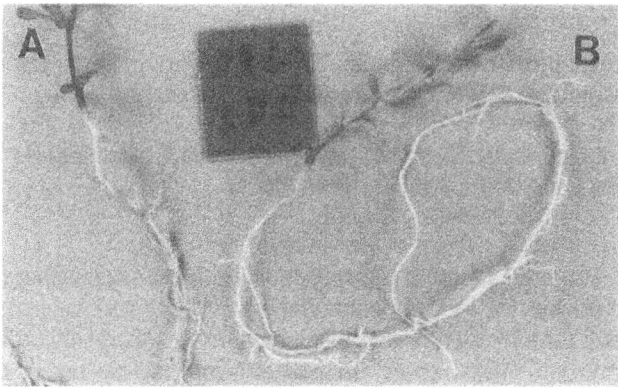

Figure 1. Phenotype of a *Lotus japonicus* mutant developing no nodules or "bumps" visible under a stereo-lup with 10x magnification. (A) Normal nodules on a wild type plant. (B) Well developed root system without nodules observed on the mutant plants.

interesting families. Subsequent lack of nodules, morphological changes or normal appearance are used to group the mutants. Increased or decreased number of nodules may not be indicated by the shoot phenotype and it is therefore necessary to look also at families without nitrogen starvation symptoms. Rescue of the mutants on medium with a normal supply of reduced nitrogen sources is used as an additional criterium for "simple" symbiotic mutations as opposed to pleiotrophic mutants influencing nodule development or function. Using these criteria we have isolated 19 symbiotic mutant lines from screening in the progeny from approximately 1000 primary transformants. This gives a frequency of symbiotic mutants at about 2 %. Preliminary investigations indicate that only one of these lines may be tagged by the T-DNA suggesting a tagging frequency of 5 % among the symbiotic mutant lines. This frequency is not too different from frequencies observed in *Arabidopsis* after tissue culture transformation and screening in the progeny (Van Lijsebettens et al. 1991). These data from *L. japonicus* indicate that T-DNA insertion mutagenesis may be possible with a reasonable frequency and that it may be worthwhile refining the transformation technology (Handberg et al. 1994, Oger et al. 1996) to minimise the manipulations needed and the time required to produce transgenic plants. A number of specialised T-DNA segments for gene trapping, promoter trapping or gene over- expression have been designed to increase the efficiency and the range of the T-DNA tagging approach (Walden et al. 1991, Koncz et al. 1992). Different transformation techniques have also been tested in *Arabidopsis* and seems to influence the efficiency. A high tagging frequency was reported after direct seed transformation (Feldman 1990, Koncz et al. 1992) while the tagging frequency appeared to be lower after tissue culture transformation (Van Lijsebettens et al. 1991). Tagged genes have however been isolated from *Arabidopsis* mutants generated by both transformation procedures. An interesting question arising from the T-DNA tagging is the cause of the untagged mutations. Are they caused by the tissue culture conditions, repair of incomplete T-DNA integrations or from the mobility of for example retroelements activated by the *in vitro* culture?

Retrotransposons activated by tissue culture have been described in both tobacco and rice (Hirochika et al. 1996). Another possibility is the activation of endogenous transposable elements by the *Ac* transposase present on some of the T-DNA segments transformed into *L. japonicus*. In *Arabidopsis* there is evidence for activation of the endogenous *Tag 1* transposable element by the *Ac* transposase (Bhatt et al. 1996).

2. Gene targeting approaches

With the increased sequence information available on plant genes, techniques to study gene function will become increasingly important. One of the methods that could contribute substantially to the understanding of the *in vivo* gene function is gene targeting or gene inactivation by homologous recombination. With this technique it would be possible to make a specific mutation in a previously cloned and characterised gene, and to investigate the resulting phenotype at the whole plant level. Unfortunately it has proven difficult to establish gene targeting in plants, apparently due to the high rate of random integration compared to homologous recombination. The construction of the negative cytosine deaminase marker gene (Stougaard, 1993) opened the possibility to try a positive negative selection scheme for enrichment of homologous recombination events in *L. japonicus*. Two target genes, glutamine synthetase and a *pzf* gene encoding a zinc finger protein were selected as chromosomal target genes for the approach. A positive marker gene conferring geneticin resistance was cloned into the coding regions of the two genes. On each side 5 to 10 kb of flanking DNA were offered for recombination events and the constructs were bordered by the cytosine deaminase negative selection marker. These T-DNA segments were transformed into *L. japonicus* using the published procedure and transgenic calli were selected with a combination of positive and negative selection. The enrichment obtained was approximately 60 fold compared to random integration estimated by positive selection on parallel control plates. The surviving calli were further tested using PCR with outside primers amplifying flanking regions starting from apparently correctly integrated positive marker sequences. On this basis five calli were taken onto regeneration medium to develop transgenic plants. The detailed PCR and Southern analysis of the transgenes has been initiated and it appears that none of the integrations are precise if integration has occurred at the correct target. Integration at the correct target has still to be proven but the positive negative selection seem to be a first step to improve the chances of selecting targeting events. Considering that the five putative targeting events were identified among 10000 transformation events the positive negative selection is an advantage. The frequency of homologous recombination events may however still limit the possibilities for gene targeting.

3. Differential display

The development of sensitive PCR based techniques has opened the possibility to identify novel differentially expressed genes. Using the differential display reverse transcription (DDRT-PCR) method (Liang and Pardee, 1992) the induction of genes during the nodulation process has been followed in *L. japonicus*. Taking advantage of the small size of the *L. japonicus* plants a high density growth regime was establish in order to obtain sufficient root material from the seedlings early after inoculation. Under these conditions seeds are germinated on a steel mesh in small plant containers. The root will grow through the mesh and can be easily harvested by sliding a scalpel along the "root side" of the mesh. From the harvested material total RNA was extracted from uninoculated roots, roots five and nine days after inoculation and finally from functional mature nodules. Short ten-mer primers and anchored oligo-T primers were used in the amplification process. The resulting fragment were separated on acrylamide sequencing gels with a resolution between 50 and 500 nucleotides. A total of 34 fragments that appeared differentially expressed during nodule development were cloned and analysed further. RNase protection studies was undertaken to confirm the expression pattern and Southerns were performed to confirm the plant origin of the identified gene fragments. Data base searches and sequence comparisons suggested a function for 7 of the isolated fragments. No similarity was found to the other fragments, which is not very surprising considering that DDRT-PCR amplifies short fragments originating mainly from the 3′untranslated regions. Full length cDNAs corresponding to the most interesting fragments are currently isolated and sequenced.

Bhatt, A.M., Page, T., Lawson, E.J.R., Lister, C. and Dean, C. 1996. The use of *Ac* as an insertional mutagen in *Arabidopsis*. Plant J. 9, 935-945.
Handberg, K. and Stougaard, J. 1992. *Lotus japonicus*, an autogamous, diploid legume species for classical and molecular genetics. Plant J. , 2, 487-496.
Handberg, K., Stiller, J., Thykjær, T. and Stougaard, J. 1994. Transgenic plants: *Agrobacterium* mediated transformation of the diploid model legume *Lotus japonicus*. in Cell Biology: A Laboratory Handbook. eds. J.E. Celis. Academic Press.
Feldman, K.A. 1991. T-DNA insertion mutagenesis in *Arabidopsis*: mutational spectrum. Plant J. 1, 71-82.
Koncz, C. Némenth, K. Rédei, G.P. and Schell J. 1992. T-DNA insertional mutagenesis in *Arabidopsis*. Plant Molec. Biol. 20, 963-976.
Liang, P. and Pardee, A.B. 1992. Differential display of eukaryotic messenger RNA by means of the polymerase chain reaction. Science, 257, 967-971.
Oger, P., Petit, A. and Dessaux, Y. 1996. A simple technique for direct transformation and regeneration of the diploid legume species *Lotus japonicus*. Plant Science, 116,159-168.
Pajuelo, E., Schauser, L., Thykjær, T., Larsen, K. and Stougaard, J. Transposon tagging in *Lotus japonicus* using the maize elements *Ac* and *Ds*. in 1996 IS-MPMI Symposium Proceedings, eds. Stacey, G., Mullin, B., Gresshoff, P.
Hirochika, H., Sugimoto, K., Otsuki, Y., Tsugawa, H. and Kanda, M. (1996) Retrotransposons of rice involved in mutations induced by tissue culture. Proc. Natl. Acad. Sci. USA, 93, 7783-7788.
Stougaard, J. 1993. Substrate-dependent negative selection in plants using a bacterial cytosine deaminase gene. Plant J, 3, 755-761.
Thykjær, T., Stiller, J., Handberg, K. Jones, J. and Stougaard, J. 1995. The maize transposable element *Ac* is mobile in the legume *Lotus japonicus*. Plant Mol. Biol. 27, 981.993.
Van Lijsebettens, M., Vanderhagen, R., Van Montagu, M. 1991. Insertional mutagenesis in *Arabidopsis thaliana*: isolation of a T-DNA linked mutation that alters leaf morphology. Theor. Appl. Genet. 81, 277-284.
Walden, R., Haryashi, H. and Schell, J. 1991. T-DNA as a gene tag. Plant J. 1, 281-288.

MOLECULAR GENETICS OF A MODEL-PLANT:
MEDICAGO TRUNCATULA

Thierry Huguet[1], Leïla Tirichine[1], Michèle Ghérardi[1], Muriel Sagan[2], Gérard Duc[2], Jean-Marie Prospéri[3]

1 Laboratoire de Biologie Moléculaire des Relations Plantes-Microorganismes, CNRS-INRA, BP27, 31326 Castanet-Tolosan Cedex, France.
2 Station de Génétique et d'Amélioration des Plantes, INRA, BV1540, 21034 Dijon Cedex, France.
3 Station de Génétique et d'Amélioration des Plantes, INRA, Domaine de Melgueil, 34130 Mauguio, France.

Introduction

Plants belonging to the genus *Medicago*, as most of the legumes, have the capacity to associate with soil bacterias of the genus *Rhizobium* to form at the surface of their roots specialized organs, the so-called nodules, in which bacterias reduce atmospheric nitrogen in ammonia which is, in turn, assimilated by the plant for its growth and development (Long, 1989). The understanding of the mechanisms which rule this symbiosis is a major objective to control the interactions between plants and soil microorganisms, pathogens or not and to reduce the inputs of fertilizers and/or to increase the forage productivity.

M.truncatula as a model-plant

Medicago.truncatula give the opportunity to combine molecular and genetical approaches. It.is an annual, diploid (2n=16) and autogamous legume having a relatively small genome (0.49-0.57 pg/1C, Blondon *et al.*, 1994); it can be transformed by *Agrobacterium tumefaciens* and regenerate *via* somatic embryogenesis (Thomas *et al.*, 1992; Chabaud *et al.*, 1996). A number of genes related to *Rhizobium*-legume symbiosis have already been isolated and sequenced (Gamas *et al.*, 1996). *M.truncatula* is nodulated by *Rhizobium meliloti* strains whose many mutants are available and their bacterial nodulation factors (sulfated lipooligosaccharides) essential for *Rhizobium*-legume symbiosis have been purified and characterized (Dénarié and Cullimore,1993).

For all these reasons, *M.truncatula* has been proposed as a model-plant (Barker *et al.*,1990)

Biodiversity

According to Lesins and Lesins (1979), annual species of *Medicago* derived from perennial ancestors at the end of the Tertiary era. This autogamous species, *M.truncatula*, can be found all around the Mediterranean basin as native populations

NATO ASI Series, Vol. G 39
Biological Fixation of Nitrogen
for Ecology and Sustainable Agriculture
Edited by A. Legocki, H. Bothe, A. Pühler
© Springer-Verlag Berlin Heidelberg 1997

which have a good tolerance to drought and salinity and grow in a wide range of soils and environmental conditions: 290 populations have been collected at the INRA laboratory of Montpellier since 1985. A high level of variability of quantitative, qualitative and molecular characters have been observed among and within natural populations (Chaulet and Prospéri,1994; Bonnin *et al.*,1996).

Genetic map

A genetic map is a powerful tool for understanding the organization and evolution of genomes. Four F1 hybrids, resulting from crosses between homozygous individuals of 4 populations (1 French, 2 Algerians and 1 cultivar Jemalong) have been selected to be the starting points of the *M.truncatula* genetic mapping program.

The genetic map of *M.truncatula* is under progress using the F2 progeny of two crosses. We mapped 70 RAPD markers, 2 isoenzymes (PGM, PGD), 6 known genes (rDNA, ENOD12, MTGSa, MTGSb, MTLb1, MTLec2) and 1 morphological marker (pod coiling) covering 514 cM.

The mapping of the genes involved in the *Rhizobium*-legume symbiosis (induced mutants and natural polymorphisms) is under progress.

Agronomically important characters such as nitrogen fixation efficiency, cold tolerance, disease resistance or osmotic stress tolerance could be mapped taking advantage of the high level of polymorphism observed in natural populations of *M.truncatula*.

Mutagenesis

Mutants are very important tools to analyse the different steps of a complex developmental process like the *Rhizobium*-legume symbiosis. The screening is done on the capacity of potential mutants to grow on nitrogen-free medium using the symbiotic nitrogen fixation as nitrogen source. After irradiation of the variety Jemalong with γ-rays, 18 symbiotic mutants were obtained at the M2 generation and were stable up to the M6 generation (Sagan *et al.*,1995). They consist of 2 [Nod⁻], 4 [Nod$^{+/-}$], 3 [Nod^{++}] and 9 [Nod$^+$Fix⁻] mutants. Non-nodulating mutants [Nod⁻] were also defective for mycorhizal symbiosis [Myc⁻]. Supernodulating mutants [Nod^{++}] were nitrate-tolerant up to 15 mM (Nts). Genetic and molecular analysis of these mutants is under way.

A mutant deficient for nitrogen fixation was studied in details; it is caracterized by the lack of bacteroid development and by defence reactions during *Rhizobium* infection (Bénaben *et al.*,1995).

Natural variants

M.truncatula shows a strong cultivar x *Rhizobium* strain specificity (Snyman and Strijdom,1980). We have screened three natural *M.truncatula* populations and one cultivar (Jemalong) with a number of wild-type *R.meliloti* strains, effective on alfalfa. From 123 genotype/strain combinations tested, 44 (36%) were deficient for nodulation or fixation. The observed polymorphism goes from the lack of visible nodulation to the formation of morphologically normal non-fixing nodules. The genetic and physiological basis of this plant and bacterial polymorphisms are not yet known.

Agronomy

M.truncatula, as all *Medicago* annual species is native from the Mediterranean basin and has an high agronomical potential. It is cultivated as a source of winter forage or to avoid soil erosion and improve soil fertility. Their capacity to reseed spontaneously from one year to another due to the presence of an high proportion of hard seeds allowing a perennial installation make them well adapted to the climatically variable Mediterranean environments. *M.truncatula* is one of the species most cultivated in dry lands, especially in Australia (ley-farming system of Cereal-Legume cultivation). Several cultivars of *M.truncatula* are commercially available. In France, the selection of annual *Medicago* to introduce them in Mediterranean agriculture is under progress since 1985 at INRA, Montpellier, France (Prospéri,1989) and four cultivars are available (1 *M.truncatula*: Salernes).

Future Prospects

M.truncatula can be used for multidisciplinary approaches linking quite different scientific and applied domains. One of our objectives is to make a genetic study of plant genes involved in symbiosis, based on symbiotic mutants and natural variants. In addition, due to the high level of genetic diversity found in natural populations, *M.truncatula* is a potential source of genes of agronomical interest to be transfered to cultivated alfalfa.

Références

BARKER D.G., BIANCHI S., BLONDON F., DATTEE Y., DUC G., FLAMENT P., GALLUSCI P., GENIER G., GUY P., MUEL X., TOURNEUR J., DENARIE J., HUGUET T.(1990) *Medicago truncatula*, a model plant for studying the molecular genetics of the *Rhizobium*-legume symbiosis. *Plant Mol. Biol. Rep.*, **8**, 40-49.

BENABEN V., DUC G., LEFEBVRE V., HUGUET T. (1995) TE7- An inefficient symbiotic mutant of *Medicago truncatula* Gaertn cv. Jemalong. *Plant Physiology*, **107**, 53-62.

BLONDON F., MARIE D., BROWN S., KONDOROSI A. (1994) Genome size and base composition in *Medicago sativa* and *M.truncatula* species. *Genome*, **37**, 264-270.

BONNIN I., HUGUET T., GHERARDI M., PROSPERI J.M., OLIVIERI I. (1996) High level of polymorphism and spatial structure in a selfing plant species *Medicago truncatula* Gaertn. using RAPDs markers. *American Journal of Botany*, **83(7)**, 843-855.

CHABAUD M., LARSONNAUD C., MARMOUGET C., HUGUET T. (1996) Transformation of Barrel Medic (*Medicago truncatula* Gaertn.) by *Agrobacterium tumefaciens* and regeneration via somatic embryogenesis of transgenic plants with the MtENOD12 nodulin promoter fused to the GUS gene. *Plant Cell Reports* **15**, 305-301..

CHAULET E., PROSPERI J.M. (1995) Genetic diversity of *Medicago truncatula* Gaertn collected in Algeria. in *Proceedings of the 7th Genetic Resources section meeting of Eucarpia*. March 1994, Clermont-Ferrand..Eds INRA.

DENARIE J., CULLIMORE J. (1993) Lipo-oligosaccharide nodulation factors: a minireview. New class of signaling molecules mediating recognition and morphogenesis. *Cell*, **74**, 951-954.

GAMAS P., DE CARVAHLO-NIEBEL F., LESCURE N., CULLIMORE J. (1996) Use of a substractive hybridization approach to identify new *Medicago truncatula* genes induced during root nodule development. *Molecular Plant Microbe Interactions* **9(4)**, 233-242.

LESINS K.A., LESINS L. (1979) Genus *Medicago*. A taxogenetic study. Dr. W.Junk bv Publishers.

LONG S.L. (1989) *Rhizobium*-legume nodulation: life together underground. *Cell*, **56**, 203-214.

PROSPERI J.M. (1989) : Selection of annual medics for French Mediterranean regions. In *Workshop on introducing the ley farming system to the Mediterranean basin*. 173-191. Perugia (Italy). Eds. S. Christiansen, L. Materon, M. Falcinelli et P. Cocks.

SAGAN M., MORANDI D., TARENGHI, DUC G. (1995) Selection of nodulation and mycorhizal mutants in the model plant *Medicago truncatula* Gaertn after gamma rays mutagenesis. *Plant Science*, **111**, 63-71.

SNYMAN C.P., STRIJDOM B.W. (1980) Symbiotic characteristics of lines and cultivars of *Medicago truncatula* inoculated with strains of *Rhizobium meliloti*. *Phytophylactica*, **12**, 173-176.

THOMAS M.R., ROSE R.J., NOLAN K.E. (1992) Genetic transformation of *Medicago truncatula* using *Agrobacterium* with genetically modified Ri and disarmed Ti plasmids. *Plant Cell Rep.*, **11**, 113-117.

Advances in Molecular Characterization of the Yellow Lupin - *Bradyrhizobium* sp. *(Lupinus)* Symbiotic Model

Andrzej B. Legocki, Wojciech M. Karłowski, Jan Podkowiński,
Michał Sikorski, Tomasz Stępkowski

Institute of Bioorganic Chemistry, Polish Academy of Sciences,
Noskowskiego 12/14, 61-704 Poznań, Poland

Keywords. *Bradyrhizobium* sp. *(Lupinus)*, *nod* genes, lupin genes

1. Introduction

Lupins belonging to tribe *Genisteae* constitute one of the most ancient collections of legumes. This monophylogenetic genus comprises of a dynamic group of species that occupy all kinds of habitats (Kass and Wink 1994). The exact number of species is unknown; roughly 500 species have been described. The real number, however, could be as high as 1700 species (Planchuelo 1994). Owing to their flexibility to adapt to adverse environmental challenges, high protein and oil content in seeds, as well as production of various alkaloids, lupins are a most useful legumes in agriculture, biotechnology and ecology.

Lupins are also interesting models for studies of symbiotic nitrogen fixation. It is presumed that they retained many ancient and unique features. However, the biology of lupin root nodule morphogenesis has not been fully recognized. There is no convincing evidence confirming the presence of primary infection threads. The way rhizobium enters the root tissues is either unknown. Finally, the infection process results in the formation of unique, collar-shaped, indeterminate nodules (Golinowski et al. 1983).

2. Results and discussion

2.1 Cloning of the nodulation genes in *Bradyrhizobium* sp. WM9 *(Lupinus)*

The selected *Bradyrhizobium* sp. WM9 nodulates lupins, seradella and *Lotus corniculatus*; three hosts belonging to distinct phylogenetic lineages of legumes. Recognition of the *nod* functions responsible for the infection of different hosts and characterization of the lipochitooligosaccharides (LCOs) are primary objectives of our work.

In plant assays, lupins were inoculated with various rhizobia whose Nod factors have already been characterized. It appeared that nodulating strains produce LCOs whose decorations are almost identical. In the cases of nodulating rhizobia the Nod factors are *N*-methylated, carbamoylated, and they carry 4-O-*acetyl*-fucose at C-6 of *N*-acetyl-glucosamine of the reducing end.

A genomic library of WM9 was constructed in pRI40 cosmid vector. The cosmids carrying the *nod* region were isolated by colony hybridization with probes specific for

NATO ASI Series, Vol. G 39
Biological Fixation of Nitrogen
for Ecology and Sustainable Agriculture
Edited by A. Legocki, H. Bothe, A. Pühler
© Springer-Verlag Berlin Heidelberg 1997

nodC and *nodZ* genes. Random sequencing led to the identification of the majority *nod* functions involved in the biosynthesis of LCOs. The genetic content was in congruence with the expected composition of the Nod factor. All *nod* genes, whose products are responsible for LCO decorations such as N-methylation (*nodS*), carbamoylation (*nodU*), fucosylation (*nodZ*) or 4-O-acetylation of fucose (*nolL*) were found (Fig. 1).

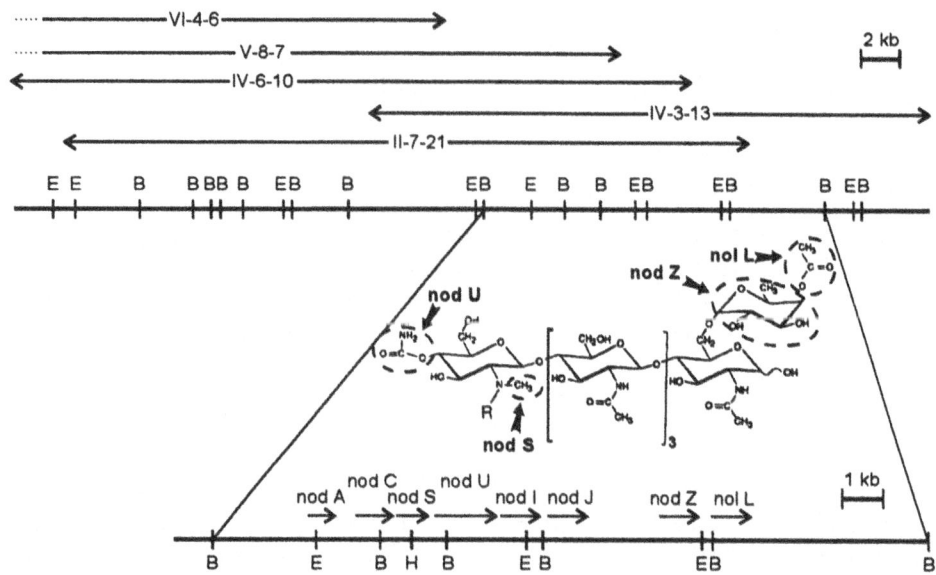

Fig. 1 Diagram of physical map of a nodulation region in *Bradyrhizobium* sp. (*Lupinus*).

Comparison of the nucleotide *nodA* and *nodC* sequences reveals surprisingly low homology to the nodulation genes in other *Bradyrhizobium* strains. Other characteristics, such as partial, 260 bp long sequence of 16S rRNA gene and presence of *nodVW* genes, resemble *B. japonicum* (Stacey 1995). The observed dissimilarity of the *nod* functions in *Bradyrhizobium* sp. WM9 to other *Bradyrhizobia* may reflect the coevolution of lupins and their microsymbionts.

2.2 Early nodulin genes

We detected and characterized two different sequences homologous to ENOD40. One sequence - LlENOD40A was found in 6-day old nodules, the other - LlENOD40B was selected from *L. luteus* genomic library. The latter sequence was highly homologous within its 12-peptide region to *G. max* ENOD40.

We have also isolated LlPRP2 gene coding for proline-rich early nodulin. The comparison of deduced amino-acid sequence coded by this gene with already published nodulin sequences shows the highest level of homology with members of the ENOD2 gene family. This protein of 103 kD is two-and-half-time bigger than other ENOD2 sequences. The analysis of the promoter region showed the presence of several interesting elements such as the sequence homologous to OSE motif and the new element characteristic for all ENOD2 promoter regions. The expression of

LlPRP2 gene, in contrast to ENOD2 genes, is located in the infected, central tissue of the root nodule. We suggest that the gene from yellow lupin represents as a homologue to ENOD2 family.

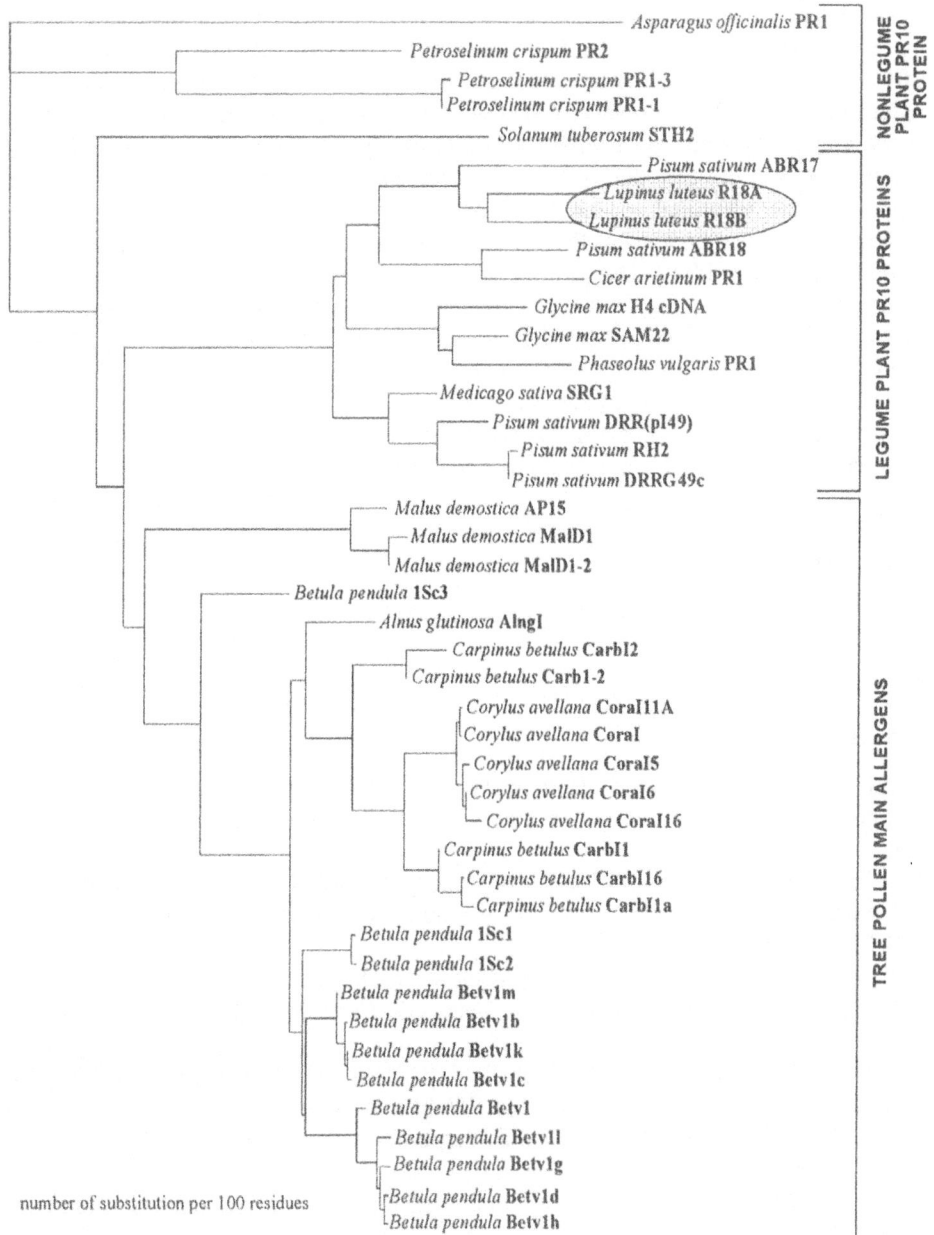

Fig. 2 The phylogenetic tree of PR10 class proteins

2.3 Symbiotically regulated PR10 genes

We identified two homologous yellow lupin proteins representing a class of small acidic polypeptides of M_r 17 000 which are constitutively expressed in roots of an uninfected plant. The expression of those genes as seen on Northern blots indicates that their transcription is down-regulated during the development of functional symbiosis with *B. lupini* (Sikorski et al. 1996). Full-length cDNA copies coding for both proteins LlPR10.1A and LlPR10.1B were selected from cDNA library. The deduced amino acid sequences showed that both genes coded for polypeptides, each composed of 156 amino acid residues. These proteins exhibit 40-80% similarity to pathogenesis-related and stress-induced proteins of other plants. Based on their amino acid composition and physical properties, the identified yellow lupin proteins could be classified as intracellular pathogenesis-related proteins of PR10 class (van Loon et al. 1994). Fig. 2 presents the phylogenetic tree of PR10 class proteins which are coded by sequences revealing similar structure and organization with one intron at comparable position. These similarities established a class of conserved defense-related proteins, which may suggest their common function in plant-microbe interactions. It has been recently suggested that nitrogen-fixing bacteria avoid induction of defense response. It is still an open question whether an active suppression of plant defense by invading microsymbiont is indispensable for the establishment of functional symbiosis.

Acknowledgments

This work was supported by the grants from State Committee for Scientific Research: 6P 204 056 06 and 6P 04B 011 10.

References

Golinowski W., Kopcińska J., Borucki W. 1987. The morphogenesis of lupine root nodules during infection by *Rhizobium lupini*. Acta Societas Botanicum Poloniae 56, 687-703.

Kass E., Wink M. 1994. Molecular phylogeny of lupins. In: "Advances in Lupin Research" (eds) Neves-Martins J.H., Beirao da Costa M.L. ISA Press, Lisboa, p. 267-270.

Planchuelo A.M. 1994. Molecular phylogeny of lupins. In: "Advances in Lupin Research" (eds) Neves-Martins J.H., Beirao da Costa M.L. ISA Press, Lisboa, p. 65-69.

Sikorski M.M., Szlagowska A.E., Legocki A.B. 1996. cDNA sequences encoding for two homologs of *Lupinus luteus* (L.) IPR-like proteins (accesssion Nos. X79974 and X79975 for LlR18A and LlR18B, respectively) (PGR 95-114). Plant Physiol. 110, 335.

Stacey G. 1995. *Bradyrhizobium japonicum* nodulation genetics. FEMS Microbiol. Lett. 127, 1-7.

van Loon L.C., Pierpoint W.S., Boller Th., Canejero V. 1994. Recommendations for naming plant pathogenesis-related proteins. Plant Mol. Biol. Rep. 12, 245-264.

Actinorhizal nodules

Katharina Pawlowski[1], Ana Ribeiro[1], Changhui Guan[1], Alison M. Berry[2] and Ton Bisseling[1]

[1] Department of Molecular Biology, Wageningen Agricultural University, 6703 HA Wageningen, The Netherlands
[2] Department of Environmental Horticulture, University of California, Davis, CA 95616, USA

Keywords. Symbiotic nitrogen fixation, actinorhiza, *Frankia*, *Alnus glutinosa*, *Casuarina glauca*, *Datisca glomerata*, glutamine synthetase

1. Introduction

Actinorhizal nodules are induced by the nitrogen-fixing actinomycete *Frankia* on the roots of several dicotyledonous plant species belonging to eight different families (Benson and Silvester, 1993). *Frankia* sp. are gram-positive soil bacteria which normally grow in hyphal form. Under nitrogen-limiting conditions and atmospheric oxygen tension, they can form spherical vesicles at the ends of hyphae or short side branches. In these vesicles, nitrogenase is formed, protected from O_2 by the vesicle envelope. In symbiosis, the shape of the vesicles and their localization in the infected cells depends on the host plant (reviewed by Baker and Mullin, 1992). Actinorhizal nodules consist of multiple lobes, each of which represents modified lateral root without root cap, with a superficial periderm and with infected cells in the expanded cortex.

While a single origin has been suggested for the predisposition of plants to enter symbioses with nitrogen-fixing bacteria (Soltis et al., 1995), several independent direct evolutionary origins of the symbiosis have been proposed among actinorhizal plants (Swensen, 1996). These independent origins may account for the considerable degree of structural variability between nodules of actinorhizal plants from different families (Fig. 1; reviewed by Silvester et al., 1990). Furthermore, they can be expected to be reflected also in differences between infection mechanisms and specialization of nodule tissues amongst actinorhizal symbioses. We have compared the expression patterns of genes showing markedly enhanced expression levels in nodules ("actinorhizal nodulin genes") compared to roots in *Alnus glutinosa*, *Datisca glomerata* and *Casuarina glauca*, three different actinorhizal plant species which are supposed to belong to distinct phylogenetic branches (Swensen, 1996).

Due to the presence of the apical meristem, the cortical cells of an actinorhizal nodule lobe are arranged in a developmental gradient (Ribeiro et al., 1995). The meristematic zone (zone 1) is followed by the infection zone (zone 2), were some of the cortical cells are invaded and progressively filled by branching *Frankia* hyphae. The nitrogen fixation zone (zone 3) is characterized by the expression of *Frankia nif* (nitrogen fixation) genes. In this zone, vesicles have been formed and *Frankia* is fixing nitrogen. In the senescence zone (zone 4), plant cytoplasm and *Frankia* material are degraded. This nomenclature will be applied in the description of actinorhizal nodulin gene expression patterns.

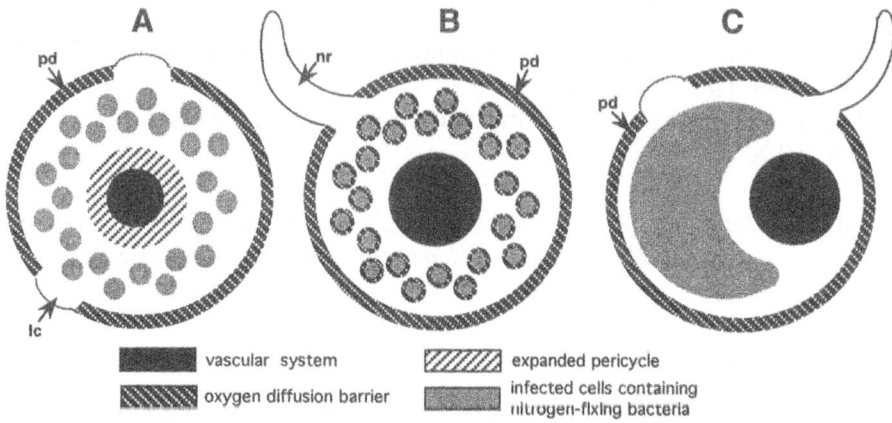

Fig. 1. Schematic cross sections of actinorhizal nodule lobes from plants of different families. (A) *Alnus glutinosa;* (B) *Casuarina glauca;* (C) *Datisca glomerata.* An O_2 diffusion barrier can be formed by cells lacking intercellular spaces as in the periderm (pd) or by lignification of the walls of infected cells as in (B) (Berg and McDowell, 1987). Nodules are aerated via lenticels (lc) or agraviotropically growing nodule roots (nr).

2. Comparing gene expression patterns

Several actinorhizal nodulin genes were found to be expressed in infected cortical cells in specific stages of development, which allows some conclusions with regard to the function of the encoded proteins. For instance, genes expressed at high levels during cortical cell infection, i.e. in zone 2, and not or at much lower levels in zone 3, probably encode products that play a role in the infection process. One of those genes encodes a putatively extracellular serine protease (*ag12;* Ribeiro et al., 1995). A homolog of *ag12* isolated from *Casuarina glauca,* termed *cg12,* shows a similar expression pattern. These data suggest that the proteins involved in the infection process of cortical cells by *Frankia* share some similarity between *Alnus* and *Casuarina.*

However, other evidence indicates that the proteins involved in the infection process can differ in nodules from different actinorhizal plant genera. cDNAs representing an actinorhizal nodulin gene family expressed in infected cells of zone 2 have been isolated from *A. glutinosa* and shown to encode small putatively extracellular metal-binding proteins. A cDNA homologous to those of *A. glutinosa* was isolated from *Datisca glomerata.* This homolog was not expressed during infection of cortical cells, but mainly in the periderm of nodules as well as roots. Thus, a clear difference exists in the proteins involved in cortical cell infection in *A. glutinosa* and *D. glomerata,* respectively. It should be noted that according to phylogenetic studies on actinorhizal plants, *Casuarina* and *Alnus* are more closely related to each other than either of them to *Datisca* (Swensen, 1996).

The expression of plant glutamine synthetase (GS), the key enzyme in the assimilation of ammonium, product of bacterial nitrogen fixation, was compared in all three nodule types. While GS expression in nodules of *Alnus* (Guan et al., 1997) and *Casuarina* was confined to the nitrogen-fixing infected cells and the pericycle of the central vascular system of the nodule lobes, in *Datisca* high levels of expression were found in the uninfected cortical cells. Thus, it seems that in actinorhizal nodules from different plant genera, tissue specialization can be strikingly different. Furthermore, the GS expression data again stress the similarity between *Alnus* and *Casuarina* nodules, and the difference of both with *Datisca* nodules.

3. Discussion

The results of molecular analysis of actinorhizal nodules support the hypothesis that there is more than one direct phylogenetic origin of the symbiosis. Based on the analysis of regarding cell types involved in metabolic specialization and of plant proteins involved in the infection of cortical cells by *Frankia*, *Datisca* seems strikingly different from *Alnus* and *Casuarina*. *Datisca* also differs from *Alnus* and *Casuarina* in the infection process (during infection, *Frankia* grows intercellularly in *Datisca*, but intracellularly in *Alnus* and *Casuarina*) and in the organization of cortical cells (Fig. 1). It remains to be examined whether the molecular differences observed are the necessitated by the different morphology and infection process, or whether they are an unrelated effect due to independent evolution of the symbiosis in the different systems.

The results of Soltis et al. (1995) indicate that based on *rbcL* sequence comparison, all plants of nitrogen-fixing root nodule symbioses are evolutionary related, although several recurrent origins of the symbiosis have been proposed among actinorhizal plants (Swensen, 1996). Interestingly, legumes seem to be as phylogenetically distant from, e.g. actinorhizal Casuarinaceae, as the latter are from actinorhizal Datiscaceae or Rosaceae. In spite of this, there is evidence for the conservation of infected cell-specific transcription factors between legumes and Casuarinaceae. It has been shown that the promoter of a nodule-specific hemoglobin gene from *Casuarina* directs expression of a GUS reporter gene specifically in the infected cells of nodules of the legume *Lotus japonicus* (Jacobsen-Lyon et al., 1995). These data seem to indicate that nodule infected cell-specific transcription factors were already present in a common ancestor of legumes and *Casuarina*, i.e. before the independent direct phylogenetic origins of Casuarinaceae- and legume symbioses.

Our results have shown that actinorhizal nodules of *Datisca* are different from actinorhizal nodules of *Alnus* with regard to tissues involved in nodule nitrogen metabolism, and with regard to genes expressed in the infected cells. Furthermore, we have preliminary evidence that in both features, *Alnus* resembles legumes. As in *Alnus* nodules, GS is expressed at high levels in the infected cells of all legume nodules examined thus far (Forde et al., 1989; Miao et al., 1991; Temple et al., 1995; Guan et al., 1997). Members of the 18Kdal protein family are expressed at high levels roots of pea and *Alnus* roots, and at low levels in nodules of both plants (RH2; Mylona et al., 1994; A.-E. Hadri, personal communication), while the *Datisca* homolog is expressed at high levels in nodules (in the infected cells) and at very low levels in roots. Thus, these aspects of nodule function seem to be more conserved between *Alnus* and legumes, than between the two actinorhizal plants *Alnus* and *Datisca*. At present it cannot be decided whether the similarity between *Alnus* and legumes is due to a common phylogenetic origin, or to parallel evolution. At any rate, these results show that further research is needed to understand the phylogeny of nitrogen-fixing symbioses, and that comparisons of the expression patterns of homologous genes in different systems, as well as studies on the conservation of symbiosis-related transcription factors, can be used in the phylogenetic analysis.

References

Baker DD, Mullin BC (1992) *In* Biological Nitrogen Fixation (Stacey G, Burris RH, Evans HJ, eds) Chapman and Hall, New York, pp. 259-292.

Benson DR, Silvester WB (1993) Microbiol Rev 57: 293-319.

Berg RH, McDowell L (1987) Can J Bot 66: 2038-2047.

Forde BG, Day HM, Turton JF, Shen WJ, Cullimore JV, Oliver JE (1989) Plant Cell 1: 391-401.

Guan C, Ribeiro A, Akkermans ADL, Jing Y, Van Kammen A, Bisseling T, Pawlowski K (1997) Plant Mol Biol, in press.

Jacobsen-Lyon K, Østergaard-Jensen E, Jørgensen J-E, Marcker KA, Peacock J, Dennis E (1995) Plant Cell 7: 213-223.

Miao GH, Hirel B, Marsolier MC, Ridge RW, Verma DPS (1991) Plant Cell 3: 11-22.

Mylona P, Moerman M, Yang WC, Gloudemans T, Van de Kerckhove J, Van Kammen A, Bisseling T, Franssen HJ (1994) Plant Mol Biol 26: 39-50.

Ribeiro A, Akkermans ADL, van Kammen A, Bisseling T, Pawlowski K (1995) Plant Cell 7: 785-794.

Silvester WB, Harris SL, Tjepkema JD (1990) *In* The Biology of Frankia and Actinorhizal Plants (Schwintzer CR, Tjepkema JD, eds) Academic Press, New York, pp. 157-176.

Soltis DE, Soltis PS, Morgan DR, Swensen SM, Mullin BC, Dowd JM, Martin PG (1995) Proc Natl Acad Sci USA 92: 2647-2651.

Swensen SM (1996) Am J Bot, in press.

Temple SJ, Heard J, Ganter G, Dunn K, Sengupta-Gopalan C (1995) Mol Plant-Microbe Interact 8: 218-227.

Molecular and Genetic Insights into Shoot Control of Nodulation in Soybean

Peter M. Gresshoff, Keishi Senoo, Jaime Padilla, Debbie Landau-Ellis,
Anatoli Filatov, Alexander Kolchinsky and Gustavo Caetano-Anollés

Plant Molecular Genetics, The Center for Legume Research,
Institute of Agriculture, The University of Tennessee,
Knoxville TN 37901-1071, USA, e-mail: JCROCKET@utk.edu

Abstract:
Mutants of soybean were used to study autoregulation of nodulation. Supernodulation resulted from mutations in the *nts-1* gene, which was mapped next to two RFLP markers. Supernodulation resulted from the absense of autoregulation nodulation and led to nitrate tolerant nodulation. Grafting experiments revealed that autoregulation functions through the shoot of the plant. Specifically, early cell division steps in the root activate a yet undefined process, which leads to the production of an inhibitory response from young leaves, which, transmitted systemically to the roots, prevents further cell divisions and nodule meristem maturation. However, meristems of sufficient size and independence progress to form nodules. Autoregulation is seen as a mechanism by which legume plants limit mitogenic stimulation from ubiquitous factors produced by rhizobia. The mechanism may have similarities to that controlling lateral root number, reflecting onto- and phylogenetic links between nodule and root formation.

1. Autoregulation of Nodulation:

Nodulation of legumes is controlled by genetic determinants within the plant and the microsymbiont. Cellular responses of the plant include an activation of plant defense genes, changes in the phenylpropanoid pathway, ion efflux, root hair deformation, and initiation of cell division events in precisely determined cells of the subepidermal cortex (for determinate nodulation legumes such as Lotus or soybean) and the pericycle next to xylem poles (Rolfe and Gresshoff, 1988). The study deals with autoregulation of nodulation which restricts the number of nodules formed on *Bradyrhizobium japonicum* infected soybean (*Glycine max* L. Merrill) roots. Chemically induced mutants defined a single gene, *nts-1*, which controls nodule number and ability to nodulate in the presence of otherwise inhibitory levels of nitrate. Recessive mutations at *nts-1* lead to supernodulation and nitrate tolerance of nodulation (Carroll et al, 1985a,b). Similar types of mutants were isolated in other soybean lines and other legumes including pea, frenchbean, *Medicago truncatula* and *Lotus japonicus*.

nts-1 is defined by several alleles, including those leading to extreme supernodulation like nts382 and nts1007 or moderate hypernodulation like nts1116. *nts-1* was mapped next to RFLP markers pUTG132a and pA381 on the USDA ARS soybean map (linkage group H; Landau-Ellis et al, 1991; Landau-Ellis and Gresshoff, 1994; Kolchinsky et al, 1997). The proximity of *nts-1* to RFLP marker pUTG132a (0.7 ± 0.5 cM), corresponding to perhaps 350 kb, may permit the cloning of this chromosomal region to facilitate the map-

NATO ASI Series, Vol. G 39
Biological Fixation of Nitrogen
for Ecology and Sustainable Agriculture
Edited by A. Legocki, H. Bothe, A. Pühler
© Springer-Verlag Berlin Heidelberg 1997

based cloning of the *nts-1* locus (Gresshoff, 1995). Physical mapping close to this region suggested that 1 cM corresponds to less than 500 kb (Funke et al, 1993). Isolation of such region is potentially possible through the selection of anchored YAC or BAC clones (Funke et al, 1994; Zhu et al, 1996). Recently we have used soybean YACs to "paint" metaphase chromosomes, valuable to detect chimerism and chromosomal locations. Probe pUTG132a detected a single region terminally located on a metaphase chromosome, consistent with its location on the distal end of linkage group H.

Marker density in the *nts-1* region was increased through placement of additional molecular markers using either arbitrary primer technology and BSA (Kolchinsky et al, 1997), DAF (Caetano-Anollés and Gresshoff, 1996), or synteny mapping using RFLP markers from *Phaseolus vulgaris* (Gresshoff and Filatov, in prep.). However, none of these new molecular markers are closer than the originally discovered pUTG132a RFLP. A range of mapping studies suggest that this region of the soybean genome is present in a single copy and that the region is strongly conserved among genotypes.

Cytological analysis demonstrated that nts382 plants, being homozygous recessive for *nts-1*, possess more successful cell division centers per unit root length than wild type Bragg (Mathews et al, 1989), suggesting that autoregulation functions by restricting meristem transition from a small cell cluster to a persistent nodule meristems perhaps 48-72 hours after infection. It is likely that the transition point from cell cluster to nodule meristem involving both pericycle and cortical cell divisions is the target for both autoregulation of nodulation and nitrate inhibition. Other environmental factors like soil acidity may also effect this critical step (Alva et al, 1988).

2. Systemic Regulation of Nodulation

That nodulation is controlled in a systemic fashion was demonstrated by a range of experiments. Research from the Bauer group in the early 1980s clearly showed that not the entire root is susceptible to infection and nodule induction. The concept of a "window of nodulatability" developed, which stated that only the cells in the region of emerging root hairs are able to be infected. Parallel work demonstrated that in a radial sense, not all root hair cells are capable of infection and nodule meristem initiation was restricted to cortical areas next to the xylem poles (see Rolfe and Gresshoff, 1988). This suggested (a) developmental gradients in an axial dimension and (b) cellular potencies produced in a radial sense by positional gradients involving putative source-sinks and unknown substances. The key conclusion was that not all root hair cells are identical and not all cortical cells behave the same way. Moreover the problem is dynamic as the root tip continuously grows, regenerating new invasion and nodulation targets.

Since the presence of rhizobia produces mitogenic nod-factors (LCOs), the plant has a requirement to control cell divisions. This is done by development and through a systemic circuit. This circuit is best demonstrated in split root experiments, in which prior inoculation of one spatially separated root portion inhibits nodulation in the second portion receiving a delayed inoculation. In soybean a 7 day delay inhibits all nodulation on the delayed side. When using the supernodulation mutants, no inhibition occurred (Olsson et al, 1989). This control suggested the involvement of the shoot, which was confirmed by reciprocal grafting to be the source of the autoregulation signal (Delves et al, 1986), who showed that a grafted plant with an nts382 shoot and a wild-type root will be supernodulating. The reverse graft (wild-type shoot:nts382 root) showed wild-type nodulation. This shoot control of nodulation has become a central feature of all supernodulation mutants, except for the nod3 mutant in pea, which is root controlled. Grafts without growing shoot apexes limited the source of the autoregulation signal to the young leaf (Delves et al, 1992).

3. Molecular Expression Studies

To investigate the molecular changes in leaves during autoregulation of nodulation, shoot RNA was in vitro translated and products were viewed by 2-D electrophoresis (Sayavedra-Soto et al, 1995). No significant and reproducible protein changes were seen related to the supernodulation and autoregulation status of the plant. however, proteins changed through development and presumably in response to new nitrogen source availability (ureide utilization, storage, etc.).

We recently started to use differential display to analyze shoot changes. mRNA was extracted from uninoculated Bragg plants, and 7 day inoculated Bragg and nts382 plants. The timing was designed to initiate autoregulation in the inoculated wild type, but to prevent interference with nodule emergence and the onset on nitrogen fixation and N-transport. Nearly 120 primer combinations were tested, yielding over 10,000 data points (presumably mRNAs, although redundancies do exist).Differentially expressed RNAs (or rather resultant DNA bands) were isolated from the gels, cloned and sequenced. Clones were used to confirm differential expression by northern blots, while DNA sequence allowed synthesis of primers for RT-PCR verification. At present 5 candidate bands have been investigated in more detail. Their expression patterns follow two main types: (1) they are absent in uninoculated wild type, but present in the inoculated material, and (2) they are absent in uninoculated wild type and inoculated nts382, but present in inoculated wild type. Class 1 expression is consistent with a general response to inoculation, without distinction of normal and supernodulation. Class 2 suggest s that the nts mutant behaves like an uninoculated plant, consistent with its phenotype of persistent nodulation.

Preliminary sequence comparisons (BLAST) showed that class 1 clone G10 is similar to sequences in genes for rh2, a pea root hair protein, and an alfalfa stress response protein. Class 2 clone G6.2 was similar to several beta-1,3 glucanases. Both findings suggest that shoots of soybean plants respond to early inoculation events through activation of stress related proteins. The fact that nts382 differs from Bragg may provide a hint towards autoregulation.

References:

Alva, A.K., Edwards, D.G., Carroll, B.J., Asher, C.J. and Gresshoff, P.M. (1988). Nodulation and early growth of soybean mutants with increased nodulation capacity under acid soil infertility factors. *Agronomy Journal* **80**: 836-841.

Caetano-Anollés, G. and Gresshoff, P.M. (1996) Generation of sequence signatures from DNA amplification fingerprints with mini-hairpin and microsatellite primers. *Bio/Techniques* **22**: 1044-1054.

Carroll, B.J., McNeil, D.L. and Gresshoff, P.M. (1985a). Isolation and properties of soybean mutants which nodulate in the presence of high nitrate concentrations. *Proc. Natl. Acad. Sci. USA* **82**: 4162-4166.

Carroll, B.J., McNeil, D.L. and Gresshoff, P.M. (1985b). A supernodulation and nitrate tolerant symbiotic (nts) soybean mutant. *Plant Physiol.* **78**: 34-40.

Delves, A.C., Mathews, A., Day, D.A., Carter, A.S., Carroll, B.J. and Gresshoff, P.M. (1986). Regulation of the soybean-*Rhizobium* symbiosis by shoot and root factors. *Plant Physiol.* **82**: 588-590.

Delves, A.C., Higgins, A. and Gresshoff, P.M. (1992) The shoot apex is not the source of the systemic signal controlling nodule numbers in soybean. *Plant, Cell and Environment* **15**: 249-254.

Funke, R., Kolchinsky, A. and Gresshoff, P.M. (1993). Physical mapping of a region in the soybean (*Glycine max*) genome containing duplicated sequences. *Plant Mol. Biol.* **22**: 437-446.

Funke, R.P., Kolchinsky, A. and Gresshoff, P.M. (1994) High EDTA concentrations cause entrapment of small DNA molecules in the compression zone of pulsed field gels, resulting in smaller than expected insert sizes in YACs prepared from size selected DNA. *Nucleic Acids Research* **22**:2708-2709.

Gresshoff, P.M. (1995) Moving closer to the positional cloning of legume nodulation genes. In: *Nitrogen Fixation: Fundamentals and Applications*. eds. I. A. Tikhonovich, V.I. Romanov, N.A. Provorov and W.E. Newton. Kluwer Academic Publishers, Dortrecht, NL., pp 416-420.

Kolchinsky, A., Landau-Ellis, D. and Gresshoff, P.M. (1997) Map order and linkage distances of molecular markers close to the supernodulation (*nts-1*) locus of soybean. *Molec. Gen. Genetics* (in press).

Landau-Ellis, D. and Gresshoff, P.M. (1994) The RFLP molecular marker closely linked to the supernodulation locus of soybean contains three inserts. *Molecular Plant Microbe Interactions* **7**:432-433.

Landau-Ellis, D., Angermüller, S. A., Shoemaker, R., and Gresshoff, P.M. (1991) The genetic locus controlling supernodulation co-segregates tightly with a cloned molecular marker. *Mol. Gen. Genetics* **228**: 221-226.

Mathews, A., Carroll, B.J., and Gresshoff, P.M. (1989) Development of *Bradyrhizobium* infections in supernodulating and non-nodulating mutants of soybean (*Glycine max* (L.) Merrill). *Protoplasma* **150**: 40-47.

Olsson, J.E., Nakao, P., Bohlool, B.B. and Gresshoff, P.M. (1989). Lack of systemic suppression of nodulation in split root systems of supernodulating soybean (*Glycine max* (L.) Merr.) mutants. *Plant Physiol.* **90**: 1347-1352.

Rolfe, B.G. and Gresshoff, P.M. (1988) Genetic analysis of legume nodule initiation. *Ann. Rev. Plant Physiol. Plant Mol. Biol.* **39**: 297-319.

Sayavedra-Soto, L.A., Angermüller, S.S., Prabhu, R. and Gresshoff, P.M. (1995) Polypeptide patterns in leaves of *Glycine max* (L.) Merr. cv. Bragg and its supernodulating mutant during early nodulation. *Physiology and Molecular Biology of Plants* **1**: 27-36.

Zhu, T., Shi, L., Funke, R.P., Gresshoff, P.M. and Keim, P. (1996) Characterization and application of soybean YACs to molecular cytogenetics. *Mol. Gen. Genetics* (in press).

Effect of nitrogen nutrition pathways on the quality of nitrogen storage compounds in legumes

G. Duc[1], D. Page[1], M. Sagan[1], G. Viroben[2], J. Gueguen[2]

[1] Institut National de la Recherche Agronomique, Station de Génétique et d'Amélioration des Plantes, BV 1540, 21034 Dijon Cedex, France
[2] Institut National de la Recherche Agronomique, Laboratoire de Biochimie et Technologie des Protéines, rue de la Géraudière, BP 527, 44026 Nantes Cedex, France

INTRODUCTION

Legumes accumulate nitrogen from two main sources : biological nitrogen fixation (BNF) resulting from plant-*Rhizobium* symbiosis and nitrate assimilation (NA). Many field experiments have measured that more than 60-70 % of legume nitrogen is supplied by BNF pathway in the absence of major limiting factor and no soil N enrichment is required in most field situations.

Legumes are famous for their yield instability and their variable composition. Pea (*Pisum sativum* L.) whose seeds are intensively used in animal feeds, provides numerous illustrations of these instabilities. Concerning 733 commercial samples harvested in France from 1987 to 1992, statistics by UNIP detected a 10 points range (19 to 29 % of dry matter) in seed protein content, the environment effect being greater than the genotypic one. Gueguen and Babot (1988) found for cv Amino an environment induced variation on the protein composition (proportion of albumin in the protein ranging from 12.6 to 38.1 % and vicilin/legumin ratio from 1.2 to 2.7). Atta *et al.* 1995 measured intra-plant variations in protein content of individual seeds which differ in amplitude with genotypes and may be explained by the level of concordance of the seed filling period with the nitrogen fixation cycle.

Since nitrogen is a main yield factor in agronomy and also a major protein constituent, it is logically incriminated in priority to explain these instabilities. Indeed, many studies have shown that BNF is highly susceptible to drought, water logging, high and low temperatures, salinity (Saxena *et al.*, 1993) and also to biotic stresses such as *Sitona* larvae which feed on the nodules. It is also well-known that soil nitrates inhibit BNF switching the plant N source toward NA. Genetic variability has been detected in several species in relation to BNF intensity, but breeding was rendered difficult by complex inheritance and heavy measurement techniques of this trait.

Whether the respective activity of the two nitrogen nutrition pathways co-existing in a legume will influence the product quality is the question considered in this paper. It is an increasingly important question for the marketing and nutritional value of agricultural products and consequently for breeders' decisions.

BNF AND NA ARE DIFFERENTIALLY REGULATED AND WILL DETERMINE DIFFERENTIAL N ACCUMULATION DURING PLANT DEVELOPMENT

BNF and NA differ in their kinetics over plant cycle and take place in distinct organs. It is generally accepted that nitrate reductase activity in annual legumes is predominant at the beginning of plant cycle and is later relayed by BNF which peaks when first pods start filling.

Main enzymes of N assimilation pathway (nitrate reductase, nitrite reductase, glutamine synthetase, asparagine synthase) are mostly located in roots and leaves. Their proportions can vary according to the genotype. In *Vicia faba* L., Caba *et al.* (1995) have measured 75 % of nitrate reductase activity in roots of one genotype and 85 % in leaves of another genotype. For BNF pathway, bacterial nitrogenase, plant glutamine synthetase and asparagine synthase are major enzymes found in the nodules. Added to this complexity, Tingey *et al.* have shown in pea that distinct glutamine synthetase polypeptides are differentially expressed in leaves, roots and nodules. In legume species which transport fixed nitrogen as ureids, an additional metabolism is found in nodules which consists in allantoin and allantoic acid synthesis. Most products of these pathways are further degraded in leaves and fruits providing remobilized N compounds for seed filling.

According to the nature of the stress applied to the legume plant and the stage at which it occurs during plant cycle, then BNF or NA will be more specifically hampered which may have different consequences on the chemical nature of plant reserves.

NATO ASI Series, Vol. G 39
Biological Fixation of Nitrogen
for Ecology and Sustainable Agriculture
Edited by A. Legocki, H. Bothe, A. Pühler
© Springer-Verlag Berlin Heidelberg 1997

WHETHER A LEGUME PLANT IS FED ON BNF OR NA INFLUENCES THE NATURE OF N PRODUCTS IN DIFFERENT PLANT TISSUES

In xylem sap of amide exporting legumes (AEL) asparagine represents more than 70 % of N compounds whereas ureids are major compounds in N_2 fixing ureid-exporting legumes (UEL). When increasing the BNF/NA activity ratio in UEL, it has been shown on cowpea (Peoples et al., 1985) and soybean (Mc Clure and Israel, 1979) that ureid content increases while nitrate and asparagine contents decrease in the xylem sap. Moreover, soybean breeders have used ureid content as a breeding criteria to enhance BNF (Herridge and Rose 1994). In AEL such as, chickpea, lentil and faba bean, Peoples et al. (1987), Rochat and Boutin (1991) have shown that nitrate content and asparagine/glutamine ratio decrease in the xylem sap while BNF/NA increases. Rochat and Boutin (1991) also noticed a lower homoserine content in the xylem sap of N fixing plants.

Contrasting with results on xylem sap, Peoples et al. (1985) on cowpea (UEL) found very low ureid contents in fruit phloem sap (ureid represent less than 5 % when glutamine becomes the major constituent) whatever the N plant source and did not detect differences on amino acids content of phloem sap between plants fed from BNF or from NA. In the fruit phloem sap of pea (AEL), Rochat and Boutin (1991) have shown that asparagine and glutamine are the main constituents. They have also shown that nitrate fed plants were poorer in asparagine and richer in homoserine and glutamine than N2 fixing plants in this sap. Atta (1995) also found a strong correlation (r = -0,89) between BNF intensity measured by acetylene reduction assay and homoserine content in the phloem sap of cv Solara.

The empty organ technique experienced on immature pea seeds measured higher homoserine and lower glutamine contents in the seed coat secretion liquid of NO_3^- fed plants in comparison with N_2 fixing plants (Rochat and Boutin, 1991).

Studies on the impact of N nutrition pathway on seed protein quality of legumes are very seldom in the literature. Zougari et al. (1995) have measured in soybean a significant decrease in the proportion of lipoxygenase in the protein between NA and BNF regimes.

In pea, this paper provides comparisons made on cv Frisson submitted to very contrasted N nutrition regimes. A field experiment (Sagan et al., 1993) has been devised using cv Frisson, two isogenic symbiotic mutants of cv Frisson (P2 : non-nodulating, P64 : super-nodulating and nitrate-tolerant) and two nitrate fertilizer treatments (0 or 500 kg N ha^{-1}). It resulted in very diverse BNF/NA ratio (Table 1). Treatment (P2 - ON) has been inducing strong N source limit and was the only one to significantly reduce seed yield and seed N content (table 1). The composition of seed nitrogen was achieved by separating the non-protein nitrogen (NPN) defined by Bell (1963) as compounds of MW lower than 10000 D (it comprises free amino acids, small peptides, ammonium salts, alkaloïds, purinic and pirimidic bases ...) and by determining the three main protein fractions (albumin, vicilin and legumin).

NPN was determined using Kjehldahl method on seed extract with 10 % (p/v) trichloracetic acid. An aliquot fraction was used to estimate free amino acids after derivatization (PITC method) according to Bildingmeyer et al. (1987). On another aliquot, total amino acids were determined after an acid hydrolysis step (HCl 6N, 24 hours, 110°C) followed by derivatization as above. The main protein fractions were separated by ion-exchange chromatography on the monoQ column as described by Baniel et al. (1992). The five major peaks were collected and analysed following the Bradford color test procedure (Bradford, 1976). The experiment was performed three times.

In all treatments, NPN represents a small proportion of the seed nitrogen (8-12 %). A tendancy of lower proportion is found in P2-ON treatment when N starvation occurs. Regardless of the type of sample, amino acid composition of NPN shows high concentrations of arginine, aspartic acid and glutamic acid specially present under free form (Table 2). The respective proportion of the different amino acids in NPN is only modified in P2-ON with a drastic decrease in free asparagine, total aspartic acid, arginine and glycine while it has no effect on glutamic acid.

Whatever sample is considered, no significant difference is detected on the albumin fraction. Only the globulin fractions are affected when nitrogen nutrition is deficient in P2-ON treatment (Figure 1). In this case, the most affected fraction is legumin which is dramaticaly reduced when vicilin rate is moderately reduced. No significant difference on globulins is detected between Frisson, P64 and P2-500N.

Table 1 : Nitrogen Regime effect on pea seed yield and quality

Genotype	N treatment kg. ha⁻¹	Seed N yield g N m⁻²	Fixed N % in seeds (2)	seed nitrogen (%)	Seed nonprotein N (% total N)
FRISSON	0	15.0 a (1)	89.3	4.0 a	10.0
FRISSON	500	18.6 a	12.3	4.2 a	11.0
P2	0	1.9 b	0	2.7 b	8.1
P2	500	20.4 a	0	4.2 a	11.2
P64	500	15.4 a	46.7	4.0 a	12.0

(1) Classification of means according to Newman-Keuls test p<0.05
(2) Measured by Soil 15 N labelling technique (Sagan *et al.* 1993)

Table 2 : Total and free main amino acids in seed Nonprotein fraction (millimoles / 100 g)

Genotype	N treatment kg. ha⁻¹	Total			Free			
		Aspartic acid	Glutamic acid	Arginine	Aspartic acid	Asparagine	Glutamic acid	Arginine
FRISSON	0	2.66	2.06	3.24	0.34	1.67	1.24	3.77
FRISSON	500	3.11	2.05	4.00	0.18	1.74	0.63	4.82
P2	0	0.98	1.96	0.61	0.05	0.31	0.52	0.68
P2	500	3.10	2.17	4.30	0.17	1.34	0.56	4.70
P64	500	3.11	2.24	4.37	0.27	2.30	1.08	6.79

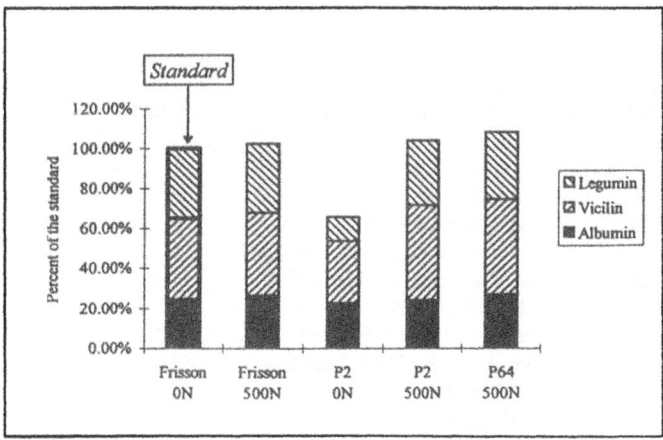

Fig 1 : Comparison to the standard (Frisson 0N) of protein content and its biochemical composition for various seed samples : Frisson grown in high nitrate conditions (Frisson 500N), P2 grown in low nitrate conditions (P2 0N), P2 grown in high nitrate conditions (P2 500N) and P64 grown in high nitrate conditions (P64 500N).

DISCUSSION - CONCLUSION

Protein content and protein composition of pea seeds strongly influence their nutritional value.

According to the nitrogen nutrition pathway used by a legume (BNF or NA), there are some qualitative differences in N compounds accumulated which are higher in vegetative tissues and in UEL and which tend to be non-significant or restricted to minor fractions in seeds (after N remobilization from vegetative tissues and embryo protein synthesis). However this seed composition aspect received little consideration in the past.

On the contrary, our data show that the level of plant N supply by BNF and NA will have a strong influence on seed quality when it is limited. Indeed Sagan et al. (1993) have shown in pea that N deficiency is firstly expressed by a dilution of N in plant and seed dry matter. Our study detected that some compounds are specifically reduced by strong N starvation (ie particular amino acids in NPN and legumins in proteins). We can explain that legumins are more responsive to N deprivation than other fractions by the fact albumins contain the main part of enzymes and proteins involved in embryo development which gives them priority, and by the fact rapidly senescing stressed plants will be mostly affected in their latest formed protein fractions which are legumins (Casey et al., 1993).

Consequently, intensity and favourable interaction of BNF and NA pathways are important aspects to consider when breeding for quality of legume products, the ultimate objective being an adequate total N supply in intensity and timing to the plant which will also be important to yield stability. Regulation and interaction between NA and BNF are therefore to be understood. In this respect, genetic nitrate tolerance of symbiosis offers a new BNF x NA interaction. Published reports indicate that nitrate tolerance occurs naturally (Herridge and Betts, 1988) but can be also induced as shown in recently selected supernodulating mutants (for example in pea, Duc and Messager 1989, Sagan et al. 1993 or soybean , Gresshoff 1993, Song et al. 1955) which maintain some N_2 fixing activity in high nitrate condition. Such a new variability presently evaluated in several laboratories may be valuable.

REFERENCES

Atta S., 1995. Thèse n° 1484 de l'Université de Rennes I, 108 pp.

Atta S., Maltese S. and Cousin R., 1995. Proceeding of 2nd European Conference on Grain Legumes. AEP (eds), 417.

Bell P.A., 1963. Anal. Biochem. 5, 443-451.

Baniel A., Gueguen J., Bertrand D., 1992. Proceeding of 1st European Conference on Grain Legumes, AEP (eds) 409-410.

Bildingmeyer B.A., Cohen S.A., Tarvin T.L., Frost B., 1987. J. Assoc. of Anal. Chem. 70, 241-247.

Bradford MM, 1976. Anal. Biochem. 72, 248-254.

Caba J.M., Lluch C., Ligero F., 1995. Phsyiologia Plantarum 93, 667-672.

Casey R., Davies D.R., 1993. In peas : Genetics, Molecular Biology and Biotechnology, R. Casey and D.R. Davies (eds), CAB (Pbs) 121-163.

Duc G., Messager A., 1989. Plant Sci. 60, 207-213.

Gresshoff P., 1993. Plant Breeding Reviews 11, 275-318.

Gueguen J., Babot J., 1988. J. Sci. Food Agric. 42, 209-224.

Herridge D.F., Betts J.H., 1988. Plant and Soil 110, 129-135.

Herridge D.F., Rose I.A., 1994. Crop Sci. 34, 360-367.

Peoples M.B., Pate J.S., Atkins C.A., 1985. J. Exp. Bot. 165, 567-582.

Peoples M.B., Sudin M.N., Herridge D.F., 1987. Exp. Bot. 189, 567-579.

Rochat C., Boutin J.P., 1991. J. Exp. Bot. 42, 207-214.

Sagan M., Ney B., Duc G., 1993. Plant and Soil 153, 33-45.

Saxena N.P., Johansen C., Saxena M.C., Silim S.N., 1993. In : Breeding for stress tolerance in Cool-Season food legumes. K.B. Singh and M.C. Saxena (eds), ICARDA and J. Wiley (Pbs), 245-270.

Song L., Carroll B.J., Gresshoff P.M., Herridge D.F., 1995. Soil Biol. Biochem. 27, 563-569.

Tingey S.V., Walker E.L., Coruzzi G.M. (1987). The EMBO Journal 6, 1-9.

Zougari A., Guy S., Planchon C., 1995. Plant Breeding 114, 313-316.

Genetic linkage map of alfalfa (*Medicago sativa*) and its use to map seed protein genes as well as genes involved in leaf morphogenesis and symbiotic nitrogen fixation

G.B. Kiss, P. Kaló, K. Felföldi, P. Kiss and G. Endre

Institute of Genetics, Biological Research Center of the Hungarian Academy of Sciences, Szeged, P. O. Box 521, H-6701, Hungary

Abstract. We have constructed a genetic map for *Medicago sativa* using morphological, isozyme, seed protein, RFLP and RAPD markers. For this we have crossed two diploid alfalfa plants (the yellow-flowered diploid *Medicago sativa* ssp. *quasifalcata* k93 and the purple-flowered *Medicago sativa* ssp. *coerulea* w2) which belong to the *Medicago sativa* complex. One F1 plant (F1/1) was selected and self pollinated to produce the F2 population which consisted of 138 individuals. Mapping was carried out in this segregating population by determining the genotype of more than 1000 genetic markers for the individual plants. The genetic map consists of eight linkage groups corresponding to the eight haploid chromosomes of alfalfa. The genetic length of the genome is approximately 550 cM, consequently the physical equivalent of 1 cM is about 1500 kb in average. The map location of the sticky leaf mutation conditioning altered leaf morphogenesis as well as an ineffective symbiotic mutation was determined. The fine mapping and the isolation of the genes in question by map-based cloning are in progress.

Keywords. genetic mapping, map-based cloning, *Medicago sativa*, RFLP, symbiotic nitrogen fixation, leaf morphogenesis

1. Introduction

In higher eukaryotic plants, the number of the genes is approximately 50,000, therefore the plant genomes are fairly complex, consequently it is very hard to correlate genes with functions. Some techniques which are applicable for prokaryotic gene isolation (e.g. complementation of mutants) can not be used for most of the plant species. In higher eukaryotes specific genes can be identified by the genetic approach. In this case a mutant individual impaired in the biological process in question has to be isolated, then the gene suffered the mutation is genetically mapped using molecular markers, and finally the gene can be cloned with the help of tightly linked molecular markers. This approach is the so called map-based cloning strategy and has a unique advantage over the ambiguous biochemical strategies, that is one can be sure that the isolated gene is really involved and necessary in the biological process studied (see Csanádi et al., 1994).

Recently genetic mapping is carried out by the use of molecular markers identified by RFLP hybridization and PCR-based amplifications. The closest molecular markers can be used as start points for chromosomal walking to reach the gene of interest. Depending on the distance between the start point and the target gene, chromosomal

NATO ASI Series, Vol. G 39
Biological Fixation of Nitrogen
for Ecology and Sustainable Agriculture
Edited by A. Legocki, H. Bothe, A. Pühler
© Springer-Verlag Berlin Heidelberg 1997

walking can be very difficult, or relatively easy if the distance is less than 100 kilobase pairs. Nowadays map-based cloning strategy is coming into general use. This approach has been applied to clone several plant genes involved in morphogenesis or plant pathogen interactions.

2. Results

Our research group started the construction of the genetic map of alfalfa several years ago in order to develop a genetic system for *Medicago sativa* which can be used for basic as well as for applied research. The existence of a genetic map of alfalfa enables that genetic techniques can be applied for breeding or map based cloning type of work. In the last five years we have constructed a genetic map for alfalfa using morphological, isozyme, seed protein, RFLP (Restriction Fragment Length Polymorphism) and RAPD (Random Amplified Polymorphic DNA) markers (Williams et al., 1990). We have crossed two diploid alfalfa plants (the yellow-flowered diploid *Medicago sativa* ssp. *quasifalcata* k93 and the purple-flowered *Medicago sativa* ssp. *coerulea* w2) which belong to the *Medicago sativa* complex (Quiros and Bauchan, 1988). One F1 plant (F1/1) was selected and self pollinated to produce the F2 population which consisted of 138 individuals. Mapping was carried out in this segregating population by determining the genotype of more than 1000 genetic markers for the individual plants. The markers mapped to eight linkage groups which correspond to the eight haploid chromosomes of alfalfa. The size of the genome is approximately 550 cM, consequently the physical equivalent of 1 cM is about 1500 kb. Among others we determined the map location of 20 nodulin genes like ENOD2, ENOD12, Nod-22, Nod-25, leghemoglobins, glutamine synthetases, etc., and two cytologically identified genes, the rDNA and the beta-tubulin. Some morphological markers like sticky leaf, dwarf and colour of the flowers, as well as several genes with known function were also mapped (Kiss et al., 1993; Kiss et al., 1994). Out of 220 RFLP loci 155, 18, 7 and 2 loci coded for single, double, triple and quadruple genes, respectively. Since the duplicated regions showed no linkage, they can be explained by multiplication of smaller regions rather than duplication of larger DNA segments.

72 individuals of the segregating population were investigated for seed protein content (12 seeds from each plant) by two dimensional gel electrophoresis. Protein patterns were compared on a pair-wise manner. Some spots were organised in "charge trains" with the same molecular mass but different isoelectric points. They represent the most abundant proteins and probably have storage function in the seed. 130 protein spots segregated representing 45 protein markers on the genetic map. 30 showed codominant and 15 showed dominant inheritance. These 2D-markers were located on all the eight linkage groups of alfalfa.

The long term goal of our continuous mapping effort is the isolation of genes involved in symbiotic function and in plant development, that is the application of the map-based gene cloning strategy for alfalfa. One important prerequisite for this work is the isolation of a mutant plant impaired in the appropriate function. In our group a leaf morphological mutant (sticky leaf) and a mutant exhibiting ineffective symbiotic nitrogen fixation are under intensive investigation.

The sticky leaf (*stl*) mutants affected in leaf morphogenesis have appeared in the mapping population used to construct the genetic map of alfalfa. After the map position of this mutation had been determined, the isolation of the *stl* gene by map-based cloning was started. The *stl* mutation was mapped to linkage group 6 on the genetic map between RFLP markers UO553 and TMS3. These molecular markers are locating at 3.1 cM and 14.1 cM distance from *stl*, respectively. However, they are too far yet to start chromosomal walking. To reduce the number of the walking steps, molecular markers in very tight linkage with the mutation are required, that is "mapping rather than walking" is recommended.

With the application of Bulked Segregant Analysis technique (Michelmore et al., 1991) we screened RAPD primers for tightly linked markers to the *stl* mutation. 610 different 10-mer RAPD primers were tested in PCR reactions, and out of them 74 primers provided 89 molecular markers locating in the vicinity of the *stl* locus in both directions. 12 RAPD markers were found to be within 1 cM distance from the *stl*. In the meantime the rRNA locus was mapped close to the *stl*. The genetic distance of the closest markers from *stl* is about 0.2 cM and we are on the way to determine their physical distance by Pulsed Field Gel Electrophoresis.

During the development of the alfalfa genetic system one of our aims was to isolate mutant plants impaired in symbiotic nitrogen fixation. Among the mutants isolated, one proved to be appropriate for further genetic analysis. This mutant plant was called McIN6 and the locus mutated was designated in_6. The McIN6 individual plant derived from the *Medicago sativa* ssp. *coerulea* seed population is incapable to form effective symbiosis in the presence of *Rhizobium meliloti* AK631. The mutant plant shows obvious nitrogen deficiency and formed many white, ineffective nodules in the absence of combined nitrogen and in the presence of *Rhizobium meliloti* AK631. In the presence of combined nitrogen (e.g. soil) this plant is able to grow normally, develops flowers and yields seeds after either self or foreign pollination. The light microscopical analysis showed that the bacteria enter the nodule cells and they develop into bacteroids, but inside the nodules few infected cells are found. The invasion process takes place normally (infection threads are visible), but the number of the bacteroids is far less than in the normal nodules. The electron-microscopical analysis revealed that the nodule cells undergo early senescence and this phenomenon is similar to the in_2 and in_3 mutants isolated from tetraploid alfalfa cultivar (Vance and Johnson, 1983).

The ineffective phenotype is maintained after either vegetative or generative propagation. From the cross between McIN6 and the wild type Mcw2 plant it turned out that the mutation is recessive, since all F1 progenies were able to fix nitrogen and the self mating progeny of the F1 plants showed that the mutation inherited as a single Mendelian marker. The McIN6 plant had also been crossed with the yellow-flowered *Medicago sativa* ssp. *quasifalcata* k93 plant and F1 plants were generated. The F2 progenies of two F1 plants (F1/19 and F1/51) were analyzed for their symbiotic effectiveness. RFLP and RAPD markers were used to map the segregating in_6 mutation on the eight linkage groups of the genetic map of *Medicago* (Kiss et al., 1993). One RAPD marker (OPE7X) which located on the linkage group 7 between two RFLP markers showed linkage with the in_6 allele.

3. Conclusions

The saturated linkage map of alfalfa containing more than 1000 markers allows us to map mutations affecting different traits of *Medicago sativa* like sticky leaf and ineffective symbiotic nitrogen fixation. Fine mapping of the mutated genes conditioning these traits is under way by identifying PCR-based molecular markers with the help of Bulked Segregant Analysis method. The closest molecular markers can be used afterwards as starting points for chromosomal walking towards the mutated genes. This map-based cloning strategy will enable us to identify and characterize genes required for leaf development, symbiotic nodule organogenesis and nodule function in alfalfa.

4. Acknowledgments

The authors thank Mrs. P. Somkúti, Mrs. L. Tóth, Mrs. K. Katona, Ms. Z. Liptay and Mr. S. Jenei for skillful technical assistance. This work was supported partly by Grants OTKA T013219, OTKA T016935, CIPA-CT93-0156, AKP 96-360/62 and by the C.N.R.S. - Hungarian Academy of Sciences collaborative program.

5. References

Csanádi, G., J. Szécsi, P. Kaló, P. Kiss, G. Endre, Á. Kondorosi, É. Kondorosi, and G. B. Kiss. 1994. ENOD12, an early nodulin gene, is not required for nodule formation and efficient nitrogen fixation in alfalfa. Plant Cell 6:201-213.

Kiss, G. B., G. Csanádi, K. Kálmán, P. Kaló, and L. Ökrész. 1993. Construction of a basic genetic map for alfalfa using RFLP, RAPD, isozyme and morphological markers. Mol. Gen. Genet. 238:129-137.

Kiss, G. B., P. Kaló, G. Endre, G. Csanádi and K. Felföldi. 1994. An improved linkage map for alfalfa. p. 169-173. In G. B. Kiss and G. Endre. (ed.), Proceedings of the "1st European Nitrogen Fixation Conference" and the Workshop on "Safe Application of Genetically Modified Microorganisms in the Environment". Officina Press, Szeged.

Michelmore, R. W., I. Paran, and R. V. Kesseli. 1991. Identification of markers linked to disease-resistance genes by bulked segregant analysis: A rapid method to detect markers in specific genomic regions by using segregating populations. Proc. Natl. Acad. Sci. U. S. A. 88:9828-9832.

Quiros, C. F. and G. R. Bauchan. 1988. The genus *Medicago* and the origin of the *Medicago sativa* complex. p. 93-124. In A. A. Hanson, D. K. Barnes, and R. R. Hill. (ed.), Alfalfa and alfalfa improvement. American Society of Agronomy, Madison, WI.

Vance, C. P. and L. E. B. Johnson. 1983. Plant determined ineffective nodules in alfalfa (*Medicago sativa*): structural and biochemical comparisons. Can. J. Bot. 61:93-106.

Williams, J. G. K., A. R. Kubelik, K. J. Livak, J. A. Rafalski, and S. Tingey. 1990. DNA polymorphism amplified by arbitrary primers are useful as genetic markers. Nucleic Acids Res. 18:6531-6135

Genetic Transformation and Regeneration of Legumes

Craig A Atkins and Penelope MC Smith

Botany Department and Centre for Legumes in Mediterranean Agriculture (CLIMA), University of Western Australia, Nedlands, WA, 6907, Australia.

Abstract. A wide range of legume species have been genetically transformed and the transformants regenerated to provide novel genotypes. The methods used for transformation include the use of *Agrobacterium tumefaciens, A. rhizogenes*, particle bombardment and electroporation. Although *Agrobacterium*-based methods using a wide variety of tissue explants have been the most widely used, generally they yield low rates of transformation (0.1-5%). *In planta* techniques, either with *Agrobacterium* or electroporation of intact meristems, hold the promise of much higher frequencies and more simple means of regeneration of transformants. Methods have been developed for soybean, pea, cowpea, moth bean, common bean, black gram, groundnut, lentil, narrow-leafed lupin, yellow lupin, winged bean, broad bean, narbon bean, pigeon pea, grass pea and chickpea among the pulses and for a number of medics and clovers, alfalfa, Townsville stylo, sainfoin and birdsfoot trefoil among pasture species. In many cases viable plants and seed have been recovered and the transformations are stable. Most commonly the reporter gene for β-glucuronidase (*gus*), or genes for antibiotic resistance (*npt, hpt*) have been used as selectable markers. More recently genes conferring tolerance to herbicides (eg *bar* or *pat* for glufosinate [Basta] or *aro* A and *cp*4 / *gox* for glyphosate [Roundup]) have proven to be more effective. These genes have also provided a basis for engineering novel herbicide-tolerant cultivars of a number of crop species. Transformations to enhance the nutritional quality of grain or herbage, or to insert genes associated with nodule function have also been generated. A detailed summary of procedures developed for the *A.tumefaciens*-mediated transformation of narrow-leafed lupin (*Lupinus angustifolius* L.) to create stable herbicide (Basta) resistant cultivars is presented.

Keywords. Genetic engineering, legumes, lupin, *Lupinus angustifolius* L., pasture, pulses, regeneration, transformation

1 Introduction

The tools of molecular genetics and recombinant DNA technology have opened the way for direct genetic manipulation (genetic engineering) of plants. While the nature of genetic engineering is really no different to what the plant breeder has always done, there are two unique features which set the acivity apart from the classical processes of crossing, backcrossing and recurrent selection. Firstly, genetic engineering offers the potential to introduce defined single genes, and secondly there

NATO ASI Series, Vol. G 39
Biological Fixation of Nitrogen
for Ecology and Sustainable Agriculture
Edited by A. Legocki, H. Bothe, A. Pühler
© Springer-Verlag Berlin Heidelberg 1997

are no genetic or reproductive barriers as to the source of introduced genes. Thus the whole of the world's gene pool is now available to alter plants in specific and controlled ways. For these reasons the introduction of a new trait is not accompanied by unwanted features which then need to be "bred out". While this may reduce the time required for a new genotype to be produced and stabilised the need for rigorous testing of trangenics and the statutory requirements which relate to their release means that there is not likely to be a significant speeding up of the release of new varieties. However this depends on the trait which being introduced. The technology has progressed to the stage where traits which require expression of a number of genes to cause the desired effect are difficult to engineer but potentially such changes will require the same time as insertion of a single gene.

Transgenic lines have been generated in a wide range of broad scale crop species, including cereals, oilseeds and pulses, in pasture species and species used for horticulture and floriculture and potentially in timber production. In the period 1986-93 there were 1025 field trials which occured in 31 countries and involved 38 species (Dale, 1995). Although a wide range of traits have been transferred more than 30% of the 1025 field releases relate to herbicide resistance and of the 20 or so subsequent applications for unrestricted use in agriculture more than 50% were for lines conferring resistance to herbicides (Dale,1995). Weed control is a significant cost component of modern agronomic practice and this, coupled with a range of more effective, less specific herbicides (such as glufosinate and glyphosate), has greatly stimulated research by their manufacturers. Coupling the manufacture of chemical products with genetic resistance to their toxicity in target crops offers the potential for considerable commercial gain. However there is expanding research in areas of disease and pest resistance, increased product shelf life, enhanced grain nutritional quality, oil content and composition and stress tolerance. More recently a diverse range of proposals and strategies to genetically engineer plants to serve as "bioreactors" have appeared (Goddijn and Pen, 1995). Substances which could be produced include specific carbohydrates, lipids and enzyme proteins for industrial processing but also pharmaceuticals, such as antigens and antibodies, growth factors and peptide hormones (Mason and Arntzen,1995) and the raw materials for biodegradable plastics based on polyhydroxyalkanoates (Poirier et al. 1992; Vanderleij and Witholt, 1995).

2 Genetic Transformation of Legumes

Legumes provide a critical component of farming systems in which ecological and economic sustainability are key criteria for adoption and development. As pulses they provide high protein grain, which in some cases also contains significant levels of high quality oil, as pasture components they provide high quality herbage while in agroforestry systems leguminous trees are fast-growing sources of fuel or timber wood and also have the capacity to contribute to the management of soil water resources. However, their value in all these guises lies in their capacity to fix atmospheric nitrogen. Fixation of nitrogen provides a significant alternative to fertilizer nitrogen and offers a means to conserve and manage soil resources more effectively. There seems little doubt that as agriculture develops and adopts further sustainable practices that leguminous crops, pastures and trees will provide an increasingly significant component of the production systems which evolve.

Genetic engineering offers a means to improve and enhance the utility of legumes by increasing their yield and harvest index, altering their architecture to more closely match mechanical harvesting practices, by changing the nature of the storage components of grain to increase their nutritional or processing value, by inserting genes to confer herbicide tolerance or resistance to viral and fungal diseases or to the many insect pests which ravage legumes crops and pastures and destroy the stored grain, to improve their agronomic features, such as reducing the tendency of legume roots to acidify soils, and perhaps to increase the effectiveness and efficiency by which their root nodules sequester nitrogen. To realise the potential which such diverse genetic manipulations might unleash requires three essential technologies.

a. An effective means to introduce novel genes and have them stably expressed-ie. *a transformation system*,
b. an efficient selectable marker gene which allows rapid and sensitive selection of putative transgenic cells or tissues at an early stage of the process-ie. *a mechanism for selection* , and
c. a highly efficient means of converting the transformed cells or explant into viable seed- ie. *a regeneration system*.

These are requirements for transforming plants of any species but in the case of legumes it is the regeneration system which has presented the greatest practical difficulty. Leguminous species are notoriously refractory to culture *in vitro* and a diversity of highly species-specific methods to overcome this limitation have been devised. Nevertheless Table 1 indicates that stable transformation and regeneration of viable plants and seed has been reported for a wide range of pulses (grain legumes) and pasture species. A recent review by Christou (1995) also provides a description of the many transformation and regeneration methods and their outcomes for a range of legumes with particular emphasis on soybean.

Most commonly transformation has been based on infection by one of the many strains of the crown-gall bacterium *Agrobacterium tumefaciens*. Most useful are strains which have been genetically disarmed to be less effective disease agents but which retain the ability to transfer the T-DNA of their Ti plasmid including one or a number of foreign transgenes to the host. *A.rhizogenes*, which induces "hairy root" disease in the host, has also been used to transfer genes along with the infection but transformation is restricted to somatic tissue of the root and so transformants can only be recovered if root cells regenerate viable plants. This has been most successful with pasture species which propagate readily from root or stolon segments and has the advantage that it avoids callus formation and possible somaclonal variation. The second type of method relies on direct transfer of DNA into plant cells or their protoplasts. This may be achieved through direct microinjection, through electroporation or electrophoresis, or by bombarding the tissue or cells with microprojectiles coated in DNA (Christou, 1992). Of the direct methods electroporation and particle bombardment have been applied to legumes (Table 1). A further set of "alternative" DNA delivery systems have been developed or proposed (Songstad et al., 1995) but these have not been applied to legumes.

It is difficult to compare the success of different methods (see Potrykus, 1991), but generally the frequency of transformation has been rather low (<1-5%). Certainly much lower than those obtained from less recalcitrant families, such as the Solanaceae, but rather similar to the cereals. Perhaps the most promising technique developed to date is the electroporation-mediated transfer system described by Chowira et al. (1995;1996). Unlike other electroporation methods which have used isolated protoplasts or cells this technique uses intact nodal meristems and relies on the plant to develop its reproductive structures from the treated apices in a more or less normal fashion. The method was successfully applied to pea, soybean, lentil and cowpea. Transient GUS expression was recovered in 13-25% of the treated plants with a lower but nevertheless quite high frequency of seed carrying the reporter gene to a second generation (7-12%, unpublished data from the authors), indicating stable integration of the transgene. Electroporation has also been used for transformation of intact embryonic explants from seeds of *Phaseolus vulgaris* which had germinated for 3 days (Dillen et al.,1995). Although limited to transient GUS expression in epicotyl and hypocotyl tissues of the explants high frequencies of transformation were obtained and the method was applicable to seven different bean lines as well as to pea, cowpea, pigeon pea, chickpea and soybean. *In planta* techniques have the advantage that regeneration does not rely on shoot production from callus or from suspensions of cells or protoplasts. Potentially these methods are more rapid, avoid the complexities of cell and tissue culture and reduce the likelihood of somaclonal variation. Oger et al. (1996) have recently described an *in planta* protocol for *Lotus japonicus* which relies on *A. tumefaciens*. The method involves inoculating the cotyledon attachment site on a seedling and takes advantage of the regenerative capability of the plant to spontaneously produce transformed shoots. A similar procedure has been developed for soybean (Chee and Slightom, 1995).

Initially the marker genes used to select and separate putative transgenics from a background of untransformed cells, explants or plants were those which conferred resistance to one of a number of antibiotics (Table 1). These include the *npt* II gene, which encodes neomycin phosphotransferase (NPT II) and provides resistance to neomycin, kanamycin and G418 (Pridmore, 1987), and the *hph*, *aph*IV or *hyg* genes for hygromycin phosphotransferase (HPT) and resistance to hygromycin (Gritz and Davies, 1983). However many plant species, including legumes, are resistant to antibiotics (Dekeyser et al., 1989) and in a number of cases their incorporation in selective media interferes with regeneration. The *gus* (or *uid*A) gene, which expresses the enzyme -glucuronidase (GUS), has been used very widely as a reporter for chimaeric gene constructs or simply to assess transient transformation in plants (Jefferson et al., 1987), including legumes (Table 1). However while GUS expression allows rapid and effective measures of transformation it does not provide an efficient screen for selection and regeneration. Replacing antibiotic resistance genes with genes for herbicide tolerance as selectable markers has greatly enhanced the efficiency of transformation both in cereals and legume crops (Vasil et al., 1993;

McElroy and Brettell, 1994; Schroeder etal., 1993). The first and most widely used herbicide for selection is Basta (glufosinate or phosphinothricin). Bacterial genes (*bar* and *pat*) which confer resistance by metabolising glufosinate to a non toxic product have been isolated from two species of *Streptomyces* (Thompson et al 1987; Strauch et al., 1988) Both encode the enzyme phosphinothricin acetyltransferase (PAT). The *bar* gene especially has provided a powerful selection tool which has been used widely in transformation of cereals and increasingly for legumes (Table 1). Like glufosinate, the herbicide glyphosate or Roundup, is a non-selective broad spectrum chemical which is extremely toxic to higher plants. Genes for resistance to glyphosate include *epsps*, a petunia sequence encoding enolpyruvylshikimate phosphate synthetase (EPSPS), a similar gene (*aro* A) from *Salmonella typhimurium* or from *Agrobacterium* (CP4), and, more recently (Barry et al., 1992), a bacterial gene encoding glyphosate oxidoreductase (GOX). Zhou et al. (1995) have shown that a combination of CP4 and GOX provides an extremely efficient selectable marker for wheat transformation and the method has been applied also to soybean (Fuchs et al., 1993). Genes for resistance to a number of other herbicides (2,4-D, atrazine, bromoxynil, sulphonylureas, sethoxydim and haloxyfop, isoxaben, 2,2-DCPA, norfluazan and phenmediphan; Chandler, 1995) have been described and it seems likely that an increasing number will be used as selectable markers.

A number of significant questions about the control needed and the extent to which transgenic crops should be used have been raised (Rogers and Parkes, 1995). One of these has arisen as the result of using herbicide tolerance genes as markers. Widespread use of the *bar* gene for example as a marker will produce unwanted resistance to Basta in a range of crops. Techniques to remove the selectable marker gene are being sought and developed in part to address this question but also to allow new genes to be added to already transgenic cultivars (ie. to allow "pyramiding" of genes). These include the use of a site-specific recombinant system from a bacteriophage (P1 Cre/ *lox*) which is functional in plant cells and has been shown to excise an *hpt* gene used as selectable marker (Dale and Ow, 1991) as well as a gene encoding resistance to sulphonylurea (Russell et al.,1992). Cloning selectable markers within transposable elements offers a further possibility for site specific excision (Goldsborough et al.,1993) as does the possibility of co-transformation where the selectable marker gene is carried in a separate plasmid or in a separate *Agrobacterium* strain to the gene of interest and is disposed of in the second transgenic generation (Shahin et al.,1986; Komari et al., 1996). It is also likely that alternative selectable markers, such as that based on adenine phosphoribosyltransferase (APRT; Schaff, 1994), which are benign in terms of phenotype and depend on substrate for expression, could allow incremental addition of new genes into already transgenic lines.

As noted earlier it is the regeneration of a viable plant and ultimately collection of its stable transgenic seed which provided a real difficulty in applying the tools of genetic engineering to legumes. The references listed in Table 1 contain a vast range of methods which, in many cases, are species-specific and frequently cultivar-specific. The high degree of specificity is usually reflected in the tissue culture or cell isolation steps and the ease with which shoots can be regenerated from these cultured materials. In a number of protocols so-called "highly regenerable" seed lines have been identified and these have provided the key to success (eg. a particular line in *cv*. Jemalong 2HA of *M. truncatula*, Wang et al.,1996; or the early line of *L. angustifolius*, Unicrop, Pigeaire et al.,1996). As the genetic base from which improvement is sought these lines may not be ideal. In fact Christou (1994) has suggested that any crop improvement program based on genetic engineering must use elite cultivars at the outset and be prepared to change to new breeding materials as they become available. Similarly a number of studies have shown that regeneration of roots in culture is also highly variable between cultivars (eg Nauerby et al.,1991). In the case of *A.tumefaciens*-mediated transformation, some strains of the bacterium are more effective with a particular species than others (Christou, 1994) and this is also the case for the hairy root -inducing plasmid of *A.rhizogenes* (Siefkes-Boer et al,1995). For methods based on the use of protoplasts the cell walls of some species are more readily digested than others. Kohler et al. (1987) noted that some mothbean cultivars produced protoplasts which were more readily transformed by DNA following heat shock than the protoplasts from other cultivars. Thus a great diversity of cell and tissue characteristics appear to determine the utility and aplicability of a particular method.

Many of the methods developed for grain legumes use tissue from developing, mature or germinating embryonic axes (Table 1). In some cases this tissue permits adventitious regeneration, in others, regeneration depends on axillary shoot growth from the axis. These embryonic explants represent a "halfway house" in the sense that although requiring initial propagation in gel media under sterile conditions they do not involve callus induction or the culture of cells or protoplasts. Both *Agrobacterium*-mediated and direct DNA transfer methods have been applied to these tissues with a high degree of success (Table 1). The most recent methods developed rely on the transformation being carried out *in planta* so that the need for tissue culture and special conditions for propagation are essentially avoided (Chowira et al., 1995; Chee and Slightom, 1995; Oger et al., 1996). Both *Agrobacterium* and electroporation have been used to transfer the DNA to cells of the shoot meristem of seedlings or young plants. The transformed meristematic cells produce transformed shoots and seed more or less in the course of normal development. It is significant perhaps that where these methods have been applied to grain legumes a wide range of cultivars and species have proved transformable (Table 1), and at high frequency (Chowira et al., 1995;1996), raising the possibility of their wide application.

All the examples in Table 1 are legumes of the Papilionoideae, possibly because this sub family contains the majority of economically useful food and forage species. However Tepfer (1990) has reported unpublished results from K Soo Ko et al. which show transformation of three species of *Cassia* (sub family Caesalpinoideae) with *A.rhizogenes*.

Table1. Protocols developed for Genetic Tranformation and Regeneration of Legumes.

Species	Method	Tissue transformed	Reporter or Marker Gene	Viable plant or Seed	Reference
Grain legumes: *Glycine max*	At[1]	cotyledon explant	NPT II		Owens and Smigocki (1988)
"	At	cotyledon explants	NPT II, GUS, EPSP	YES	Hinchee et al (1988)
"	PB	callus	NPT II	YES	Christou et al. (1988); McCabe et al. (1988)
"	EP	protoplasts	NPT II, GUS		Christou et al.(1987)
"	At	plumule,and cotyledonary node of intact germinated seed	NPT II	YES	Chee et al (1989)
"	EP	intact nodal meristems	GUS	YES	Chowira et al. (1995; 1996)
"	PB	embryogenic suspension culture tissue	GUS, HPT		Hadi et al. (1996)
"	PB	embryo, cell suspension culture	NPTII,GUS	YES	Sato et al. (1993)
"	Ar	seedling hypocotyl	sRAB1,vRAB7 with Lbc3 promoter		Cheon et al. (1993)
"	At	cotyledonary nodes of germinating seed	coat protein precursor (CP-P) of BPMV	YES	Di et al. (1996)
"	PB	callus	Corynebacterium dapA and E.coli lysC,to enhance seed lysine	YES	Falco et al. (1995)
"	EP	intact embryonic axis	GUS		Dillen et al. (1995)

G. argyrea	Ar	hypocotyl	NPT II	YES	Kumar etal. (1991)
G. canescens	Ar	hypocotyl	pRi TL	YES	Rech et al. (1988)
Pisum sativum	At	embryonic axis ,hypocotyl segments	HPT		Puonti-Kaerlas et al. (1989); (1990)
"	At	embryonic axis of developing seed	NPT II and BAR	YES	Schroeder et al. (1993)
"	At	stem explants	NPT II,HPT		Lulsdorf et al. (1991)
"	At	nodal explants	NPT II,GUS HPT	YES	De Kathen and Jacobsen (1990); Nauerby et al. (1991)
"	At,Ar	root explants, protoplasts	NPT II		Schaerer and Pilet (1991)
"	EP	intact nodal meristems	GUS	YES	Chowira et al. (1995;1996)
"	At	developing cotyledon segments	NPT II, BAR	YES	Grant et al. (1995)
"	EP	intact embryonic axis	GUS		Dillen et al. (1995)
Vigna unguiculata	At	mature embryo	GUS	YES	Penza et al. (1991)
"	EP	intact nodal meristems	GUS	YES	Chowira et al. (1995; 1996)
"	EP	intact embryonic axis	GUS		Dillen et al. (1995)
V. aconitifolia	Ar	seedling hypocotyl	nodulin-35		Lee et al. (1993)
"	HS/PEG	protoplasts	NPT II	YES	Kohler et al. (1987)
V. mungo	At	primary leaf segments	NPT II		Karthikeyan et al. (1996)
Vicia narbonensis	At	epicotyl explants via callus	Brazil nut 2S albumin	YES	Pickardt et al. (1991) Saalbach et al. (1994 and 1995)
V. sativa	Ar	stem sections	pRi		Mugnier (1988)

V. hirsuta	Ar	seedling epicotyl	GUS; Vf Lhb promoter; VfENOD-GRP3; PsENOD12 A and B		Quandt et al. (1993); Kuster et al. (1995); Vijn et al. (1995)
V. faba	Ar	stem	10 different pRi		Siefkes-Boer et al. (1995)
Arachis hypogaea	PB	shoot meristem of mature embryonic axis	GUS, BAR, and tswv-np		Brar et al. (1994); Schnall and Weissinger (1992)
"	PB	embryogenic tissue culture	HPT	YES	Oziasakins et al. (1993)
"	At	leaf section explants	GUS, NPT II	YES	Cheng et al. (1996)
Phaseolus vulgaris	PB	seed meristems	GUS, BAR, BGMV coat protein		Russell et al. (1993)
"	PB	shoot apex of embryonic axis of germinating seed	GUS / ConA promoter	YES	Kim and Minamikawa (1996)
"	Ar		pRi		Hamill et al. (1987)
"	EP	intact embryonic axis	GUS		Dillen et al. (1995)
Lens culinaris	EP	intact nodal meristems	GUS	YES	Chowira et al. (1995;1996)
"	At	shoot apex of germinating embryonic axis	GUS, BAR		Barton et al. (1996)
Lupinus angustifolius	At	shoot apex of germinating embryonic axis	GUS, BAR	YES	Atkins et al (1994);Pigeai re etal. (1996)
"	At	slices of maturing seed	GUS, BAR and S-rich sunflower seed albumin	YES	Molvig et al.(1994)
L. albus , polylhyllus	Ar	stem sections	pRi		Mugnier (1988)
L. luteus	At	germinating embryo	GUS, BAR		Somsap et al. (1994)

Psophocarpus tetragonolobus	Ar	root	pRi	YES	Tepfer (1990)
Cicer arietinum	At	embryonic axis	NPTII, GUS	YES	Fontana et al. (1993)
"	Ar	stems	10 different pRi		Siefkes-Boer et al. (1995)
"	EP	intact embryonic axis	GUS		Dillen et al. (1995)
Cajanus cajan	EP	intact embryonic axis	GUS		Dillen et al. (1995)
Lathyrus sativus	PB, At	immature leaflets and nodal segments	GUS		Barna and Mehta (1995)
Pasture legumes:					
Stylosanthes humilis	At, Ar	leaf sections, stolons	NPT II, nos	YES	Manners (1988); Manners and Way (1989)
S. guianensis	At	leaf sections via callus	NPT II, BAR	YES	Sarria et al. (1994)
Medicago sativa	At	stem cuttings	NPT II,	YES	Deak et al (1986);
"	At	leaf segments	chicken ovalbumin	YES	Schroeder et al (1991)
"	At	mature stem segments	nos/NPT II	YES	Shahin et al. (1986)
"	At	stem and petiole discs	BAR, NPTII, HPT	YES	D'Halluin et al. (1990)
"	Ar	root	pRi / nos	YES	Spano et al. (1987); Sukhapinda et al. (1987)
"	PB	calli from petiole and stem sections	NPTII	YES	Pereira and Erickson (1995)
M. truncatula	At	leaf	MtENOD 12/GUS	YES	Chabaud et al. (1996)
"	At, Ar	embryogenic calli	GUS	YES	Thomas et al. (1992) Rose and Nolan (1995)
"	At	embryogenic calli from leaf explants	NPTII, GUS	YES	Wang et al. (1996)
M. varia	At	stem cuttings	NPTII	YES	Deak et al. (1986)

M. arborea	Ar	stems of intact plants	pRi / HPT	YES	Damioni and Arcioni (1991)
Lotus corniculatus	Ar	stolon segments	pRi 15834	YES	Webb et al (1990)
"	Ar	seedling stolons	soybean Lhb	YES	Stougaard et al. (1986); Petit et al. (1987)
"	Ar	stem	GUS, P.vulgaris GS	YES	Forde et al. (1989)
"	Ar	hypocotyl of germinating seed	ACE 1,nod45 (prom), AAT-P2		Mett et al. (1996)
L. japonicus	At	cotyledonary node in planta	NPT II	YES	Oger et al. (1996)
Lotononis bainesii	At	leaf disc/ callus	NPT II	YES	Wier et al. (1988)
Trifolium repens	At	stolon internode segments	NPT II, nos	YES	White and Greenwood (1987)
"	Ar	stolon segments	pRi 15834		Webb et al. (1990)
"	At	cotyledonary axil	NPT II, GUS	YES	Voisey et al. (1994)
"	At	internodal stolon explants	pea albumin 1 (PA1)	YES	Ealing et al. (1994)
"	At	cotyledons of mature imbibed seed	GH3-GUS, BAR	YES	Larkin et al. (1996)
T. subterraneum	At	hypocotyl segment explants from seed	GUS, BAR	YES	Khan et al. (1994)
T. pratense	Ar	stolon segments	pRi 15834		Webb et al. (1990)
Macroptilium atropurpureum	Ar	seedlings	pRi		Beach and Gresshoff (1988)
Onobrychis viciifolia	At	seedling tissue explants	pRi	YES	Golds et al. (1991)

[1] **Key to Abbreviations**: At = *Agrobacterium tumefaciens*; Ar = *Agrobacterium rhizogenes*; EP = electroporation; PB = particle bombardment; GUS = -glucuronidase; BAR = resistance to the herbicide Basta (glufosinate); GS = glutamine synthetase; HPT = hygromycin

phosphotransferase; NPT II = neomycin phosphotransferase; HS/PEG = direct uptake of DNA by protoplasts following heat shock and treatment with polyethylene glycol; AAT = aspartate aminotransferase; ACE1 = yeast metalloregulatory transcription factor; Lhb = leghemoglobin; *nos* = nopaline synthase; pRi = wild type or engineered plasmids of *A. rhizogenes*; EPSP = 5-enolpyruvylshikimic acid 3-phosphate synhetase (tolerance to the herbicide glyphosate) ; Con A = Concanavalin A; GH3 = auxin responsive promoter from soybean; BPMV = bean pod mottle virus

---

3 Transformation and Regeneration of *Lupinus angustifolius*

As an example of a successful approach to developing novel transgenic cultivars of an economically important legume crop the details for transformation and regeneration of two cultivars of *Lupinus angustifolius* follow. This summary is based on the work of Pigeaire et al. (1996) and has been extended to the subsequent analysis and testing of transgenic generations. The method described is reproducible and is in routine use to genetically engineer lines with a range of novel traits. These include alterations to growth regulator synthesis and sensitivity, changes to the nutritional quality of the grain, tolerance to herbicides and resistance to viruses.

3.1 Preliminary Screening of Lupin Cultivar and *Agrobacterium* Strain. Ninteen different cultivars were tested with four different disarmed *Agrobacterium tumefaciens* strains each carrying the same binary vector (pCGP83; Janssen and Gardner, 1989) which contained a *gus*-CaMV*35S* construct. Transient GUS expression was greatest when *A.tumefaciens* strain AgL0 (Lazo et al., 1991) was used to infect *cv*. Unicrop. For this reason the transformation protocol was developed with *cv*. Unicrop and later applied to a current commercial cultivar, Merrit.

The *Agrobacterium* strain subsequently used was AgL0 containing the binary vector pCGP 963 or pCGP 1258. The plasmid pCGP 963 was derived from pGA 492 (An, 1986) and contained the *bar* gene controlled by the CaMV-35S promoter and an *ocs* (octopine synthase) terminator. The plasmid pCGP 1258 was the same except that a *gus* gene from pKIWI101 (Janssen and Gardner, 1989), also under control of the 35S promoter, was added.

3.2 Explants and Inoculation. Seed was surface sterilized and germinated for ca. 20 h on wet sterile filter paper in the dark at 25C or until the root was 2-10 mm long. After removing the seed coat the cotyledons and the two pairs of leaves were excised under a dissecting microscope to leave the plumule with an apical dome and the third pair of leaf primordia. These were stabbed 5-10 times with a 30G needle and plated upright (ie with the root in solid medium) on a co-cultivation medium containing MS salts (Murashige and Skoog, 1962) together with B5 vitamins, 10 mg BAP and 1 mg NAA per litre, adjusted to pH5.8 and solidified with 0.25 %(w/v) gelrite (Sigma).

A drop of *Agrobacterium* (with $5x10^8$ cells per ml in MS salts) was placed on top of the wounded apex. The explants were co-cultivated with the bacterium for 2 d in low light at 25C, after which they were transferred to MS medium supplemented with one tenth the level of growth regulators and 150 mg per litre of timentin (to inhibit growth of the *Agrobacterium*). After 2 d a thin film of a solution containing glufosinate (2 mg per ml) and 0.1% Tween 20 was applied to the apex and the explants grown on under low light and 25C with an 8 h photoperiod.

3.3 Selection and Regeneration of Transgenic Axillary Shoots.

The glufosinate solution rapidly killed explants which were co-cultivated with AgL0 (without *bar*) but in almost 100% of cases in those with the *bar*-containing plasmid explants grew as controls, producing green shoots (10-15 mm) after 2 wk in culture. The axillary shoots were excised and placed on a "micropropagation" medium (as above but with a further tenth dilution of the growth regulators, 150 mg timentin, 20 mg glufosinate per litre and solidified with 0.9% agar) which encouraged growth of transformed axillary buds. This transfer was repeated every 2 wk until some of the surviving shoots developed axillary shoots on more than one side. Control shoots of lupin on this medium (minus glufosinate) normally develop with a decussate phyllotaxis and exhibit axillary shoots on four sides. However, glufosinate-tolerant shoots ("primary") from the initial excision of the explant frequently developed a single axillary shoot ("secondary") on one side with the others being non-transformed and dying. The excised secondary shoot in turn formed resistant "tertiary" shoots on two sides and when excised again and cultured with selection pressure formed four shoots ("quaternary") on each of four sides. The "tertiary" or "quaternary" shoots were collected as the transformed material and cultured on in the absence of herbicide.

3.4 Grafting Transgenic Shoots.

In a series of preliminary studies transformd lupin shoots did not form roots reliably in culture, despite changes to the levels of growth regulators to encourage rooting. For this reason the protocols described here involve grafting to a non-transgenic root stock. More recently much higher frequencies of rooting and adequate development of roots have been obtained (PMC Smith, N. Fletcher, CA Atkins, unpublished data).

Several shoots (clones) representing each transgenic event were grafted separately onto the cotyledonary node of 10-14 day old seedlings of their corresponding cultivar growing in sand with a complete nutrient solution (equivalent to Hoagland's solution with 5 mM KNO_3), or potting mix supplemented with nutrients. The tissue above the cotyledonary node was excised and the remaining 5 mm long piece of stem split longitudinally so that the small cultured shoot could be inserted and held securely. The grafts with their root stock were covered in a plastic vial wrapped in aluminium foil for 3-4 d and then the foil was removed and finally the plastic vial. In this time the graft union formed and new shoot tissue was evident. After a further 1-2 d the vial was removed and the plants grown on under normal glasshouse conditions to obtain viable seed. The extent to which the transgenic shoots grew depended on the time to flowering. If shoots were held in petridish culture for extended periods before grafting they flowered soon after grafting and seed was not set. This could be overcome to some extent if the floral organs on the main axis were removed once the graft was established leaving the axillary shoots to flower later when sufficient vegetation had been formed to support seed set and filling.

Table 2. Protocol for transformation and regeneration of *Lupinus angustifolius* L. with a *bar*-CaMV 35S-*nos* construct to confer tolerance to the herbicide Basta (glufosinate).

Step	Number	Months
Embryonic explants infected with AgL0/pCGP963 and the primary Basta selection applied at the sites of infection(2 mg/ml glufosinate)	2463	0
Shoots into culture for micropropogation with 0.2 mg/ml glufosinate in the medium		
T_0-herbicide resistant shoots	9	5
Grafts made to root stocks		
T_0-seed	57	8-10
T_1-seedlings	28	
T_1 resistant plants selected with a leaf spray of 0.1 mg/ml glufosinate	15	14
T_1-seed	8 plants from only one accession (3 other accessions held in reserve)	
T_2-seedlings	39	
T_2 resistant plants selected with a leaf spray of 0.1 mg/ml glufosinate	33	
T_2-seed	2522	20
T_3-seed from GMAC-approved outside screenhouse trials	ca. 0.5×10^6	26
T_4-seed from GMAC-approved field trial plots	ca. 40×10^6	38

3.5 Assays for Genetic Transformation. Transformed axillary shoots which developed on the glufosinate medium (20 mg per ml) were tolerant of a 2 mg per ml glufosinate solution (in 0.1% Tween 20) painted on
the leaflets. Grafted shoots and foliage which expressed the *bar* gene in subsequent generations were tolerant to application of 0.1 mg glufosinate per ml in a leaf paint assay and for a whole plant spray treatment of a commercial formulation of Basta containing 0.1 mg glufosinate per ml. Resistance of leaflets or whole plants was assessed after 6-14 d. Stable integration of the *bar* gene was also confirmed by southern analysis of genomic DNA, by northern analysis of RNA isolated from leaf tissue and by the assay of leaf extracts for activity of phosphinothricin-N-acetyltransferase (PAT).

Inheritance patterns in the first generation of transgenic seed (T_1) equated roughly to a 3:1 Mendelian ratio, ie. a segregation expected for a dominant gene at a single locus in a self-pollinating species.

3.6 Transformation Frequency and Time Required to Make a New Genotype.
Table 2 indicates the outcome and time required for successful transformation of *cv.* Merrit, a current commercial lupin cultivar, to express tolerance to Basta at the field trial stage of development. Overall the transformation frequency was about 0.4%. However this varies considerably with cultivar and frequencies of 2.8% have been found for the older variety, *cv.* Unicrop. Confirming the initial transformation required about 5 months at the end of which a series of transformed shoots were available for grafting. The T_1 seed was available after a further 5 months and sufficient seed was generated in the T_3 generation for screenhouse and field trials. Similar periods have been described for transformation of peas and peanuts using comparable protocols for regeneration (Grant et al. 1995; Cheng et al. 1996). The time required from isolating the initial transgenic lupin explant to generating sufficient seed for controlled field site trialling at three locations was just over three years. Although the regulatory process has not yet been applied to the transgenic lupin, and similarly, the herbicide Basta is not yet approved for broad scale use on lupin crops in Western Australia, potentially the new technology could be available to the farmer some six years after the initial transformations.

Since the initial experiments which produced Basta-tolerant lines the transformation system has been improved to give frequencies of 2.5 % for cultivar Merrit (N. Fletcher, R. Walker, CA Atkins, unpublished). The methodology is currently being applied to a wide range of manipulations to alter yield potential and harvest index, grain composition, resistance to virus and susceptibility to stress.

4 Genetic Transformation and Nitrogen Fixation

As well as providing a means to alter legumes so that they express novel features which could increase yield or enhance their quality and value to downstream processing industry, genetic engineering also offers a means to study mechanisms of gene expression and the regulation of physiological and developmental processes. This latter is especially relevant to the functioning of nodules and nitrogen fixation as the genomes of both partners to the symbiosis interact and contribute to the effectiveness and efficiency of the process. Consequently both have been engineered to alter the course of nodule initiation and development so that new insights into the symbiosis have been gained. It is only through a detailed understanding of the genetic interaction of the partners, and how this generates the special structural and biochemical features of a nodule, that the important regulatory mechanisms which allow symbiotic development to occur will be able to be manipulated. Such manipulations no doubt hold the keys for a rational approach to expressing nitrogen fixation at high levels in non-legumes.

M. truncatula has been suggested to be a useful model for studying the molecular genetics of the legume-*Rhizobium* symbiosis (Barker et al.,1990) and successful transformation and expression of foreign genes has been achieved (see Table 1). Similarly Petit et al.(1987) have suggested *Lotus corniculatus* could fulfil such a role, and successful insertions of plant genes which alter nodule function have been reported (see Table 1). In most cases transformation has involved *A. rhizogenes* and the development of nodulated "hairy roots". This system has been used with a number of legumes (*M.sativa, M. truncatula, L. corniculatus V.aconitifolia and V. hirsuta*) to investigate nodule-specific expression using reporter gene fusions with promoters of both early and late nodulin genes (Lauridsen et al., 1993; Journet et al., 1993; Pichon et al., 1992; 1994; Quandt et al., 1993; Kuster et al., 1995; Lee et al., 1993; Mett et al., 1996). Recently Oger et al. (1996) have demonstrated the stable transformation and regeneration of the diploid species *Lotus japonicus*. The method used *in planta* inoculation at the cotyledon attachment site and so avoided the need for regeneration via callus. Transgenic T1 plants were produced from seed in 15 weeks. Larkin et al.,(1996) have described an *A. tumefaciens* -based method for white clover in which the cotyledons of mature , imbibed seed are used and which yielded a transformation frequency of 50% with the bar gene as the selectable marker. No doubt these "model plants" will continue to be used to examine expression of genes in nodules and quite possibly they will provide new insights into nodule development.

However there is a need for similar model species for the pulses. A major problem is that the biochemistry of nitrogen metabolism in nodules is not the same in all species; those of temperate origin form asparagine as the principal product of ammonia assimilation in the infected cells of the nodule, while members of the more tropical tribe, the Phaseoleae, form the ureides, allantoin and allantoic acid (Atkins 1991). The nature of the two pathways and their intra- and intercellular compartmentalisation is very different. Furthermore there is a large difference in the efficiency by which the two nodule types use plant carbon (sugars) to produce and translocate nitrogenous solutes to the host plant (Atkins, 1991). Among the amide-forming symbioses pea probably offers the most useful model. An improved technique has been described which allows transformation and regeneration to seed in seven months (Grant et al., 1995) and the *in planta* electroporation method described by Chowira et al. (1995) has also been applied to peas. Narrow-leafed lupin offers a readily transformed model plant. However nodules on lupin appear to differ from those of most other amide species in their mode of initiation (Tang et al., 1992), their morphology (Corby et al., 1983) and physiology (Witty et al., 1994). On the other hand pea forms indeterminate nodules which are similar in their infection/initiation processes, morphology and physiology to nodules on a wide range of pulse and pasture species. Among the ureide-formers both cowpea and soybean have been extensively studied (Atkins, 1991) and each has been successfully transformed (Table 1). Both form nodules through similar processes of infection and initiation, show determinate, spherical morphology and appear to be regulated by O_2 supply in a very similar way. However some of the other members of the genus *Vigna*, such as mothbean (*V.aconitifolia*) or mungbean (*V.radiata*), could offer useful alternatives. Some cultivars of mungbean are very small, produce mature seed in about eight weeks from planting (CA Atkins, unpublished) and translocate essentially all their fixed nitrogen as allantoin (RM Rainbird and CA Atkins, unpublished). Methods for their efficient and high frequency transformation have not been described, but in view of the success of the *in planta* electroporation technique employed by Chowira et al. (1995; 1996) for cowpea, transformation of mungbean through this means seems very likely. Similarly other *in planta* transformation protocols developed for *Lotus* spp (Oger et al., 1996) or for soybean (Chee and Slightom, 1995) using *A.tumefaciens* show that the process of generating transgenic plants can be greatly simplified and speeded up if the compex and slow regeneration of shoots from explants, cells or callus is avoided.

There is little doubt that the tools of recombinant DNA technology coupled with efficient transformation and regeneration systems have the potential to provide insights into genetic regulation of nodulation and nodule functioning which can be exploited to develop more effective legume-*Rhizobium* symbioses. The sorts of alterations to symbioses which might increase their effectiveness relate to the carbon cost of fixation and nitrogenous solute synthesis, susceptibility of nodules to stress, such as temperature extremes, drought and restricted O_2 supply and the premature senescence which results from such environmental variations, their sensitivity to combined forms of nitrogen in the soil, to pesticides and to changes in the source/sink relationships of the host. Whether the same insights also open the door for transforming non-legumes to fix atmospheric nitrogen remains an intriguing and inviting question.

Acknowledgement. The methodology developed for transformation and regeneration of lupin was supported by grants from the Legume Research Council and the Grains Research and Development Corporation of Australia.

Literature Cited.

An G (1986) Plant Physiology 81: 86-91.

Atkins CA (1991) In. Biology and Biochemistry of Nitrogen Fixation. MJ Dilworth and AR Glenn (eds) Elsevier Amsterdam. pp 293-319.

Atkins CA, Smith PMC and Pigeaire A (1994) In. Proceedings of the First Australian Lupin Technical Symposium. M Dracup and J Palta (eds) Department of Agriculture Western Australia. pp 119-122.

Barker DG, Bianchi S, Blondon F, Dattee Y, Duc G, Essad S, Flament P, Genier G, Guy P, Muel X, Tourneur J, Denarie J and Huguet T (1990) Plant Molecular Biology Reporter 8: 40-49.

Barna KS and Mehta SL (1995) Journal of Biochemistry and Biotechnology 4: 67-71.

Barton J, Smith PMC, Klyne AM and Atkins CA (1996) unpublished results.

Barry G, Kishore G, Padgette S, Taylor M, Kolacz K, Weldon M, Re D, Eicholtz D, Fincher K, and Hallas L (1992) In. Biosynthesis and Molecular Regulation of

Amino Acids in Plants. BK Singh, HE Flores and JC Shannon (eds). American Society of Plant Physiologists. pp 139-145.

Beach KH and Gresshoff PM (1988) Plant Science 57: 73-81.

Brar GS, Cohen BA, Vick CL and Johnson GW (1994) The Plant Journal 5: 745-753.

Chandler SF (1995) In. Herbicide Resistant Crops and Pastures in Australian Farming Systems. GD McLean and G Evans (eds) Bureau of Resource Sciences Canberra. pp 229-240.

Chabaud M, Larsonneau C, Marmouget C and Huguet T (1996) Plant Cell Reports 15: 305-310.

Chee PP and Slightom JL (1995) Methods in Molecular Biology 44: 101-119.

Chee PP, Fober KA and Slightom JL (1989) Plant Physiology 91: 1212-1218.

Cheng M, Jarret RL, Li Z, Xing A and Demski JW (1996) Plant Cell Reports 15: 563-657.

Cheon C-III, Lee N-G, Siddique A-BM, Bal AK and Verma DPS (19993) The EMBO Journal 12: 4125-4135.

Chowira GM, Akella V, Fuerst PE and Lurquin PF (1996) Molecular Biotechnology 5: 85-96.

Chowira GM, Akella V and Lurquin PF (1995) Molecular Biotechnology 3: 17-23.

Christou P (1992) The Plant Journal 2: 275-281.

Christou P (1994) Euphytica 74: 165-185.

Christou P, Murphy J and Swain WF (1987) Proceedings of the National Academy of Science. USA 84: 3962-3966.

Christou P, McCabe DE and Swain WF (1988) Plant Physiology 87: 671-674.

Corby HDL, Polhill RM and Sprent JI (1983) In. Nitrogen Fixation Vol3: Legumes. WJ Broughton (ed). Clarendon Press, Oxford. pp 1-35.

Damiani F and Arcioni S (1991) Plant Cell Reports 10: 300-303.

Dale PJ (1995) Trends in Biotechnology 13: 398-402.

Dale EC and Ow DW (1991) Proceedings of the National Academy of Science. USA 88: 10558-10562.

Deak M, Kiss GB, Koncz C and Dudits D (1986) Plant Cell Reports 5: 97-100.

De Kathen A and Jacobsen HJ (1990) Plant Cell Reports 9: 276-279.

Dekeyser R, Claes B, Marichal M, Van Montagu M and Caplan A (1989) Plant Physiology 90: 217-223.

D'Halluin K, Botterman J and de Greef W (1990) Crop Science 30: 866-871.

Di R, Purcell V, Collins GB and Ghabrial, SA (1996) Plant Cell Reports 15: 746-750.

Dillen W, Engler G, Van Montagu M and Angenon G (1995) Plant Cell Reports 15: 119-124.

Ealing PM, Hancock KR and White DWR (1994) Transgenic Research 3: 344-354.

Falco SC, Gulda T, Locke M, Mauvais J, Sanders C, Ward RT and Webber P (1995) Biotechnology 13: 577-582.

Fontana GS, Santini L, Caretto S, Frugis G and Mariotti D (1993) Plant Cell Reports 12: 194-198.

Forde BG, Day HM, Turton JF, Wen-jun F, Cullimore JV and Oliver JE (1989) Plant Cell 1: 391-401.

Fuchs RL, Re DB, Rogers SG, Hammond BG and Padgette SR (1995) In. Proceedings 3rd International Symposium on the Biosafety Results of Field Tests of Genetically-Modified Plants and Micro-organisms. Monterey (Nov 1994).

Goddijn OJM and Pen J (1995) Trends in Biotechnology 13: 379-387.

Golds TJ, Lee JY, Husnain T, Ghose TK and Davey MR (1991) Journal of Experimental Botany 42: 1147-1157.

Goldsborough AP, Lastrella CN and Yoder JI (1993) Bio/Technology 11: 1286-1292.

Grant JE, Cooper PA, McAra AE and Frew TJ (1995) Plant Cell Reports 15: 254-258.

Gritz L and Davies J (1983) Gene 25: 179-188.

Hadi MZ, McMullen MD and Finer JJ (1996) Plant Cell Reports 15: 500-505.

Hamill J, Parr J, Rhodes R, Robins R and Walton N (1987) Biotechnology 5: 800-804.

Hinchee MAW, Connor-Ward DV, Newell CA, McDonnell RE, Sato SJ, Gasser CS, Fischoff DA, Re DB, Fraley RT and Horsch RB (1988) Bio/Technology 6: 915-922.

Janssen BJ and Gardner RC (1989) Plant Molecular Biology 14: 61-72.

Jefferson RA, Kavanagh TA and Bevan MW (1987) The EMBO Journal 6: 3901-3907.

Journet E-P, Pichon M, Dedieu A, de Billy F, Truchet G and Barker DG (1993) The Plant Journal 6: 241-249.

Karthekeyan AS, Sarma KS and Veluthambi K (1996) Plant Cell Reports 15: 328-331.

Khan MRI, Tabe LM, Heath LC, Spencer D and Higgins TJV (1994) Plant Physiology 105: 81-88.

Kim JW and Minamikawa T (1996) Plant Science 117: 131-138.

Kohler F, Golz C, Eapen S, Kohn H and Schieder O (1987) Plant Cell Reports 6: 313-317.

Komari T, Hiei Y, Saito Y, Murai N and Kumashiro T (1996) The Plant Journal 10: 165-174.

Kumar V, Jones B and Davey MR (1991) Plant Cell Reports 10: 135-138.

Kuster H, Quandt H-J, Broer I, Perlick AM and Puhler A (1995) Plant Molecular Biology 29: 759-772.

Larkin PJ, Gibson JM, Mathesius U, Wienman JJ, Gartner E, Hall E, Tanner GJ, Rolfe BG and Djordjevic MA (1996) Transgenic Research 5: 1-11.

Lauridsen P, Franssen H, Stougaard J, Bisseling T and Marcker KA (1993) The Plant Journal 3: 483-492.

Lazo GR, Stein PA and Ludwig RA (1991) Plant Molecular Bio/Technology 9: 963-967.

Lee N-G, Stein B, Suzuki H and Verma DPS (1993) The Plant Journal 3: 599-606.

Lulsdorf MM, Rempel H, Jackson JA, Baliski DS and Hobbs SLA (1991) Plant Cell Reports 9: 479-483.

Manners JM (1988) Plant Science 55: 61-68.

Manners JM and Way H (1989) Plant Cell Reports 8: 341-345.

Mason HS and Arntzen CJ (1995) Trends in Biotechnology 13: 388- 392.

McCabe DE, Swain WF, Martinell BJ and Christou P (1988) Bio/Technology 6: 923-926.

McElroy D and Brettell RIS (1994) Trends in Biotechnology 12: 62-68.

Mett VL, Podivinsky E, Tennant AM, Lochhead LP, Jones WT and Reynolds HS (1996) Transgenic Research 5: 105-113.

Molvig L, Schroeder HE, Tabe LM, Moore A, Kortt AA, Spencer D and Higgins TJV (1994) In. Proceenings af the First Australian Lupin Technical Symposium. M Dracup and J Palta (eds) Department of Agriculture Western Australia. pp 285.

Mugnier J (1988) Plant Cell Reports 7: 9-12.

Murashige T and Skoog FJ (1962) Physiologia Plantarum 15: 473-497.

Nauerby B, Madsen M, Christiansen J and Wyndaele R (1991) Plant Cell Reports 9: 676-679.

Oger P, Petit A and Dessaux Y (1996) Plant Science 116: 159-168.

Owens LD and Smigocki AC (1988) Plant Physiology 88: 570-573.

Oziasakins P, Schnall JA, Anderson WF, Singsit C, Clemente TE, Adang MJ and Weissinger AK (1993) Plant Science 93: 185-194.

Penza R, Lurquin PF and Filippone E (1991) Journal of Plant Physiology 138: 39-43.

Pereira LF and Erickson L (1995) Plant Cell Reports 14: 290-293.

Petit A, Stougaard J, Kuhle A, Marcker KA and Tempe J (1987) Molecular and General Genetics 207: 245-250.

Pichon M, Journet E-P, Dedieu A, de Billy F, Truchet G and Barker DG (1992) Plant Cell 4: 1199-1211.

Pichon M, Journet E-P, de Billy F, Dedieu A, Huguet T, Truchet G and Barker DG (1994) Molecular Plant-Microbe Interactions 7: 740-747.

Pickardt T, Meixner M, SchadeV and Schieder O (1991) Plant Cell Reports 9: 535-538.

Pigeaire A, Abernethy D, Smith PMC, Simpson K, Fletcher N, Lu C-Y, Atkins CA and Cornish E (1996) Molecular Breeding (accepted for publication).

Poirier Y, Nawrath C and Somerville C (1995) Bio/Technology 13: 142-150.

Potrykus I (1991) Annual Review of Plant Physiology and Plant Molecular Biology 42: 205-225.

Pridmore RD (1987) Gene 56: 309-312.

Puonti-Kaerlas J, Stabel P and Ericsson T (1989) Plant Cell Reports 8: 321-324.

Puonti-Kaerlas J, Ericsson T and Engstrom P (1990) Theroetical and Applied Genetics 80: 246-252.

Quandt H-J, Puhler A and Broer I (1993) Molecular Plant-Microbe Interactions 6: 699-703.

Rech EL, Golds TJ, Hammatt N, Mulligan BJ and Davey MR (1988) Journal of Experimental Botany 39: 1275-1286.

Rogers HJ and Parkes HC (1995) Journal of Experimental Botany 46: 467-488.

Rose RJ and Nolan KE (1995) Plant Cell Reports 14: 349-353.

Russell SH, Hoopes JL and Odell JT (1992) Molecular and General Genetics 234: 49-59.

Russell DR, Wallace KM, Bathe JH, Martinell BJ and McCabe DE (1993) Plant Cell Reports 12:165-169.

Saalbach I, Pickardt T, Machemehl F, Saalbach G, Schieder O and Muntz K (1994) Molecular and General Genetics 242: 226-236.

Saalbach I, Pickardt T, Waddell DR, Hillmer S, Schieder O and Muntz K (1995) Euphytica 4: 1-12.

Sarria R, Calderon A, Thro AM, Torres E, Mayer JE and Roca WM (1994) Plant Science 96: 119-127.

Sato S, Newell C, Kolacz K, Tredo L, Finer J and Hinchee M (1993) Plant Cell Reports 12: 408-413.

Schaurer S and Pilet P-E (1991) Plant Science 78: 247-258.

Schaff DA (1994) Plant Science 101: 3-9.

Schnall JA and Weissinger AK (1992) In Vitro 28: 122A.

Schroeder HE, Khan MRI, Knibb WR, Spencer D and Higgins TJV (1991) Australian Journal of Plant Physiology 18: 495-505.

Schroeder HE, Schotz AH, Wardley-Richardson T, Spencer D and Higgins TJV (1993) Plant Physiology 101: 751-757.

Seifkesboer HJ, Noonan MJ, Bullock DW and Conner AJ (1995) Israel Journal of Plant Sciences 43: 1-5.

Shahin E, Sukhapinda K, Simpson R and Spivey R (1986) Theoretical and Applied Genetics 72: 770-777.

Somsap V Cooper JJ, Li D and Jones MGK (1994) In. Proceedings of the First Australian Lupin Symposium. M Dracup and J Palta (eds) Department of Agriculture Western Australia. pp 312.

Songstad DD, Somers DA and Griesbach RJ (1995) Plant Cell, Tissue and Organ Culture 40: 1-15.

Spano L, Mariotti D, Pezotti M, Damiani F and Arcioni S (1987) Thoeretical and Applied Genetics 73: 523-530.

Stougaard J, Marcker KA, Otten L and Schell J (1986) Nature 321: 669-674.

Strauch E, Wohlleben W and Puhler A (1988) Gene 63: 65-74.

Sukapinda K, Spivey R and Shahin EA (1987) Plant Molecular Biology 8: 209-216.

Tang C, Robson AD, Dilworth MJ and Kuo J (1992) New Phytologist 121: 457-467.

Tepfer D (1990) Physiologia Plantarum 79: 140-146.

Thomas MR, Rose RJ and Nolan KE (1992) Plant Cell Reports 11: 113-117.

Thompson CJ, Movva RN, Tizard R, Crameri R, Davies JE, Lauwereys M and Bottermann J (1987) The EMBO Journal 9: 2519-2523.

Vanderleij FR and Witholt B (1995) Canadian Journal of Microbiology 41: 222-238.

Vasil V, Srivastrava V, Castillo AM, Fromm ME and Vasil IK (1993) Bio/Technology 11: 1553-1558.

Vijn I, Christiansen H, Lauridsen P, Kardailsky I, Quandt HJ, Broer I, Drenth J, Jensen EO, Vankammen A and Bisseling T (1995) Plant Molecular Biology 28: 1103-1110.

Voisey CR, White DWR, Dudas B, Appleby RD, Ealing PM and Scott AG (1994) Plant Cell Reports 13: 309-314.

Wang JH, Rose RJ and Donaldson BI (1996) Australian Journal of Plant Physiology 23: 265-270.

Webb KJ, Jones S, Robbins MP and Minchin FR (1990) Plant Science 70: 243-254.

White DWR and Greenwood D (1987) Plant Molecular Biology 8: 461-469.

Wier AT, Thro AM, Flores HE and Janyes JM (1988) International Journal of Experimental Botany 48: 123-131.

Witty JF and Minchin FR (1994) Journal of Experimental Botany 45: 967-978.

Zhou H, Arrowsmith JW, Fromm ME, Hironaka CM, Taylor ML, Rodriguez D, Pajeau ME, Brown SM, Santino CG and Fry JE (1995) Plant Cell Reports 15: 159-163.

Part 11

Coevolution of Symbiotic Systems

Phylogenetic Perspectives on the Origins and Evolution of Nodulation in the Legumes and Allies

Jeff J. Doyle and Jane L. Doyle
L. H. Bailey Hortorium
Cornell University
Ithaca, New York 14853, USA

Despite the importance of the Leguminosae, the third largest family of flowering plants with some 670 genera and 18,000 species (Polhill et al. 1981; Polhill 1994), many questions remain concerning relationships both within the family and between it and other families. Recent advances in legume phylogeny have been spurred by the availability of molecular data, and are thus part of the more general explosion of phylogenetic studies in the flowering plants (Doyle 1995).

1 Familial Relationships: a Predisposition for Symbiosis?

In phylogenetic studies of the chloroplast gene *rbcL*, sequences from the family Polygalaceae (milkwort family) have consistently and unexpectedly appeared either as the closest relatives (sister group) to sequences from Leguminosae (e.g., Chase et al. 1993), or share this distinction with sequences from Surianaceae (Soltis et al. 1995). Neither Polygalaceae or Surianaceae are known to nodulate. Intriguingly, however, *rbcL* sequences from all eight of the flowering plant families known to participate in rhizobial or actinorhizal nitrogen-fixing symbioses belong to the same large group (clade), suggesting that all of these families share a common ancestry (Soltis et al. 1995). There are many other families in this clade that do not participate in nitrogen-fixing symbioses, however, so it is perhaps less likely that there was a single origin of nodulation in these disparate groups than that some key unknown innovation arose in their common ancestor that permitted different members of this large clade to form symbioses (Soltis et al. 1995).

2 Relationships within Leguminosae: Multiple Origins of Nodulation?

An analysis of *rbcL* sequences from ca. 100 genera of legumes (Doyle et al. accepted; Fig. 1) has many similarities with the comparably comprehensive (albeit preliminary) non-molecular phylogenetic analysis of Chappill (1995). Phylogenetic considerations indicate that the simplest hypothesis to account for the distribution of nodulation in Leguminosae involves multiple origins (Fig. 1): 1) in an ancestor of subfamily Papilionoideae; 2) in the ancestor of a lineage that includes subfamily Mimosoideae and members of the *Sclerolobium* and *Dimorphandra* groups of Caesalpinieae, the groups that include most nodulating members of subfamily Caesalpinioideae; and 3) in *Chamaecrista*, the only remaining genus of Caesalpinioideae known definitively to

NATO ASI Series, Vol. G 39
Biological Fixation of Nitrogen
for Ecology and Sustainable Agriculture
Edited by A. Legocki, H. Bothe, A. Pühler
© Springer-Verlag Berlin Heidelberg 1997

308

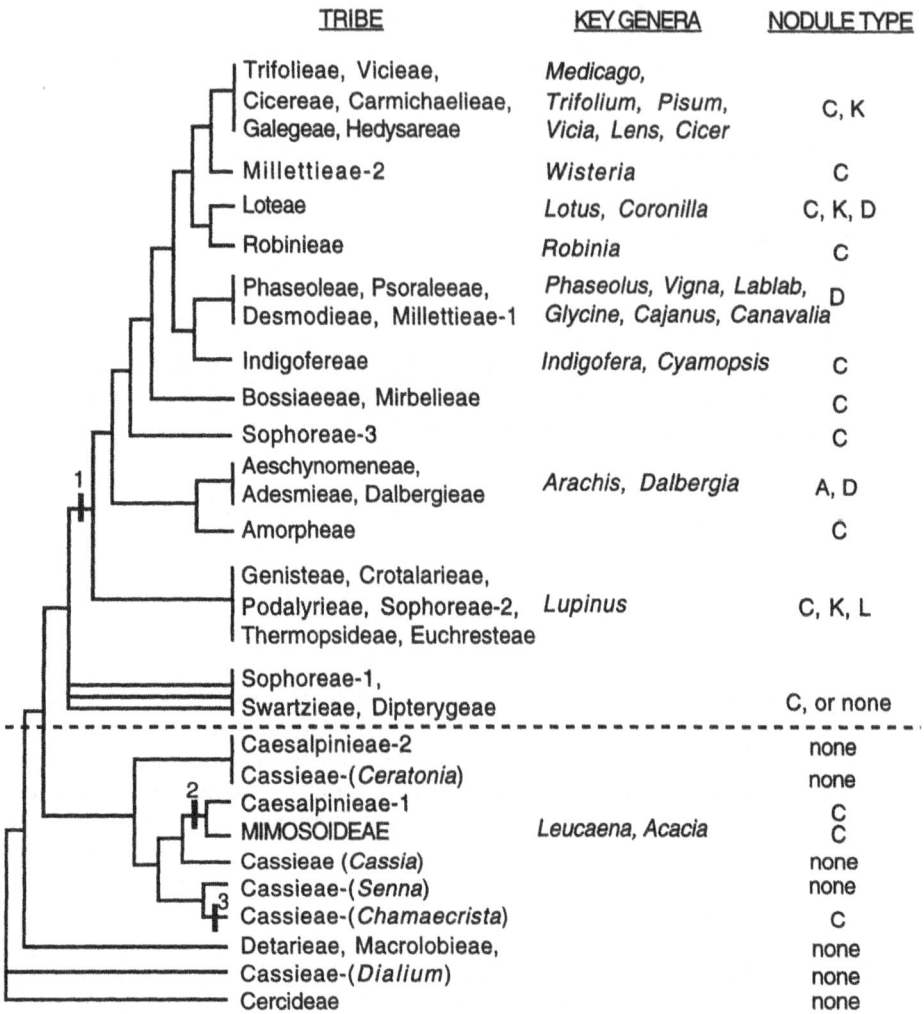

TRIBE	KEY GENERA	NODULE TYPE

Trifolieae, Vicieae, Cicereae, Carmichaelieae, Galegeae, Hedysareae	Medicago, Trifolium, Pisum, Vicia, Lens, Cicer	C, K
Millettieae-2	Wisteria	C
Loteae	Lotus, Coronilla	C, K, D
Robinieae	Robinia	C
Phaseoleae, Psoraleeae, Desmodieae, Millettieae-1	Phaseolus, Vigna, Lablab, Glycine, Cajanus, Canavalia	D
Indigofereae	Indigofera, Cyamopsis	C
Bossiaeeae, Mirbelieae		C
Sophoreae-3		C
Aeschynomeneae, Adesmieae, Dalbergieae	Arachis, Dalbergia	A, D
Amorpheae		C
Genisteae, Crotalarieae, Podalyrieae, Sophoreae-2, Thermopsideae, Euchresteae	Lupinus	C, K, L
Sophoreae-1, Swartzieae, Dipterygeae		C, or none
Caesalpinieae-2		none
Cassieae-(Ceratonia)		none
Caesalpinieae-1		C
MIMOSOIDEAE	Leucaena, Acacia	C
Cassieae (Cassia)		none
Cassieae-(Senna)		none
Cassieae-(Chamaecrista)		C
Detarieae, Macrolobieae,		none
Cassieae-(Dialium)		none
Cercideae		none

Fig. 1. Phylogeny of Leguminosae based primarily on *rbcL* sequence data (Doyle et al. accepted). Tribal names are used; para- or polyphyletic tribes (e.g. Sophoreae) occur in more than one place on the tree; tribes above the dotted line are Papilionoideae, while those below the line are Caesalpinioideae, with Mimosoideae indicated in capital letters. Numbered rectangles indicate points beyond which most species are known to nodulate. "Key genera" are generally economically or scientifically important taxa, or are mentioned in the text; all are known to nodulate. "Nodule type" is the predominant type (Corby 1981, 1988) for the tribe or tribes listed: A and D are determinate types (aeschynomenoid and desmodioid, respectively), while C and K are indeterminate types (caesalpinioid, crotalarioid), and L is the lupinoid type unique to *Lupinus*; Corby's "mucunoid" type can be either determinate or indeterminate and is not listed here, with the assumption that it is either a specialized type of desmodioid (if determinate) or caesalpinioid (if indeterminate) nodule.

nodulate, and which is closely related to the non-nodulating genus *Senna*. Such a hypothesis is bolstered by the finding that nodulation likely has arisen numerous times in more distantly related families (see above). McKey (1994) has argued that the high nitrogen "lifestyle" of legumes predates nodulation, and that this has imposed a need for large amounts of nitrogen on legumes. This need has been met by the independent evolutionary development of alternative nitrogen acquisition systems in members of the family, among which could be independent origins of bacterial symbiosis (McKey 1994; Sprent 1994). On the other hand, nodulation is a complex syndrome, whose evolutionary innovations include a novel organ, the nodule, and a novel organelle, the symbiosome (Verma et al. 1992). Given this complexity, a single origin of nodulation followed by multiple independent losses, though not most parsimonious, might be more likely than multiple origins, particularly given the cost to the plant of maintaining a symbiosis (see Sprent 1994).

Testing these competing hypotheses requires an assessment of homology (similarity due to common descent) in nodules of these different lineages. Nonhomology hypothesized on the basis of the phylogeny could be corroborated by showing that nodules in the different groups differed from one another in essential features; however, the basic "caesalpinioid" type of indeterminate nodules found in Mimosoideae, Caesalpinioideae, and in basal Papilionoideae (Corby 1988) meet the similarity criterion. Nodule homologies could also be tested by assessing the phylogenetic relationships and expression patterns of genes encoding nodule proteins (Doyle 1994b). Nodule nonhomology would be supported by the finding that nonhomologous genes perform similar functions in the nodules of different groups. Nonhomologous genes performing identical functions could either be truly unrelated, having separate *de novo* origins, or paralogous (Fitch 1970), being related by ancient gene duplication. In the latter case, "nodulin" genes would belong to different gene subfamilies. Independent recruitment during separate origins of nodulation could result in a situation in which, for example, paralogous genes were found to encode "nodulins" in *Pisum* and *Chamaecrista*; a single origin should initially only involve orthologous genes. Such tests are in progress using glutamine synthetase (GS), which consists of several phylogenetically-defined cytosolically expressed gene subfamilies (Doyle 1991, 1994b). Analysis of sequences from several previously unstudied legumes, as well as nodule cDNAs from *Chamaecrista* (JJD and JLD, unpublished; Fig. 2) suggests that the cytosolic family is quite complex; moreover, expression patterns of cytosolic GS are complicated (e.g., Temple et al. 1995). It is therefore difficult to identify a clade of putatively orthologous nodule-enhanced sequences, and current results do not discriminate between single or multiple origins of nodulation. Other gene families may prove more useful for this approach, for example the MADS-box family of transcriptional regulators, where gene evolutionary patterns are relatively well-defined (Doyle 1994a; Purugganan et al. 1995) and at least one gene is known to be expressed in nodules of *Medicago* (Heard and Dunn 1995).

3 The Evolution of Nodulation in Papilionoideae: Shifts from Indeterminate to Determinate Nodules

Basal members of Papilionoideae either do not nodulate or possess the indeterminate caesalpinioid (formerly "astragaloid") type of nodule (Corby 1981, 1988). More derived elements of the subfamily include several different lineages. Members of the

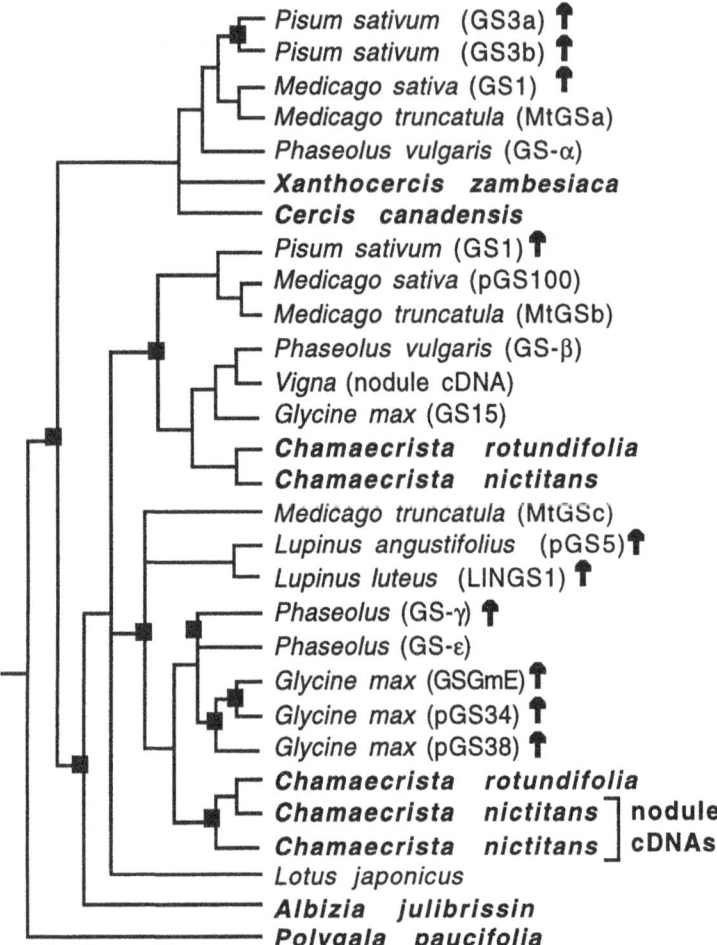

Fig. 2. Glutamine synthetase phylogenetic hypothesis, based on several overlapping parsimony analyses. Boldface sequences are partial genomic sequences from Caesalpinioideae (*Cercis, Chamaecrista*), Mimosoideae (*Albixia*), basal Papilionoiideae (*Xanthocercis*), or the related family Polygalaceae; two nearly full-length nodule cDNAs from *Chamaecrista* are also included (JJD and JLD unpublished). Some sequences, such as those of *Medicago truncatula*, are quite fragmentary, and were not used in full phylogenetic analyses. Black squares denote possible gene duplication events; upward arrows indicate genes showing nodule-enhanced expression patterns.

"Genistoid Alliance" (Genisteae, Crotalarieae, etc.; Fig. 1), like basal Papilionoideae, have indeterminate nodules, generally either to the caesalpinioid or crotalarioid types. Among the remaining derived members of the subfamily, indeterminate nodules of the same two types also typify most members of the clade that includes the "temperate herbaceous" tribes, to which such genera as *Pisum*, *Vicia*, *Medicago*, and *Trifolium* belong. Relatives of the temperate herbaceous tribes suggested by cpDNA data include the closely related genera *Coronilla* and *Lotus* (Polhill 1994); *Coronilla* has indeterminate nodules typical of this entire group, while *Lotus* has determinate nodules of the desmodioid type (Corby 1981, 1988). Determinate desmodioid (and related mucunoid) nodules predominate in Phaseoleae and its allies; Phaseoleae is the largest tribe of the Leguminosae with some 83 genera, including *Glycine*, *Phaseolus*, and *Vigna*. A second type of determinate nodule, the smaller, clustered aeschynomenoid type (Corby 1981, 1988), occurs in *Arachis* (peanut) and other members of its tribe (Aeschynomenae) as well as the closely related Adesmieae. Interspersed with Aeschynomeneae in the *rbcL* tree are members of Dalbergieae that have desmodioid nodules (Corby 1988). Based on the *rbcL* tree, it would appear that determinate nodules have evolved at least three times from indeterminate ancestors: once in *Lotus* and once each in the Phaseoleae and Aeschynomeneae clades. The degree to which parallelism might include other differences known to exist between well-studied determinate and indeterminate nodules remains to be explored.

4 Conclusion: The Value of Phylogenetic Hypotheses

Phylogenetic hypotheses provide an objective framework for addressing a diversity of questions concerning the origin(s) and evolution of nodulation in the flowering plants generally and in the Leguminosae specifically. The *rbcL* hypothesis discussed here is a first effort to provide such a framework for the family using molecular data, but much more needs to be done. Additional sampling, even for *rbcL*, is clearly needed, given the enormous size of the family. Furthermore, *rbcL* provides too little variation in some parts of the family to provide a robust hypothesis. Finally, it is now widely recognized that a tree constructed from a single gene can be misleading with regard to the relationships of the organisms from which the genes are sampled (e.g., Nei 1987; reviewed in Doyle 1992). Consequently, additional data, both molecular and morphological, are needed to produce a comprehensive hypothesis for this important plant family. Additionally, efforts to understand the origins and evolution of nodulation in the legumes will be furthered by the extension of biochemical and molecular studies beyond the handful of economically important or "model system" species that have received most attention to date.

Acknowledgments

Several US NSF-DEB and NSF-BSR grants have supported our legume phylogenetic research.

References

Chappill JA (1995) Cladistic analysis of the Leguminosae: The development of an explicit hypothesis. *In* Crisp MD, Doyle JJ (*eds*) Advances in legume systematics, part 7: phylogeny, Royal Botanic Gardens, Kew, *pp* 1-9

Chase MW, Soltis DE, Olmstead RG, *et al.* (1993) Phylogenetics of seed plants: an analysis of nucleotide sequences from the plastid gene *rbcL*. Ann Missouri Bot Gard 80: 528-580

Corby HDL (1981) The systematic value of leguminous root nodules. *In* Polhill RM, Raven PH (*eds*), Advances in legume systematics, part 2, Royal Botanic Gardens, Kew, *pp* 657-670

Corby HDL (1988) Types of rhizobial nodules and their distribution among the Leguminosae. Kirkia 13: 53-123

Doyle JJ (1991) Evolution of higher plant glutamine synthetase genes: tissue specificity as a criterion for predicting orthology. Mol Biol Evol 8: 366-377

Doyle JJ (1992) Gene trees and species trees: molecular systematics as one-character taxonomy. Syst Bot 17: 144-163

Doyle JJ (1994a) Evolution of a plant homeotic multigene family: toward connecting molecular systematics and molecular developmental genetics. Syst Biol 43: 307-328

Doyle JJ (1994b) Phylogeny of the legume family: an approach to understanding the origins of nodulation. Ann Rev Ecol Syst 25: 325-349

Doyle JJ (1995) DNA data and legume phylogeny: a progress report. *In* Crisp MD, Doyle JJ (*eds*) Advances in legume systematics, part 7: phylogeny, Royal Botanic Gardens, Kew, *pp* 11-30

Doyle JJ, Doyle JL, Ballenger JA, Dickson EE, Kajita T, Ohashi H. A phylogeny of the chloroplast gene *rbcL* in the Leguminosae: Taxonomic correlations and insights into the evolution of nodulation. Amer J Bot (accepted)

Faria SM. de, Lewis GP, Sprent JI, and Sutherland JM (1989) Occurrence of nodulation in the Leguminosae. New Phytol 111: 607-19

Fitch WM (1970) Distinguishing homologous from analogous proteins. Syst Zool 19: 99-113

Heard J, Dunn K (1995) Symbiotic induction of a MADS-box gene during development of alfalfa root nodules. Proc Natl Acad Sci USA 92: 5273-5277

McKey D (1994) Legumes and nitrogen: The evolutionary ecology of a nitrogen-demanding lifestyle. *In* Sprent JI, McKey D(*eds*) Advances in legume systematics, part 5: nitrogen economy, Royal Botanic Gardens, Kew *pp* 211-228

Nei M. (1987) Molecular evolutionary genetics. Columbia, New York

Polhill RM (1994) Classification of the Leguminosae. *In* Bisby FA, Buckingham J, Harborne JB (*eds*), Phytochemical dictionary of the Leguminosae, Chapman and Hall, New York, *pp* xxxv-lvii

Polhill RM, Raven PH, Stirton CH (1981) Evolution and systematics of the Leguminosae. *In* Polhill RM, Raven PH (*eds*), Advances in legume systematics, part 1, Royal Botanic Gardens, Kew, *pp* 1-26

Purugganan MD, Rounsley SD, Schmidt RJ, Yanofsky MF (1995) Molecular evolution of flower development: Diversification of the plant MADS-box regulatory gene family. Genetics 140: 345-356

Soltis DE, Soltis PS, Morgan DR, Swensen SM, Mullin BC, Dowd JM, Martin PG (1995) Chloroplast gene sequence data suggest a single origin of the predisposition for symbiotic nitrogen fixation in angiosperms. Proc Natl Acad Sci USA 92: 2647-2651

Sprent JI (1994) Nitrogen acquisition systems in the Leguminosae. *In* Sprent JI, McKey D(*eds*) Advances in legume systematics, part 5: nitrogen economy, Royal Botanic Gardens, Kew *pp* 1-16

Temple SJ, Heard J, Ganter G, Dunn K, Sengupta-Gopalan C (1995) Characterization of a nodule-enhanced gluatmine synthetase from alfalfa: nucleotide sequence, *in situ* localization, and transcript analysis. Mol Plant-Microbe Interact 8: 218-227

Verma DPS, Hu C-A, Zhang M (1992) Root nodule development: origin, function and regulation of nodulin genes. Physiol Plant 85: 253-265

CO-EVOLUTION OF LEGUME–RHIZOBIAL SYMBIOSES: IS IT ESSENTIAL FOR EITHER PARTNER?

Janet I Sprent

Department of Biological Sciences, University of Dundee, Dundee, DD1 4HN, Scotland, UK

Keywords Evolution, legumes, rhizobia, nodules

1. Introduction. In the last few years much new information has been published, principally using molecular methods, on the taxonomy and phylogeny of both legumes and their nodulating bacteria. Much of this new information comes from tropical areas, where the biodiversity of both legumes and rhizobia is greater than in more temperate areas. With respect to co-evolution, things now seem much more complicated than before.

2. Legumes. The Leguminosae is a large and complex family of dicotyledonous plants, divided into three sub-families which vary greatly in their ability to nodulate (Faria *et al*, 1989). It is likely that several distinct nodulation events occurred, an aspect which is considered by Doyle in this volume. Based on *rbcL* sequences, the legumes have been shown to be part of a clade, Rosid I, which also includes all actinorhizal plant families and several which have no known nodulating members (Soltis *et al*, 1995). These workers suggested that members of this clade may have a predisposition to form nodules. What is this predisposition? At the simplest level it must relate to the ability to produce outgrowths which have vascular connections to the subtending root. The origin of lateral roots as we now know them is obscure, but probably dates back to the carboniferous, about 300 Ma BP (Thomas and Spicer, 1987). Recent work with *Arabidopsis* is consistent with some genetic control of lateral root formation which is separate from that of primary root development (Scheres *et al*, 1996; Celenza *et al*, 1995). A series of mutants has been isolated with modified lateral root development, usually coupled to altered auxin metabolism. All the various 'para-nodules' and similar structures described on a wide range of flowering plants might be explained on the basis of altered hormone metabolism producing structures similar to those on mutated *Arabidopsis*.

The major next step is entry of the microsymbiont into the outgrowth. In the non-nodulating soybean line T201, this two stage process can be clearly seen. Treatment with 2,4-D led to formation of outgrowths which had features of modified roots, but when rhizobia were also added functional nodules, with a peripheral vascular system were produced (Akao *et al*, 1991). This latter feature remains unique to legume

NATO ASI Series, Vol. G 39
Biological Fixation of Nitrogen
for Ecology and Sustainable Agriculture
Edited by A. Legocki, H. Bothe, A. Pühler
© Springer-Verlag Berlin Heidelberg 1997

nodules. Assuming that legumes were nodulated fairly early on in their evolution (probably in the late Cretaceous, about 80 Ma BP (Herendeen *et al*, 1992)) what other predisposing factors have been acquired? One possibility is the acquisition of genetic information from an ancestor of *Agrobacterium* which exploited wound-induced cell divisions (Baron and Zambryski, 1995). The latter do not occur in monocots, so this step might have occurred after their separation. Fundamental to the overlap between pathogenesis and symbiosis is a signal exchange system. Similarities between signals involved in mycorrhizal and nodule formation will be covered elsewhere in this volume. Endo-mycorrhizas are a very ancient phenomenon, suggesting that signalling systems are also ancient. Does this mean that the occasional non mycorrhizal nodulated legume (e.g. many *Lupinus* spp) have lost another component of mycorrhizal formation? What is the position of the essentially non-mycorrhizal families Proteaceae and Cruciferae (Brassicaceae)? The latter is in a clade, Rosid II, close to the N_2-fixing clade of Soltis *et al* (1995). The former are in a more basal section of the dicots (Chase *et al*, 1993). A search for signalling molecules in these groups could be informative. Examination of non-nodulating legumes for nodulation genes is another possible avenue. As Soltis *et al* (1995) point out, some nodulins are really 'housekeeping' proteins and at least one, *ENOD* 12, is not essential for nodulation in alfalfa (Csanádi *et al*,1994). We have recently looked for common features at the morphological and molecular level in non-nodulating African species of *Acacia* (Harrier *et al*, 1996) and find them to be a very closely related group of species. Preliminary work using a probe for *GmENOD* 2 has shown that they may contain this gene although other non-nodulating mimosoid legumes may not (Harrier, 1995). Have nodulated legumes acquired the full complement of *nod* factors in a series of separate events or have non-nodulating spp lost critical components? Techniques now available make these questions realistic and hold out prospects of extending genuine nodulation to new species by introducing a few vital missing components.

We have argued (e.g. Sprent, 1994) that when supplies of abiotically generated combined N and that cycled from free living N_2-fixing organisms becomes insufficient to support the burgeoning land flora, symbiotic fixation became advantageous for plants as well as providing a desiccation-resistant, nutrient rich niche for micro-organsisms. An alternative suggestion is that legumes adopted a high–N lifestyle to optimize carbon-fixation and that nitrogen fixation evolved to support this lifestyle (McKey, 1994). Both suggestions require that legumes are good scavengers for soil N and that nodules are an optional extra rather than essential, except when N is the *major* growth-limiting factor. Thus it is perhaps not surprising that not all legumes *can* nodulate and of those that *can*, not all *do*, especially under stressed conditions. All evidence to date (although little has been published on non-crop plants) suggests that legumes grown on mineral N are more stress tolerant than those which are nodulated.

3. Rhizobia. In a recent review of the diversity and phylogeny of rhizobia, Young and Haukka (1996) point out that the ability to induce nodules was probably not present in the common ancestor of all rhizobia and that nodulation genes may have been transferred between phylogenetically different bacteria. A consequence of this, as Dobert *et al* (1994) have shown, is that the phylogeny of *nod* genes may be more related to the phylogeny of host plants than is 16 *s* RNA or *nif* H genes. Perhaps the older idea of more-or-less fixed strains of rhizobia should be replaced by one of a dynamic population of soil organisms varying as necessary to cope with particular sets of circumstances. I have argued elsewhere (Sprent, 1994) that, under some circumstances it may be more important for rhizobia to be good saprophytes than (good or bad) symbionts. With a few exceptions remarkably little recent attention has been given to what rhizobia may live on in soil. Bradyrhizobia may use a wider range of aromatic components than *Rhizobium* spp. (Parke and Ornston, 1984). Ability to metabolise aromatic compounds may be important not only in soil, but in modulating the host–rhizobial interaction. Cooper and Rao (1995) have shown, for example, that *R meliloti* can degrade the *nod* gene inducing flavonoid luteolin and *B. japonicum* the isoflavonoid inducers diadzein and genistein.

A further complication could be that "in the soil environment, bacteria do not exhibit a growth curve like that observed in laboratory cultures" (Trevors, 1996). In particular, stressed (including starved) gram negative bacteria may be more competent to incorporate DNA from soil, for example, that complexed with clay particles. In addition, it has recently been shown that non-symbiotic rhizobia in the soil can transfer genetic material to the chromosome of symbiotic *R. loti* used as an inoculant (Sullivan *et al*, 1995). In soil (and not just in the rhizosphere) rhizobia are part of a complex *dynamic* system which can be independent of host legumes.

4. Symbioses. In terms of crop species grown under reasonably predictable conditions a high level of specificity and effectiveness can be advantageous. Is this also true of natural ecosystems? In the longer term the answer must be no because of lack of adaptability to environmental fluctuations. Variation is necessary in genetic make-up of both partners. There are various ways in which this may be achieved. At one extreme we may have the situation recently described for *Amphicarpaea bracteata* and *Bradyrhizobium* sp. The host is an annual legume, largely self-fertile, native to the eastern part of N. America. Populations tend to consist of two or more lineages with largely distinct bacterial endosymbionts. In demonstrating this, Spoerke *et al* (1996) argued that "non-random association . . . imply that genetic diversity of hosts is an important factor in the maintenance of polymorphism within the symbiont population". Implicit in this is that to be within a nodule is of advantage to the rhizobia and following from that, in the absence of host plants, populations of rhizobia would decline. Further, the authors suggest that if only one host species is available rhizobia will, by chance, gradually lose the ability to nodulate other species.

At the other extreme we have the broadest known host range in *Rhizobium* strain NGR 234, which nodulates over 70 legume genera, although not always effectively (Relic *et al*, 1993). This broad range largely results from the production of over 18 chemically distinct nod factors (lipochitin oligosaccharides). NGR 234 was isolated from New Guinea by Trinick (1980). The humid tropics contain a large diversity of legume genera whose nodules contain a wide range of rhizobia with little obvious relationship between the two as shown for the Brazilian Amazon by Moreiro and Franco (1994). However, there is also great diversity in arid and semi-arid tropical Africa where the number of legume species is lower. A combined study of strains from the Amazon, Africa and elsewhere led to the formulation of new genera and species (de Lajudie *et al*, 1994), to which more will undoubtedly be added. This apparently chaotic situation is clarified if we return to the suggestion of Young and Haukka that *nod* genes arose relatively late and have been transferred laterally amongst rhizobia. A late origin of *nod* genes is also consistent with a large pool of saprophytic rhizobia. We have no idea how large this pool is but it should be present in soils which have never been invaded by legumes, for example, those now in the vicinity of the Antarctic, but which in former times were nearer the equator.

5. References.

Akao, S., Nakata, S. and Yoneyama, T. (1991). *Plant and Soil*, **138**, 207–212.

Baron, C. and Zambryski, P.C. (1995). *Annual Review of Genetics*, **29**, 107–129.

Celenza, J.L., Grisafi, P.L. and Fink, G.R. (1995). *Genes and Development*, **9**, 2131–2142.

Chase, M.W. *et al.* (1993). *Annals of Missouri Botanic Gardens*, **80**, 528–580.

Cooper, J.E. and Rao, J.R. (1995). *Symbiosis*, **19**, 91–98.

Csanádi, G. *et al.* (1994). *The Plant Cell*, **6**, 201–213.

De Lajudie, P. *et al.* (1994). *International Journal of Systematic Bacteriology*, **44**, 713–733.

Dobert, R.C., Breil, B.T. and Triplett, E.W. (1994) *Molecular Plant–Microbe Interactions*, **7**, 564–572.

Faria, S.M. *et al.* (1989). *New Phytologist*. **111**, 607–619.

Harrier, L.A. (1995). *Ph. D. Thesis*, University of Dundee, Scotland.

Harrier *et al.* (1996). *Plant Systematics and Evolution* (in press).

Hereendeen, P.K.S,, Crepet, W.L. and Dilcher, D.L. (1992), In *Advances in Legume Systematics 4. The Fossil Record*. Royal Botanic Gardens, Kew, 303–316.

McKey, D. (1994). In *Advances in Legume Systematics 5. The Nitrogen Factor*. Royal Botanic Gardens, Kew, 211–228.

Moreira, F.M.S. and Franco, A.A. (1994).In *Advances in Legume Systematics 5. The Nitrogen Factor*. Royal Botanic Gardens, Kew, 63–74.

Parke, D. and Ornston, N. (1984). *Journal of General Microbiology*, **130**, 1743–1750.

Relic, B. *et al.* (1993). In *New Horizons in Nitrogen Fixation*. Kluwer, Dordrecht, 183–189.

Scheres, B. *et al.* (1996). *New Phytologist*, **111**, 959–964.

Soltis, D.E. *et al.* (1995), *Proceedings of the National Academy of Sciences, U.S.A.*. **92**, 2647–2651.

Spoerke, J.M., Wilkinson, H.H. and Parker, M.A. (1996). *Evolution.* **50**, 146–154.

Sprent, J.I. (1989). *New Phytologist*, **111**, 129–153.

Sprent, J.I. (1994). *Plant and Soil*, **161**, 1–10.

Sprent, J.I. (1994). In *Advances in Legume Systematics 5. The Nitrogen Factor*. Royal Botanic Gardens, Kew, 1–16.

Sullivan, J.T. *et al.* (1995). *Proccedings of the National Academy of Sciences, USA*, **92**, 8985–8989.

Thomas, B.A. and Spicer, R.A. (1986). *The Evolution and Paleobiology of Land Plants*. Croom Helm: London.

Trevors, J.T. (1996). *Antonie van Leeuwenhoek.* **70**, 1–10.

Trinick, M.J. (1980), *Journal of Applied Bacteriology*, **49**, 39–53.

Young, J.P. and Haukka, K.E. (1996). *New Phytologist*, **133**, 87–94.

INTERACTIONS BETWEEN DIAZOTROPHS AND GRASSES

BARABARA REINHOLD-HUREK* and THOMAS HUREK
Max-Planck-Institut für terrestrische Mikrobiologie, Arbeitsgruppe Symbiose-
forschung, Karl-von-Frisch-Str., D-35043 Marburg, Germany

CO-EVOLUTION OF MICROORGANISMS AND THEIR HOSTS

Molecular phylogenetic studies based on comparative sequence analysis of genes or protein sequences has gained increasing importance in understanding evolutionary relationships of organisms. Moreover, these techniques allowed an insight in development of intimate interactions between organisms in some cases. Microrganisms (bacteria or fungi) which interact very closely with a host in a mutual interaction (symbiosis) may have developed in a synchronous manner. This is obvious especially for microorganisms which are living in an endosymbiosis in cells of an eukaryotic host, as intracellular associates. A co-evolution was evident *e.g.* for specialized prokaryotic lineages that are essential for the survival and reproduction of insects: Bacteria of the group *Buchnera aphidicola* are closely related to the *Escherichia coli/Proteus vulgaris* branch within the gamma group of the *Proteobacteria* according to 16S rDNA sequences (Moran *et al.*, 1993), and provide essential amino acids and possibly vitamins to their hosts, aphids (Douglas and Prosser, 1992). This intimate interaction of the maternally transmitted endosymbionts with aphids is reflected in the concordance of the phylogenetic trees of the symbionts with the respective aphid host species, indicating that an aphid and bacterial ancestor formed an ancient association followed by similar levels of divergence (Moran and Baumann, 1994). Examples for co-evolution in plant-microbe symbioses will be discussed in other contributions to this volume. As an example, it has been calculated from 16S rRNA gene phylogeny of obligate endosymbionts of most land plants, arbuscular mycorrhizal fungi, that their origin correlated with appearance of structures which resemble vesicles and spores formed by these fungi in the first land plants (Simon et al., 1993).

Most examples of co-evolution indicate that the interaction must be very intimate (endosymbiosis), if not essential for survival of one of the partners in order to be reflected in an organismal co-evolution. In the case of nitrogen-fixing bacteria associated with grass roots, this type of co-evolution is not evident. Most grass-associated diazotrophs occur in or on roots of a wide variety of hosts, which can harbor several different taxa of diazotrophs: *Azospirillum brasilense/lipoferum e.g.* can be isolated from many tropical, subtropical and temperate grasses and cereals (Döbereiner and Pedrosa, 1987). *Herbaspirillum* spp. has also been isolated from various plants such as sugar cane, maize, *Pennisetum*, weeds or leguminous plants , whereas *Acetobacter diazotrophicus* was found in sugar cane and sweet potato (Döbereiner *et al.*, 1993). Since sugar beet and sugar cane are only distantly related, whereas *A. diazotrophicus* comprises of one species only, this interaction may not have co-evolved but reflect physiological similarities of the hosts (sucrose content).

NATO ASI Series, Vol. G 39
Biological Fixation of Nitrogen
for Ecology and Sustainable Agriculture
Edited by A. Legocki, H. Bothe, A. Pühler
© Springer-Verlag Berlin Heidelberg 1997

However, a more detailed study on divergence of *Acetobacter diazotrophicus* strains and their respective sugar cane hosts might allow some conclusions about synchronous evolution. *Azoarcus* spp. was so far found only in roots and stems of Kallar grass in Pakistan (Reinhold-Hurek *et al.*, 1993; Hurek and Reinhold, 1995); however this might be not an exception of the rule but merely a constraint of isolation or detection techniques (see below). Moreover, five species of *Azoarcus* were detected in Kallar grass (Reinhold-Hurek *et al.*, 1993), making speculations on organismal co-evolution most doubtful.

Perhaps the concept of organismal co-evolution is less likely to be applicable to grass-associated diazotrophs, since the interactions might not be sufficiently intimate and essential for both partners. However, more information will be needed on ecological distribution and EXACT location of diazotrophs in their hosts as well as on their functions, in order to distinguish loosely from intimately associated bacteria, for which the principle of co-evolution still might apply.

MOLECULAR DIAGNOSTIC TESTS BASED ON 16S rRNA AND NITROGENASE GENES

16S rDNA based phylogeny is a widely accepted tool to study organismal phylogeny in bacteria (Olsen and Woese, 1993). Application of molecular diagnostic tests on environmental DNA samples may allow detection and identification of bacteria without isolation of the strains. The experimental approach is as following: DNA is isolated (*e.g.* from plant roots), which serves as template for PCR primers specifically targeted to 16S rDNA sequences; when these primers are specific for certain organisms or genera (Hurek *et al.*, 1993), PCR products do not need to be cloned but may be subjected directly to sequencing (Hurek and Reinhold-Hurek, 1995) in order to identfy the microorganism by comparison with known 16S rDNA sequences.

In various habitats, as yet uncultivated microorganisms have been detected by this approach (Amann *et al.*, 1991; Giovannoni *et al.*, 1990). This observation was also made for plant-associated diazotrophs of the genus *Azoarcus*: from roots of Kallar grass plants an as yet unknow sequence was retrieved which clustered within sequences of the genus *Azoarcus*, although no bacteria of this type could be isolated (Hurek and Reinhold-Hurek, 1995). This has to be taken into account when a population of diazotrophs is described by traditional methods of cultivation; numbers or even identity might be different due to strains difficult to cultivate.

Another molecule which is considered to be usefull for phylogenetic studies is nitrogenase, since it is highly conserved in function and structure (Young, 1992). When primers targeted to structural genes such as *nifH* are used in the above-mentioned approach, a clone library of PCR products may contain the information of the diazotrophs present in a given habitat. This has been applied for field-grown rice by Ueda *et al.* (1995) and revealed several clusters of nitrogenase sequences from so far unknown organisms.

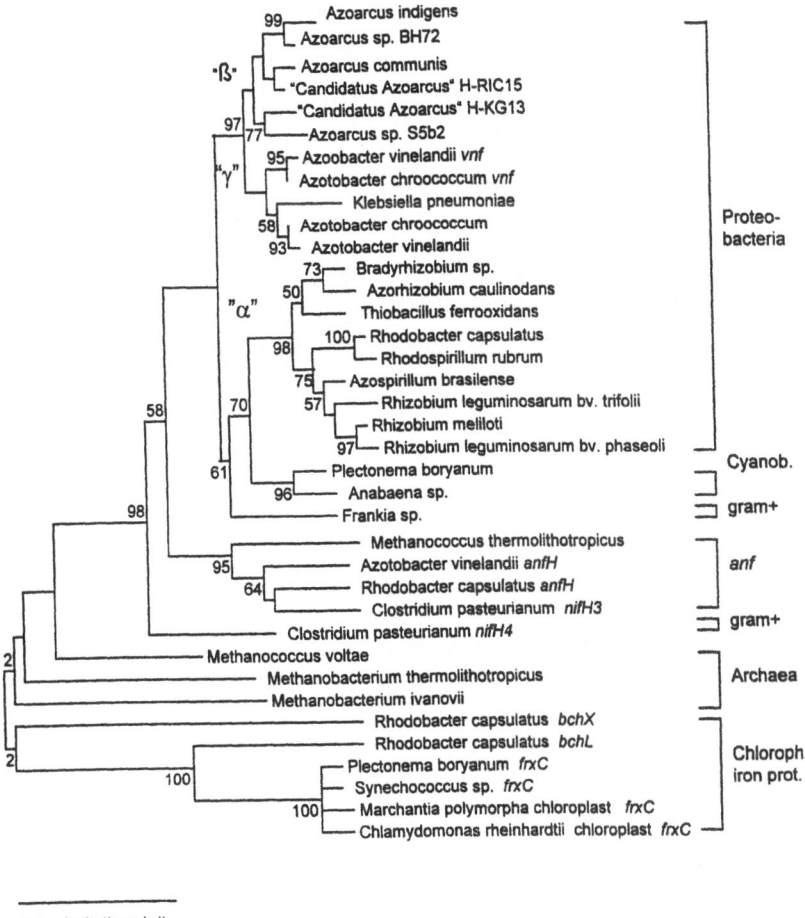

0.4 substitutions / site

Fig. 1. Phylogeny of partial NifH protein sequences. *Azoarcus* spp. sequences and "Candidatus *Azoarcus* H-KG13" obtained by us, the latter from Kallar grass roots. All other sequences from databases, including "Candidatus *Azoarcus* H-RIC15" from Ueda *et al.* (1995) from field-grown rice in Japan.

We applied this technique to pure cultures of *Azoarcus* in order to obtain sequences of several species for comparison, and to roots of Kallar grass. Phylogenetic tree inference from partial NifH protein sequences was carried out using the neighbor joining (Mega package) (Fig. 1) and the parsimony (Phylip 3.1c, tree not shown) method. Topologies of trees were almost identical, indicating that tree inference was robust. Nitrogenase phylogeny followed largely the organismal phylogeny based on 16S rDNA sequences. Plant-associated *Azoarcus* species clustered together, being most closely related to *Azotobacteraceae* (Fig. 1). The only sequence of another diazotroph of the beta group of *Proteobacteria*, from *Herbaspirillum seropedicae*, was reported to be related to *Thiobacillus ferrooxidans* and *Azospirillum brasilense* within the alpha group of *Proteobacteria* (Pedrosa *et al.*, 1995), indicating that nitrogenase sequences are divergent within the beta *Proteobacteria* and perhaps of different phylogenetic origin.

Surprisingly, within the *Azoarcus* cluster a sequence obtained from rice grown in Japan (H-RIC15, Ueda *et al.*) was located, being related to the *A. communis* sequence. This indicated that *Azoarcus* spp. might also be present in field-grow rice in Asia, which would for the first time extend the natural host range. In gnotobiotic cultures, strain BH72 was already shown to grow endophytically in rice (Hurek *et al.* 1994), demonstrating that the potential of colonizing this host exists. Prompted by these results we will carry out an extensive study on natural occurrence of *Azoarcus* spp.

LITERATURE

Amann, R., *et al.* (1991) *Nature* **351**:161-164

Döbereiner, J., and F. Pedrosa (1987). *Nitrogen-fixing bacteria in non-leguminous crop plants*. Springer Verlag, Berlin

Döbereiner, J., V. M. Reis, M. A. Paula, and F. Olivares (1993) in *New Horizons in Nitrogen Fixation*, R. Palacios, J. Mora, and W. E. Newton (eds.), Kluwer Academic Publishers, Dordrecht, pp. 671-676

Douglas, A. E., and W. A. Prosser (1992) *J. Insect Physiol.* **38**:565-568.

Giovannoni, S. J., T. B. Britschgi, C. L. Moyer, and K. G. Field (1990) *Nature* **345**:60-63

Hurek, T., S. Burggraf, C. R. Woese, and B. Reinhold-Hurek (1993) *Appl. Environ. Microbiol.* **59**:3816-3824

Hurek, T., and B. Reinhold-Hurek (1995) *Appl. Environ. Microbiol.* **61**:2257-2261.

Hurek, T., B. Reinhold-Hurek, M. Van Montagu, and E. Kellenberger (1994) *J. Bacteriol.* **176**:1913-1923

Moran, N., and P. Baumann (1994) *TREE* **9**:15-20

Moran, N. A., M. A. Munson, P. Baumann, and H. Ishikawa (1993) *Proc. R. Soc. London Ser. B.***253**:167-171

Olsen, G. J., and C. R. Woese (1993) *FASEB J.* **7**:113-123

Pedrosa, F. O., *et al.* (1995) presented at the Int. Symp. on Sustainable Agriculture for the Tropics, Angra dos Reis, Rio de Janeiro, Brazil.

Reinhold-Hurek, B, *et al.* (1993) *Int. J. Syst. Bacteriol.* **170**:1445-1451

Simon, L., J. Bousquet, R. C. Lévesque, and M. Lalonde (1993) *Nature* **363**:67-69.

Ueda, T., Y. Suga, N. Yahiro, and T. Matsuguchi (1995) *J. Bacteriol.* **177**:1414-1417

Young, J. P. W. (1992) In *Biological Nitrogen Fixation* G. Stacey, R. H. Burris, and H. J. Evans (eds.), Chapman and Hall, New York, pp. 43-86

Have Common Plant Systems Co-evolved in Fungal and Bacterial Root Symbioses ?

V. Gianinazzi-Pearson

Laboratoire de Phytoparasitologie INRA/CNRS, Station de Génétique et d'Amélioration des Plantes, INRA, BV 1540, 21034 Dijon Cedex, France

1. Introduction

The ability of root systems to establish beneficial symbiotic relationships with soil microorganisms represents one of the most successful strategies that land plants have developed to cope with abiotic and biotic stresses imposed by poor soil conditions during their colonisation of terrestrial ecosystems. The two main categories of root symbioses found in extant plant taxa, arbuscular mycorrhiza and nodules, are both formed by legumes. Evidence from molecular clocks and fossil records is consistent with the appearance of arbuscular mycorrhiza very early in the history of ancestral land plants in the Devonian period (about 400 Myr ago) (1, 2). The eukaryotic fungal partners have successfully adapted during evolution to recent plant taxa, so that the symbiosis is now widespread throughout over 80% of families in the plant kingdom. Nodulation with prokaryotic soil bacteria originated much later in evolutionary terms, after the appearance of ancestral legumes probably some 60-70 Myr ago in the late Cretaceous-early Tertiary period (3). At first sight the two root symbioses appear to share little in common apart from the cortical parenchyma being the same target tissue for mycorrhiza development and nodule organogenesis, and the formation of an extended interface for reciprocal nutrient exchange between symbiont cells where both microorganisms are surrounded by a plant-derived membrane: the peribacteroid membrane in nodules, and the periarbuscular membrane in mycorrhizal tissues. However, consideration of genetic and molecular events has revealed unexpectedly similar features which suggest that the two symbioses may be partly regulated by common plant functions.

2. Symbiosis-related genes

Studies of different legumes have shown that non-nodulating genotypes can often be defective in their mycorrhiza-forming abilities (4, 5, 6). Genetic analyses of EMS-induced isogenic pea mutants have provided evidence that some early infection stages are modulated by common plant genes in both symbioses (4, 7). The most frequent mutant phenotype, Myc^{-1}Nod^{-} found in isogenic mutants induced in different pea varieties, is characterised by complete resistance to both arbuscular mycorrhizal fungi and rhizobia (8). Three recessive mutated alleles determine this

NATO ASI Series, Vol. G 39
Biological Fixation of Nitrogen
for Ecology and Sustainable Agriculture
Edited by A. Legocki, H. Bothe, A. Pühler
© Springer-Verlag Berlin Heidelberg 1997

double resistance, and these are now known to correspond to the *sym8*, *sym19* and *sym30* genes in pea (Sagan and Duc, unpublished data). The three genes appear specific to the regulation of the mycorrhiza and nodule symbioses in roots, since their mutation has no known effect on plant interactions with other root-infecting microorganisms, whether these be fungi, nematodes or bacteria (9), or with a biotrophic aerial pathogen like *Peronospora viciae* (unpublished data). These *sym* genes intervene in the early stages of arbuscular mycorrhiza formation, after appressorium induction at the root surface. Resistance to fungal symbionts in the $Myc^{-1}Nod^-$ mutants is associated with elicitation of important cell wall appositions which are encrusted with phenolics and callose, and never observed during normal symbiotic interactions (10). The likelihood is that the mutated genes are somehow involved in the regulation of plant defence responses to the fungal symbionts (11). Another, rarer, mutation has been found to alter later symbiotic events in pea roots. In the resulting mutants, designated $Myc^{-2}Nod^{+/-}$, arbuscules do not differentiate from hyphae penetrating the host cells (12) and infection threads abort or nodulation is retarded (13), which may imply that developmental signals are defective in these plants. Furthermore, fungal and plant ATPase activities characteristic of functional arbuscules are absent from the symbiotic membranes of the poorly-developed intracellular interface, so that the mutation appears to affect also processes leading to the metabolic specialisation of the symbiotic interface. Only one mutated allele is so far known to give this symbiosis-defective phenotype and the corresponding plant gene has not yet been identified.

3. Molecular events in symbiosis development

The modulation of arbuscular mycorrhiza and nodule development by the same plant genes indicates that the infection processes in the two symbioses involve common elements. In the early stages of symbiont interactions, both arbuscular mycorrhizal fungi and rhizobia respond to flavonoids and isoflavonoids, metabolites of the phenylpropanoid pathway and exuded by host roots. Such molecules enhance spore germination, hyphal growth and root colonization by the fungal symbionts (for references see 14), whilst they serve to specifically activate nodulation genes of the symbiotic bacteria in response to a host legume (15). Rhizobial Nod-factors can stimulate arbuscular mycorrhiza formation (14) which strengthens the hypothesis that analogies may exist in early signaling events between host plants and the two microsymbionts.

Although plant-microbe interactions leading to arbuscular mycorrhiza and nodule formation result in two different morphological entities, similar cellular events can accompany some of the developmental processes in the two symbioses. One common feature is that invasion of host cells and tissues by arbuscular mycorrhizal fungi and rhizobia occurs without triggering a strong plant defence response, whilst these are induced in defective symbioses (11, 16). Another is that a number of host wall or membrane components linked to the nodulation process are associated with comparable plant-derived cell structures in host-fungal interfaces in mycorrhizal tissues. Monoclonal antibodies detect glycoproteins with common epitopes in both the infection thread matrix and wall deposits around arbuscule hyphae penetrating host cells, and likewise they have revealed that certain glycoconjugates are shared by the peribacteroid and periarbuscular membranes produced by the host plant (8). Some of these molecules are typical of the plant glycocalyx whilst others are related to other developmental events in plant tissues, suggesting that they are linked to common structural responses triggered by different stimuli.

4. Nodulin gene expression

Although nodulins have been considered to be products of genes performing unique functions in nodules, several early nodulin (ENOD) genes have been found to be expressed outside nodules, and it has been suggested that many nodulin genes could have been recruited from genes already functioning in the plant (17). These considerations, together with the similarities so far observed between fungal and bacterial root symbioses, have prompted recent research into the possible implication of nodulin-related genes in arbuscular mycorrhiza. Expression analysis has shown that four nodule genes encoding proline-rich and undescribed proteins, and the nodulin gene *ENOD40*, are enhanced in mycorrhizal roots of broadbean (18) and *Medicago* (A.M. Hirsch, pers. comm.), respectively. Likewise, the proline-rich protein encoding genes *ENOD2*, *ENOD11* and *ENOD12* appear to be activated in arbuscule-containing cells of *Medicago* roots (A.M. Hirsch, pers. comm. ; V. Gianinazzi-Pearson, E.P. Journet, D. Barker, unpublished). This may not be surprising since proline-rich proteins are also common cell wall components and may therefore be part of the host reaction to the invading fungus. More unexpected is the novel induction of a late nodule leghemoglobin gene (*VfLb29*) in arbuscular mycorrhizal tissues in fababean roots (Frühling et al., submitted). This gene has an unusually low homology with known leghemoglobins and its role in symbiotic interactions has yet to be determined.

5. Conclusions

There are increasing examples of similar molecular events in arbuscular mycorrhiza and nodule symbioses, and legume mutants provide evidence that some of these are determined by the same plant genes. Whilst some genes may be specific to symbiosis development and functioning, it is likely that others represent existing plant genes that perform their traditional tasks in a new compartment and context. Some symbiotic events are no doubt related to normal developmental processes in plants, and phytohormones may be good candidates as signal molecules in arbuscular mycorrhiza, as proposed for nodulation (15). Furthermore chitooligosaccharides released from fungal walls, perhaps by mycorrhiza-induced plant chitinases (19), may behave as elicitors of plant reactions in the symbiosis, in analogy to Nod factors during nodule induction.

　　Whilst differences do obviously exist between the two symbiotic systems, more information about the extent to which they are regulated by common plant functions will greatly contribute to understanding mechanisms involved. Moreover, seeing the ancestral nature of arbuscular mycorrhiza, the speculation that part of the plant processes leading to nodulation may have evolved from those already established for the fungal symbiosis provides an exciting challenge to future research into root-microbe interactions.

6. References

1. Simon L., Bousquet J., Levesque R.C., Lalonde M., 1993. Nature 363, 67-69.
2. Remy W., Taylor T.N., Hass H., Kerp H., 1994. Proceedings of the National Academy of Science 91, 11841-11843.
3. Sprent J.I., 1994. Plant and Soil 161, 1-10.
4. Duc G., Trouvelot A., Gianinazzi-Pearson V., Gianinazzi S., 1989. Plant Science 60, 215-222.
5. Bradbury S.M., Peterson R.L., Bowley S.R., 1993. New Phytologist 124, 665-673.
6. Sagan M., Morandi D., Tarenghi E., Duc G., 1995. Plant Science 111, 63-71.
7. Gianinazzi-Pearson V., Gianinazzi S., Guillemin J.P., Trouvelot A., Duc G., 1991. In Advances in Molecular Genetics of Plant-Microbe Interactions 1, Kluwer Academic Publishers, 336-342.
8. Gollotte A., Lemoine M.C., Gianinazzi-Pearson V., 1996. In Concepts in Mycorrhizal Research, Kluwer Academic Publishers, Dordrecht, 91-111.
9. Gianinazzi-Pearson V., Gollotte A., Dumas-Gaudot E., Franken P., Gianinazzi S., 1994. In Advances in Molecular Genetics of Plant-Microbe Interactions 3, Kluwer Academic Publishers, Dordrecht, 179-186.
10. Gianinazzi-Pearson V., Dumas-Gaudot E., Gollotte A., Tahiri-Alaoui A., Gianinazzi S., 1996. New Phytologist 133, 45-57.
11. Gollotte A., Gianinazzi-Pearson V., Giovannetti M., Sbrana C., Avio L., Gianinazzi S., 1993. Planta 191, 112-122.
12. Gianinazzi-Pearson V., Gollotte A., Lherminier J., Tisserant B., Franken P., Dumas-Gaudot E., Lemoine M.C., van Tuinen D., Gianinazzi S., 1995. Canadian Journal of Botany 73, S526-S532.
13. Sagan M., 1992. La symbiose Rhizobium-légumineuses. Analyse de mutants symbiotiques de pois Pisum sativum L. Ph.D. Thesis, University of Paris-Sud - Centre d'Orsay, F, 158 p.
14. Xie Z.P., Staehelin C., Vierheilig H., Wiemken A., Jabbouri S., Broughton W.J., Vögeli-Lange R., Boller T., 1995. Plant Physiology 108, 1519-1525.
15. Hirsch A.M., 1992. New Phytologist 122, 211-237.
16. Long S.R., Staskawicz B.J., 1993. Cell 73, 921-935.
17. Nap J.P., Bisseling T., 1990. Science 250, 948-954.
18. Frühling M., Perlick A., Roussel H., Gianinazzi-Pearson V., Puhler A., 1996. 2nd European Nitrogen Fixation Conference and NATO Advanced Workshop, Poznan, PL, Abstract.
19. Slezack S., Dassi B. and Dumas-Gaudot E., 1996. In Chitin Enzymology Vol. 2, Atec Edizioni, Grottamare, 339-347.

Nitrogen Fixing Systems and Evolution of Plant Hemoglobins

Paweł M. Stróżycki and Wojciech M. Karłowski

Institute of Bioorganic Chemistry Polish Academy of Sciences
Noskowskiego 12/14, 60-337 POZNAŃ,
POLAND

Keywords. Plant hemoglobins, evolution, *Lupinus luteus*, symbiosis.

Nitrogenase is the main bacterial enzyme responsible for the reduction of atmospheric nitrogen to its biologically active form. The enzyme is extremely sensitive to oxygen: even low oxygen concentrations may inactivate nitrogenase. Therefore, there must exist a system protecting nitrogenase from the oxygen. On the other hand, reduction of nitrogen is a highly energy consuming process which means that bacterial respiration has to be very intense. The plant protein which combines two functions: facilitating low oxygen tension and providing the necessary oxygen for respiration is leghemoglobin. For a long time leghemoglobins - hemoglobins from the nodules of leguminous plants, were thought to be the only plant hemoglobins. Now we know that there exist not only hemoglobins connected with the symbiotic nitrogen fixation, but also hemoglobins expressed in tissues other than nodules. Through its high oxygen affinity, leghemoglobin forms a kind of oxygen gradient from the surface of the infected cell to the bacteroid. Oxygen is then shifted to the terminal oxydases of the nitrogen fixing bacteria, supporting its respiration. In this way, leghemoglobin becomes a key element controlling the efficiency of nitrogen reduction.

The plant - bacterium model we use in our studies on nitrogen fixation is lupin (*Lupinus luteus*) as a plant host and *Bradyrhizobium lupini* as a microbial symbiont. Nodules which are formed on lupin main root as a result of bacterial infection are „undetermined" type of nodules so called „collar" type: they grow tightly surrounding the root (Golinowski et. al 1987; Golinowski et. al 1991).

There are two leghemoglobins synthesized in lupin nodules. They show 82% of amino acid sequence identity. Using probes specific to each cDNA (based on the 3' nontranslated sequences) we found out that both leghemoglobins are expressed at about the same time, around the 12th day after the infection with *B. lupini*. Besides both are probably coded by single copy genes (Stróżycki et. al 1995).

When we first compared the amino acid sequences of all known plant hemoglobins, we came up with a „similarity tree", which we arbitrarily rooted in-between the legume and non legume families. The „similarity tree" confirmed the suggestion that lupin was one of the oldest plants in the legume family. Then lupin peptides appeared to be the longest leghemoglobins known, with the size just in-between legume and non legume hemoglobins. A following finding was a significant loss in the length of hemoglobin polypeptide in the course of evolution. It was also clear that in the case of *Casuarina*, where there are two hemoglobins known,

NATO ASI Series, Vol. G 39
Biological Fixation of Nitrogen
for Ecology and Sustainable Agriculture
Edited by A. Legocki, H. Bothe, A. Pühler
© Springer-Verlag Berlin Heidelberg 1997

a symbiotic and nonsymbiotic one (Jacobsen-Lyon et.al 1995), the former is much shorter: its size is closer to that of legume leghemoglobins. We drew a conclusion that this may be a general trend. It was confirmed by the latest finding of soybean nonsymbiotic hemoglobin, which is much longer than the symbiotic one (Andersson et. al 1996). When we introduced new sequences published (soybean nonsymbiotic and rice), a new „similarity tree was not much different than the first one. The main difference was that soybean nonsymbiotic hemoglobin grouped together with *Casuarina* nonsymbiotic hemoglobin, which made rooting of the tree in-between legume and non legume families not possible. It appeared that the most proper way is to root this tree in-between the symbiotic and nonsymbiotic hemoglobins. The result of such rooting clearly confirms the hemoglobin gene duplication theory (Fig.1).

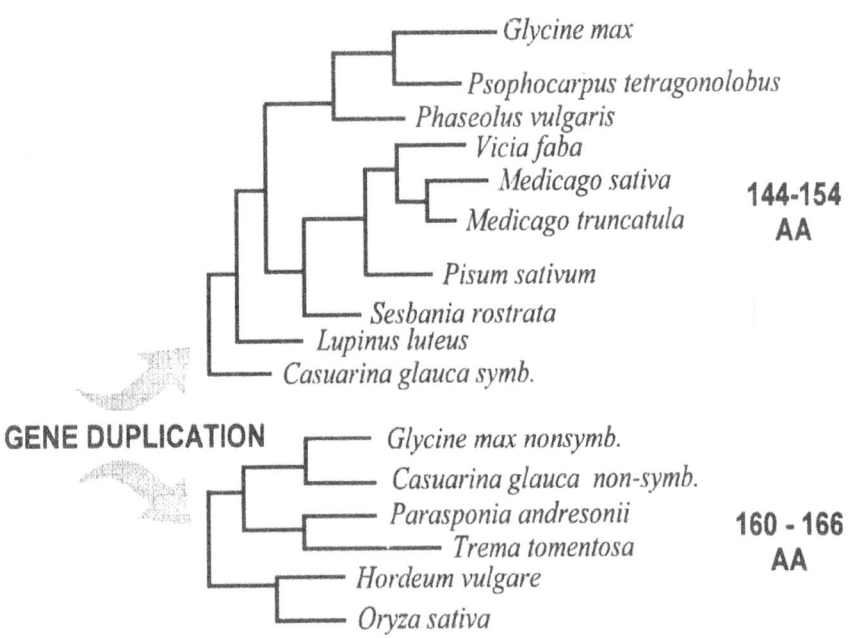

Fig.1 The „similarity tree" of plant hemoglobin polypeptides.

The length difference between symbiotic and nonsymbiotic hemoglobins is partly due to the size of the N terminus of polypeptides. All nonsymbiotic hemoglobins consist of five to ten additional amino acids - the so called "extension polypeptide" (Fig.2). The length of the exons of genes coding for different hemoglobins also shows a clear tendency towards shortening. The most stable parts are exons two and three, which code for elements of the polypeptide forming the so called "hem pocket", in which the oxygen is transported (Fig.3). It needs to be investigated why all hemoglobins of the symbiotic type are shorter. It is known for example, that soybean symbiotic hemoglobin has much higher oxygen affinity than *Parasponia* hemoglobin (Appleby et.al 1988). Is it only due to the amino acid composition ?

Glycine max lbc3	GAFTDKQEAL
Psophocarpus tetragonolobus LB	GGFTEKQEAL
Phaseolus vulgaris lb1	GAFTEKQEAL
Vicia faba lbI	GFTEKQEAL
Medicago sativa lb3	GFTDKQEAL
Medicago truncatula lb2	GFTEKQEAL
Pisum sativum lbI	GFTDKQEAL
Sesbania rostrata LB3	GFTEKQEAL
Lupinus luteus lbI	**GVLTDVQVAL**
Lupinus luteus lbII	**GALTESQAAL**
Casuarina glauca hb-sym	ALTEKQEAL

Casuarina glauca hb-nonsym	STLEG	RGFTEEQEAL
Glycine max hb-nonsym	TTTLE	RGFSEEQEAL
Parasponia andersonii hb	SSSEVN	KVFTEEQEAL
Trema tomentosa hb	SSSEVD	KVFTEEQEAL
Hordeum vulgare hb	SAAEGA	VVFSEEKEAL
Oryza sativa	ALVEDNNAVA	VSFSEEQEAL

Fig.2 The alignment of N terminal amino acid sequences of symbiotic and nonsymbiotic (gray background) hemoglobins.

PLANT	EXON 1	EXON 2	EXON 3	EXON 4	LENGTH (AA)
Glycine max lbc3	98	109	105	126	145
Phaseolus vulgaris lb	98	109	105	129	146
Medicago sativa MsI	92	112	108	129	146
Medicago trunculata I	95	112	108	129	147
Sesbania rostrata lb2	95	115	108	129	148
Lupinus luteus	98	115	114	135	154
Casaurina glauca sym	95	115	117	132	152
Glycine max hb	113	115	117	138	160
Casaurina glauca hb	113	115	117	138	161
Parasponia andresonii hb	116	115	117	141	162
Trema tomentosa hb	116	115	117	138	161

Fig.3 Comparison of the length of exons of genes coding for symbiotic and nonsymbiotic types of plant hemoglobins.

The analysis of leghemoglobin gene promoters, especially of soybean and *Sesbania* (She et. al 1993; Szczygłowski et. al 1994), revealed several structural elements which play a major role in the regulation of nodule specific expression. The so called "Organ Specific Element" (OSE) is the most conservative element present in all leghemoglobin gene promoters known (with only slight modifications). It consists of two nodulin motifs AAAGAT and CTCTT, particularly important for the level of nodule specific expression (Ramlov et. al 1993).

The sequence of *Lupinus luteus* leghemoglobin I gene promoter shows a very low similarity to other leghemoglobin promoter sequences. Despite the lack of typical nodulin elements such as OSE, that promoter directs the expression of the reporter gene into the central zone of the nodule of *Lotus* transgenic plants. Surprisingly the same promoter also shows some activity in root tissue out of the nodule (paper in preparation). A different structure of the promoter and a pattern of expression similar to that of nonlegume *Parasponia*, make lupin leghemoglobin I exceptional among hemoglobins of the symbiotic type. On the other hand, like other symbiotic type hemoglobins, it shows low amino acid sequence similarity to the nonsymbiotic type hemoglobins (Fig.4) and like all of them lcks on extension peptide (Fig.2).

lupin ⇔ symbiotic	59 - 66%
lupin ⇔ nonsymbiotic	52 - 56%
symbiotic ⇔ symbiotic	59 - 66%
nonsymbiotic ⇔ nonsymbiotic	80 - 86%

Fig. 4 Amino acid sequence similarity of plant hemoglobins of different types.

The nature of the expression of lupin hemoglobins with reference to their evolutionary position is currently under investigation.

Andersson C.R., Jensen E.O., Llewellyn D.J., Dennis E.S., Peacock W.J. 1996. A new hemoglobin gene from soybean: A role for hemoglobin in all plants. Proc. Natl. Acad. Sci.93:5682-5687.

Appleby C.A., Bogusz D., Dennis E.S., Peacock J.W. 1988. A role for haemoglobin in all plant roots. Plant,Cell and Environment 11:359-67.

Golinowski W., Kopcinska J., Borucki W. 1987. The morphogenesis of lupine root nodules during infection by Rhizobium lupini. Acta Soc Bot Pol 56:687-703.

Golinowski W., Lotocka B., Kopcinska J., Borucki W. 1991. Structure and Development of Lupine Root Nodules. Folia Histochem Cytobiol 29:157

Jacobsenlyon K., Jensen E.O., Jorgensen J.E., Marcker K.A., Peacock W.J., Dennis E.S. 1995. Symbiotic and nonsymbiotic hemoglobin genes of Casuarina glauca. Plant Cell 7(2):213-23.

Ramlov K.B., Laursen N.B., Stougaard J., Marcker K.A. 1993. Site-Directed Mutagenesis of the Organ-Specific Element in the Soybean Leghemoglobin Ibc3 Gene Promoter. Plant J 4:577-80.

She Q.X., Lauridsen P., Stougaard J., Marcker K.A. 1993. Minimal Enhancer Elements of the Leghemoglobin Iba and Ibc3 Gene Promoters from Glycine max L Have Different Properties. Plant Mol Biol 22:945-56.

Stróżycki P.M., Legocki A.B. 1995. Leghemoglobins from an evolutionarily old legume, Lupinus luteus. Plant Sci 110(1):83-93.

Szczygłowski K., Szabados L., Fujimoto S.Y., Silver D., Debruijn F.J. 1994. Site-Specific Mutagenesis of the Nodule-Infected Cell Expression (NICE) Element and the at-Rich Element ATRE- BS2* of the Sesbania rostrata Leghemoglobin glb3 Promoter. Plant Cell 6:317-32.

NATO ASI Series G

NATO ASI Series G

The manufacturer's authorised representative in the EU is Springer
Nature Customer Service Centre GmbH, Europaplatz 3, 69115 Heidelberg,
Germany. If you have any concerns regarding our products, please
contact ProductSafety@springernature.com

Printed and bound by CPI Group (UK) Ltd, Croydon, CR0 4YY

28/04/2026

02098453-0006